Animal Health and Welfare

Animal Health and Welfare

Editor: Dennis Cameron

R CALLISTO REFERENCE

www.callistoreference.com

Callisto Reference,
118-35 Queens Blvd., Suite 400,
Forest Hills, NY 11375, USA

Visit us on the World Wide Web at:
www.callistoreference.com

ISBN: 978-1-64116-642-3 (Hardback)

Cataloging-in-Publication Data

Animal health and welfare / edited by Dennis Cameron.
    p. cm.
Includes bibliographical references and index.
ISBN 978-1-64116-642-3
1. Animal health. 2. Animal welfare. 3. Animals. 4. Veterinary medicine. I. Cameron, Dennis.
SF740 .A55 2022
636.089 6--dc23

Table of Contents

Preface

Animal health refers to a concept in agriculture which is involved in ensuring that farm animals are healthy, cared for and free from diseases. It also deals with the implementation of government policies regarding the management and prevention of the outbreak of animal diseases. The well-being of nonhuman animals falls under the domain of animal welfare. It is involved in issues related to how the animals are slaughtered, how they are used in scientific research and how they are kept as pets. The effect of human activities upon the welfare and survival of wild species is also studied within animal welfare. This book contains some path-breaking studies in the field of animal health and welfare. Also discussed herein is a detailed explanation of the various concepts and applications of this field. This book will serve as a valuable source of reference for graduate and post graduate students.

The researches compiled throughout the book are authentic and of high quality, combining several disciplines and from very diverse regions from around the world. Drawing on the contributions of many researchers from diverse countries, the book's objective is to provide the readers with the latest achievements in the area of research. This book will surely be a source of knowledge to all interested and researching the field.

In the end, I would like to express my deep sense of gratitude to all the authors for meeting the set deadlines in completing and submitting their research chapters. I would also like to thank the publisher for the support offered to us throughout the course of the book. Finally, I extend my sincere thanks to my family for being a constant source of inspiration and encouragement.

Editor

Nutritional Characteristics of Forage Grown in South of Benin

Nadia Musco, Ivan B. Koura[1], Raffaella Tudisco, Ghislain Awadjihè[1], Sebastien Adjolohoun[1], Monica I. Cutrignelli*, Maria Pina Mollica[2], Marcel Houinato[1], Federico Infascelli, and Serena Calabrò

Department of Veterinary Medicine and Animal Production, University of Napoli Federico II, Napoli 80137, Italy

ABSTRACT: In order to provide recommendations on the most useful forage species to smallholder farmers, eleven grass and eleven legume forages grown in Abomey-Calavi in Republic of Benin were investigated for nutritive value (i.e. chemical composition and energy content) and fermentation characteristics (i.e. gas and volatile fatty acid production, organic matter degradability). The *in vitro* gas production technique was used, incubating the forages for 120 h under anaerobic condition with buffalo rumen fluid. Compared to legume, tropical grass forages showed lower energy (8.07 vs 10.57 MJ/kg dry matter [DM]) and crude protein level (16.10% vs 19.91% DM) and higher cell wall content (neutral detergent fiber: 63.8% vs 40.45% DM), respectively. In grass forages, the chemical composition showed a quite high crude protein content; the *in vitro* degradability was slightly lower than the range of tropical pasture. The woody legumes were richer in protein and energy and lower in structural carbohydrates than herbaceous plants, however, their *in vitro* results are influenced by the presence of complex compounds (i.e. tannins). Significant correlations were found between chemical composition and *in vitro* fermentation characteristics. The *in vitro* gas production method appears to be a suitable technique for the evaluation of the nutritive value of forages in developing countries. (**Key Words:** Grass, Legume, *In vitro* Gas Production, Nutritive Value, Degradability)

INTRODUCTION

In Benin agriculture contributes 32.5% to the national total Gross Domestic Product (GDP) (INSAE, 2012) and the livestock sub-sector represents 6.2% of the total GDP (FAO, 2006). Small ruminants are essentially bred by smallholders and this activity generates an important household income during lean periods (Babatoundé et al., 2010). In all Sub-Saharan African area, feed shortages remained the main constraint to ruminant breeding (Adjolohoun et al., 2008; Koura et al., 2015). This is the result of agricultural pressure, urbanization and climate variability on the extensive production system, which now results in poor grazing areas (Koura et al., 2015). According to this author, in some areas of Southern-Benin, the most palatable species such as *Panicum maximum* and *Pennisetum* spp. are disappearing. The cropping strategies of herders can no longer resist the increasing changes. The intensification of the production system through forage cultivation is therefore necessary (Adjolohoun et al., 2008) to ensure animal products availability. However, little information is available on the nutritive value of these forages under cultivated conditions that could facilitate their integration in farms. The *in vitro* gas production technique (IVGPT) has been proposed for several years as a valid method to determine the nutritive value of feedstuffs, since rate and extent of degradation and rumen fermentation can be easily determined by measurement of cumulative gas production (Calabrò et al., 2005a). The IVGPT is to be considered useful especially in developing countries, because the manual system does not require large financial resources and is not a time-consuming method to obtain dynamic descriptions of nutritive value of feedstuffs (Calabrò et al., 2007). Moreover, IVGPT is highly reproducible and allows analyzing many samples

* Corresponding Author: Monica I. Cutrignelli.
E-mail: monica.cutrignelli@unina.it
[1] Department of Animal Sciences, Faculty of Agricultural Sciences, University of Abomey-Calavi, Cotonou 526, Benin.
[2] Department of Biology, University of Napoli Federico II, Napoli 80126, Italy.

simultaneously using small amounts of material. The aim of this investigation was to estimate the nutritive value of some grass and legume forages grown in Abomey-Calavi in Republic of Benin, in order to provide to the smallholder farmers useful recommendations on the best forage species to utilize based upon their nutritional characteristics.

MATERIALS AND METHODS

Study area

The study was carried out on grass grown in the experimental field of the Faculty of Agricultural Sciences of the University of Abomey-Calavi. The University is situated in the Abomey-Calavi common of the Republic of Benin that is located between 6°21' to 6°42' North latitude and 2°13' to 2°25' East longitude, 11 m.a.s.l. The area has a sub-equatorial climate with two rain seasons alternated with two dry seasons of unequal duration. The rainfall amounts recorded by the Agency for Aerial Navigation Safety in Africa and Madagascar (ASECNA) between 1981 and 2012 are between 739.6 mm and 2,203.3 mm with an average of 1,305.95 mm. The soil is sandy and ferrallitic types. The vegetation consists of shrubs, grassland swamps, swamp forest and mangrove forest on the coastal belt and of semi-deciduous dense forests on bar land area.

Species installation method

The experimental area was parceled into two main blocks, grasses and legumes. These two blocks were separated by a main path of 1.5 m and each block being separated by 1.0 m side aisles; each plot had an area of 5×5 m. Eleven grasses (*Andropogon gayanus*, *Andropogon tectorum*, *Brachiaria ruziziensis*, *Cynodon datylon*, *Echinochloa stagnina*, *Hyparrhenia diplandra*, *Panicum maximum* var. C1, *Panicum maximum* var. 673, *Panicum maximum* (local), *Pennisetum purpureum*, *Vetiveria zizanoïdes*) and eleven legumes (*Aeschynomene histrix*, *Cajanus cajan*, *Centrosema pubescens*, *Chamaecrista rotundifolia*, *Gliricedia sepium*, *Leucaena leucocephala*, *Moringa oleifera*, *Mucuna utilis*, *Stylosanthes hamata*, *S. scabra*, *Tephrosia pedicellata*) were planted in June 2011. Grasses were installed by bursts of strains, whereas herbaceous legumes either by seed (*C. cajan*, *C pubescens*, *T. pedicellata*, *A. histrix*, *C. rotundifolia*, *S. hamata*, *S. scabra*, *L. leucocephala*) or by cuttings (*G. sepium*, *M. oleifera*). Some plants (*A. histrix*, *C. rotundifolia*, *S. hamata*, *S. scabra*) were sown after seed dormancy breaking by hot water. After removing seed dormancy, *L. leucocephala* was established in plastic bags and later seedlings were transplanted in the field. Planting was done in a continuous line. Spacings were 40 cm (between plants)×40 cm (between lines) for most species. However, it was 80 cm× 80 cm for *Pennisetum purpureum* and *Echinochloa stagnina*,

50 cm×50 cm for *Mucuna utilis* and 1 m×1 m for *Cajanus cajan*. The sample cuts were made in the upper part of the plants (leaves and stems), during the consolidation phase of forages.

Chemical composition

All the plants were oven-dried for 48 to 72 h at 60°C for dry matter (DM) quantification; the samples were ground to pass a 1 mm screen (Brabender Wiley mill, Brabender OHG, Duisburg, Germany) and analyzed for residual DM, crude protein (CP), ether extract (EE) and ash as suggested by AOAC (2000) procedures (ID number: 2001.12, 978.04, 920.39 and 930.05 for DM, CP, EE, and ash, respectively).

In vitro gas production

The fermentation characteristics and kinetics were studied using the IVGPT, by incubating all the forages at 39°C under anaerobic conditions with buffered rumen fluid (Calabrò et al., 2005b). The test substrates were weighed (1.0005 g±0.0003) in triplicate in 120 mL serum flasks, to which 74 mL of anaerobic medium were added. The rumen fluid was collected in a pre-warmed thermos at a slaughterhouse authorized according to EU legislation (2004) from six male buffalo bulls (*Bubalus bubalis*) fed a standard diet (neutral detergent fiber [NDF] 45.5% DM and CP 12% DM). The collected material was rapidly transported to the laboratory, where it was pooled, flushed with CO_2, filtered through a cheesecloth and added to each flask (10 mL). Three flasks with no substrate were incubated as blanks to correct for the organic matter (OM) disappearance, and gas and end products production.

Gas production of the fermenting cultures was recorded 21 times (at 2 to 24 h intervals) during the period of incubation (120 h) using a manual pressure transducer (Cole and Palmer Instrument Co, Vernon Hills, IL, USA).

The fermentation was stopped at 120 h and the fermentation liquor was analyzed for pH with a pH-meter (model 3030 Alessandrini Instrument glass electrode, Jenway, Dunmow, UK) and sampled for end product analysis. The extent of sample disappearance, expressed as organic matter degradability (dOM, %), was determined by weight difference of the incubated OM and the undegraded filtered (sintered glass crucibles; Schott Duran, Mainz, Germany, porosity # 2) residue burned at 550°C for 5 h. Cumulative volume of gas produced after 120 of incubation was related to incubated OM (OMCV, mL/g) and to degraded OM (Yield, mL/g).

For volatile fatty acids (VFA) determination, fermenting liquors were centrifuged at 12,000 g for 10 min at 4°C (Universal 32R centrifuge, Hettich FurnTech Division DIY, Melle-Neuenkirchen, Germany). One milliliter of supernatant was then mixed with 1 mL of oxalic acid (0.06 mol). Volatile fatty acids were measured by gas chromatography (ThermoQuest 8000top Italia SpA, Rodano, Milan, Italy; fused silica capillary column 30 m, 0.25 mm ID, 0.25 µm film thickness), using external standard solution composed of acetic, propionic, butyric, isobutyric, valeric and isovaleric acids (Zicarelli et al., 2011; Calabrò et al., 2015).

Data processing

The nutritive value of forages was estimated as net energy for lactation (NE$_l$, MJ/kg DM) using the equation proposed by Menke and Steingass (1988):

$$NE_l = 0.54+0.0959GP+0.0038CP+0.0001733CP^2$$

where, GP is the gas obtained *in vitro* (mL/200 mg incubated DM) after 24 h of incubation and CP is the content (g/kg DM) of crude protein. For each flask the gas production profiles were fitted to the sigmoid model described by Groot et al. (1996):

$$G = A/(1+B/t)^C$$

where, G is the total gas produced (mL/g of OM) at time t (h), A is the asymptotic gas production (mL/g of OM), B (h) is the time at which one-half of the asymptote is reached, and C is the switching characteristic of the curve. Maximum fermentation rate (R$_{max}$, mL/h) and the time at which it occurred (T$_{max}$, h) were also calculated according to the following formulas (Bauer et al., 2001):

$$R_{max} = (A \cdot C^B) \cdot B \cdot [T_{max}^{(-B-1)}]/[(1+C^B) \cdot (T_{max}^{-B})^2]$$

$$T_{max} = C \cdot [(B-1)/(B+1)]^{(1/B)}$$

Fermentation characteristics (OMCV, Yield, Dom, and pH) and the model parameters (A, B, C, t$_{max}$, R$_{max}$) were subjected, separately for grass and legume, to analysis of variance (PROC GLM, SAS 2000) according to the model:

$$y_{ij} = \mu+F_i+\varepsilon_{ij}$$

where, y is the single data, µ is the mean, F is the forage effect (i = 11 for grass and i = 11 for legume) and ε the error term. The minimum significant difference (p<0.01 and p<0.05), was used to verify the differences between means using the Tukey test.

The correlation between the chemical parameters and the *in vitro* fermentation data were also studied (PROC CORR, SAS, 2000) separately for grass and legume samples.

RESULTS

Table 1 and 2 show the chemical composition and the nutritive value of grass and legume forages, respectively. As regards grasses, the local variety of *Panicum* and *Pennisetum purpureum* showed a very high CP (25.86% and 27.33% DM) and the lowest structural carbohydrates (NDF: 50.80% and 50.30% DM) content, consequently, energy content (15.14 and 16.10 MJ/kg DM) was very high, though ash level was quite high too (16.42% and 19.68% DM), respectively. On the contrary, *V. zizanoïdes* showed the lowest CP and energy level (6.69% DM and 2.531 MJ/kg DM, respectively) and the highest NDF (79.66% DM) content. The other forages showed a similar chemical composition, on average equal to: CP 14.7%±3.00% DM, NDF 63.4%±4.41% DM and NE$_l$ 6.89±1.89 MJ/kg DM. Observing legume data, *M. oleifera* was the most interesting forage that had the highest CP, EE, and NE$_l$ content (34.45% DM, 6.22% DM and 24.60 MJ/kg DM, respectively) and the lowest NDF (14.03% DM) and acid detergent lignin (ADL) (5.94% DM) contents. Also *G. sepium* and *L. leucocephala* showed high values for CP (22.08% and 28.04% DM, respectively) and energy (12.34

Table 1. Chemical composition and nutritive value of grass forages

Graminae	DM	Ash	CP	EE	NDF	ADF	ADL	NE$_l$
	---- % ----	--- % DM ---						---- MJ/kg DM ----
Andropogon gayanus	71.10	10.06	12.71	1.80	64.67	35.31	8.66	5.376
Andropogon tectorum	79.00	7.52	14.32	1.81	62.62	40.59	8.92	6.464
Brachiaria ruziziensis	71.16	13.79	16.36	3.29	53.80	38.24	15.90	8.165
Cynodon dactylon L.	83.57	10.86	12.63	2.13	68.86	43.02	14.02	5.630
Echinochloa stagnina	53.49	9.50	16.38	2.75	63.73	41.10	9.29	8.124
Hyparrhenia diplandra	78.62	10.99	14.79	1.87	65.90	48.23	19.73	6.441
Panicum maximum var. C1	81.16	14.10	10.09	2.36	65.60	47.83	14.57	4.482
Panicum maximum var. 673	67.56	13.17	19.99	2.71	62.37	40.85	11.75	10.32
Panicum maximum (local)	52.60	16.42	25.86	3.56	50.80	38.85	13.77	15.14
Pennisetum purpureum	57.20	19.68	27.33	2.72	50.30	30.43	9.62	16.10
Vetiveria zizanoïdes	85.31	8.17	6.69	1.41	79.66	52.17	9.64	2.531

DM, dry matter; CP, crude protein; EE, ether extract; NDF, neutral detergent fiber; ADF, acid detergent fiber; ADL, acid detergent lignin; NE$_l$, net energy for lactation calculated as suggested by Menke and Staingass (1988).

Table 2. Chemical composition and nutritive value of legume forages

Leguminae	DM	Ash	CP	EE	NDF	ADF	ADL	NE_l
	----- % -----				----- % DM -----			--- MJ/kg DM ---
Aechynomene histrix	86.15	10.06	18.57	2.34	44.08	34.47	14.52	9.190
Cajanus cajan	83.36	5.86	19.09	5.41	43.81	39.74	23.82	9.478
Centrosema pubescens	90.81	9.10	20.51	2.55	52.70	37.96	16.73	9.474
Chamaecrista rotundifolia	84.83	6.83	19.51	3.88	44.17	36.91	18.49	9.723
Gliricedia sepium	71.60	12.66	22.08	1.88	29.30	23.76	12.08	12.34
Leucaena leucocephala	82.58	7.83	28.04	4.92	25.24	17.66	11.08	16.94
Moringa oleifera	80.01	12.80	34.45	6.22	14.03	14.49	5.94	24.60
Mucuna utilis	78.11	9.01	14.89	2.67	47.27	41.07	18.84	6.769
Stylosanthes hamata	87.41	10.53	12.62	1.88	49.91	42.13	11.74	6.122
Stylosanthes scabra	85.27	7.97	12.24	3.81	49.62	48.18	18.07	5.478
Tephrosia pedicellata	85.98	6.93	16.96	2.44	44.79	40.09	16.04	6.169

DM, dry matter; CP, crude protein; EE, ether extract; NDF, neutral detergent fiber; ADF, acid detergent fiber; ADL, acid detergent lignin; NE_l, net energy for lactation calculated as suggested by Menke and Staingass (1988).

and 16.94 MJ/kg DM, respectively) whereas both *Stylosanthes* spp. showed the lowest CP (12.43% DM) and high structural carbohydrates contents (49.77% DM) and low energy level (5.80 MJ/kg DM). All the other legume forages showed on average the following values: 18.26%±2.02% DM (CP), 46.14%±3.46% DM (NDF) and 8.47±1.57 MJ/kg DM (NE_l).

Table 3 and 4 show the *in vitro* fermentation parameters of grass and legume forages, respectively. Both for grass and legume forages, all parameters were highly (p<0.01) influenced by the forage type. For all forages, data related to gas production (real_OMCV and potential_A) showed similar trend. In grass forages, the average value of OM degradability was 68.41±10.89 and of OMCV 188.45±31.0

mL/g. In particular, in most grass forages (*Panicum* spp., *Pennisetum*, *Brachiaria*, and *Echinochloa*) dOM values exceeded 70%, whereas *Vetiveria* showed the lowest dOM value (40.6%; p<0.01). Regarding gas production, the highest value was found in *Echinochloa* (241 mL/g; p<0.01) and the lowest in *Vetiveria* (123 mL/g; p<0.01). In legume forages, the average values were 56.48%±12.45% and 155.45%±33.6%, for dOM and OMCV, respectively. For OM degradability, *Moringa* showed the highest (79.0%; p<0.01) value, and in all the others the values ranged between 27.9% and 64.6%. Regarding gas production, *Stylosanthes* spp. showed the highest values (195 mL/g; p<0.01). *Cajanus* showed the lowest (p<0.01) dOM (27.9%) and OMCV (70.0 mL/g) values.

Table 3. *In vitro* fermentation characteristics of grasses

Graminae	dOM	OMCV	Yield[1]	A[1]	B[1]	C[1]	t_{max}[1]	R_{max}[1]
	%	mL/g	mL/g	mL/g	h		h	mL/h
Andropogon gayanus	64.8	186	288	204	29.4	1.94	16.4	4.45
Andropogon tectorum	62.3	195	313	224	27.7	1.62	12.2	4.95
Brachiaria ruziziensis	72.4	197	272	204	19.6	2.01	11.3	6.81
Cynodon datylon	68.5	203	296	220	26.8	1.94	14.9	5.25
Echinochloa stagnina	78.5	241	307	244	24.2	2.18	15.3	6.85
Hyparrhenia diplandra	63.2	170	269	184	27.2	1.94	15.1	4.35
Panicum maximum var. C1	77.2	206	268	216	24.1	2.31	17.4	5.41
Panicum maximum var. 673	73.1	213	291	230	28.4	2.11	16.1	6.31
Panicum maximum (local)	77.3	181	235	189	21.9	2.16	13.8	5.81
Pennisetum purpureum	74.6	158	212	168	24.8	2.01	14.4	4.43
Vetiveria zizanoïdes	40.6	123	302	150	40.0	1.43	11.8	2.31
Significance P	***	***	***	***	***	*	***	***
MSD[2]	2.88	21.1	31.6	33.8	8.55	0.77	2.28	0.91
MSD[3]	3.48	25.5	38.1	40.7	10.3	0.92	2.78	1.11
MSE[4]	0.979	52.3	117	134	8.58	0.068	0.494	0.079

dOM, organic matter degradability (% of incubated OM); OMCV, cumulative volume of gas related to incubated OM (mL/g).
[1] Yield = cumulative volume of gas related to degraded OM; A = potential gas production (mL/g); B = time at which A/2 was formed (h); C = constant determining the curve sharpness; t_{max} = time at which maximum rate was reached (h); R_{max} = maximum fermentation rate (mL/h).
[2] Minimun significant differences for p<0.05. [3] Minimun significant differences for p<0.01. [4] MSE, mean square error. *, ***: p<0.05, p<0.001, respectively.

Table 4. *In vitro* fermentation characteristics of legumes

Leguminae	dOM %	OMCV mL/g	Yield[1] mL/g	A[1] mL/g	B[1] h	C[1]	t_{max}[1] h	R_{max}[1] mL/h
Aechynomene histrix	58.9	154	262	172	19.9	1.38	5.21	5.44
Cajanus cajan	27.9	70	250	101	32.7	0.77	-	-
Centrosema pubescens	47.8	136	285	154	18.7	1.26	3.31	5.37
Chamaecrista rotundifolia	52.4	142	271	154	16.3	1.42	4.70	5.85
Gliricedia sepium	64.6	161	250	177	18.9	1.52	6.69	5.73
Leucaena leucocephala	62.6	165	263	180	21.6	1.63	6.94	5.11
Moringa oleifera	79.0	160	203	166	12.3	1.59	5.28	7.84
Mucuna utilis	55.8	170	306	187	18.5	1.50	6.34	6.17
Stylosanthes hamata	60.9	195	320	205	16.7	1.70	7.55	7.58
Stylosanthes scabra	59.3	193	326	208	18.7	1.64	7.87	6.81
Tephrosia pedicellata	52.1	164	315	184	22.4	1.46	6.92	5.05
Significance P	***	***	***	***	***	***	***	***
MSD[2]	3.55	21.2	42.63	23.75	7.71	0.21	2.22	1.36
MSD[3]	4.27	25.6	51.39	28.63	9.29	0.25	2.70	1.65
MSE[4]	1.475	52.7	213.3	66.19	6.97	0.005	0.554	0.207

dOM, organic matter degradability (% of incubated OM); OMCV, cumulative volume of gas related to incubated OM (mL/g); -, not possible to calculate.
[1] Yield = cumulative volume of gas related to degraded OM; A = potential gas production (mL/g); B = time at which A/2 was formed (h); C = constant determining the curve sharpness; t_{max} = time at which maximum rate was reached (h); R_{max} = maximum fermentation rate (mL/h).
[2] Minimun significant differences for $p<0.05$; [3] Minimun significant differences for $p<0.01$. [4] MSE, mean square error. *** $p<0.001$.

Table 5 and 6 picture pH and VFA produced after 120 hours of incubation for grass and legume forages, respectively. For grass, pH values range between 6.61 and 6.85, whereas for legumes between 6.72 and 6.85. All the parameters reported in both tables, were highly different ($p<0.001$) between substrates. Grass forage had a higher production of total VFA compared to legume (73.3 vs 59.3 mM/g), however in both families the proportion among the most representative acids are similar (65%, 20%, and 7% for acetic, propionic and butyric, respectively). Regarding grass, the highest value of tVFA was found in *Panicum maximum* var. 673 (83.91 mM/g) and the lowest in *Vetiveria zizanoïdes* (53.90 mM/g), although the differences were not statistically significant. In all the cases the acetic acid is the major VFA responsible for this result. Regarding the branched chain fatty acids (BCFA: isobutyrate, isovalerate, valerate), *Brachiaria*, *Panicum maximum* local and var. 673, and *Pennisetum* showed the highest proportion on tVFA (6.53, 6.81, 6.30, and 7.28, respectively).

The legume *Stylosanthes scabra* had the highest tVFA

Table 5. Volatile fatty acids and pH at 120 h for grass

Graminae	pH	Acetate	Propionate	Isobutyrate	Butyrate	Isovalerate	Valerate	tVFA
					mM/g iOM			
Andropogon gayanus	6.67	46.55	14.41	0.77	5.69	1.45	0.78	69.64
Andropogon tectorum	6.69	48.97	14.62	0.88	5.38	1.66	0.84	72.35
Brachiaria ruziziensis	6.71	45.91	15.07	1.14	5.18	2.10	1.39	70.79
Cynodon datylon	6.63	54.54	17.17	0.93	6.12	1.62	1.14	81.52
Echinochloa stagnina	6.65	54.28	18.47	0.99	6.10	1.88	1.16	82.87
Hyparrhenia diplandra	6.69	42.58	13.86	0.79	5.18	1.46	0.78	64.65
Panicum maximum var. C1	6.75	52.91	18.02	0.84	5.91	1.56	1.00	80.25
Panicum maximum var. 673	6.61	54.54	18.26	1.34	5.82	2.48	1.46	83.91
Panicum maximum (local)	6.83	51.84	16.68	1.38	5.53	2.47	1.56	79.46
Pennisetum purpureum	6.85	44.20	13.24	1.31	4.42	2.36	1.19	66.72
Vetiveria zizanoïdes	6.67	35.35	12.28	0.60	4.05	1.12	0.50	53.90
Significance P	***	***	***	***	***	***	***	***
MSD[1]	0.04	8.03	2.66	0.30	1.08	0.49	0.19	11.6
MSD[2]	0.048	9.68	3.20	0.36	1.30	0.59	0.23	14.0
MSE[3]	0.0002	7.57	0.83	0.011	0.14	0.03	0.04	1.58

tVFA, total volatile fatty acids.
[1] Minimun significant differences for $p<0.05$. [2] Minimun significant differences for $p<0.01$. [3] MSE, mean square error. *** $p<0.001$, respectively.

Table 6. Volatile fatty acids and-pH at 120 h for legume

Leguminae	pH	Acetate	Propionate	Isobutyrate	Butyrate	Isovalerate	Valerate	tVFA
		--- mM/g iOM ---						
Aechynomene histrix	6.80	37.14	11.59	0.93	5.14	1.78	1.65	58.23
Cajanus cajan	6.83	24.21	8.32	0.19	1.94	0.31	0.27	35.24
Centrosema pubescens	6.82	39.88	12.10	0.92	4.89	1.85	1.24	60.87
Chamaecrista rotundifolia	6.78	30.06	9.99	0.81	4.33	1.55	1.37	48.10
Gliricedia sepium	6.85	38.83	12.94	1.23	4.95	2.78	1.39	62.13
Leucaena leucocephala	6.75	43.81	14.02	1.01	4.81	1.96	1.27	66.88
Moringa oleifera	6.85	44.99	13.74	1.64	5.61	3.08	1.82	70.89
Mucuna utilis	6.73	34.18	9.71	0.62	4.74	1.23	0.98	51.47
Stylosanthes hamata	6.73	43.50	13.27	0.90	5.23	1.77	1.24	65.91
Stylosanthes scabra	6.72	47.74	15.35	0.91	5.54	1.79	1.18	72.51
Tephrosia pedicellata	6.78	39.16	12.98	0.78	4.11	1.59	1.28	59.90
Significance	***	***	***	***	***	***	***	***
MSD[1]	0.05	10.6	3.39	0.20	1.80	0.41	0.31	15.9
MSD[2]	0.06	12.7	4.08	0.25	2.17	0.49	0.38	19.1
MSE[3]	0.0003	13.2	1.35	0.005	0.38	0.019	0.012	29.5

tVFA, total volatile fatty acids.
[1] Minimun significant differences for p<0.05. [2] Minimun significant differences for p<0.01. [3] MSE, mean square error. *** p<0.001.

value (72.5 mM/g) and *Cajanus cajan* the lowest (35.2 mM/g), although the differences were not statistically significant. Except for *Cajanus*, where the mean value of BCFA was 6.89.

Regarding fermentation kinetics, the parameters presented in Table 3 and 4 (B, C, t_{max}, R_{max}), can be better explained by observing the Figure 1 and 2 (Panel A and B). In particular, for grass forages it is clear that the

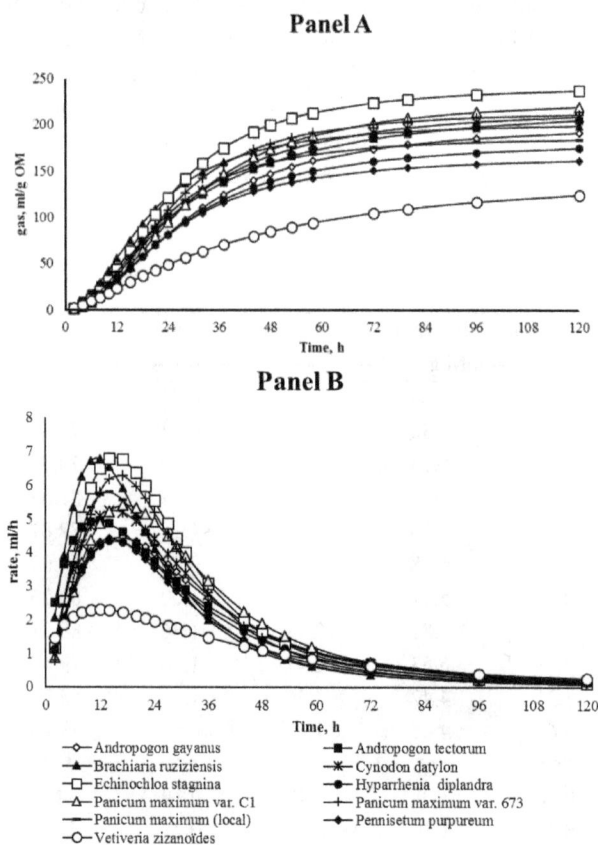

Figure 1. *In vitro* cumulative gas production (Panel A) and fermentation rate (Panel B) over time for grass.

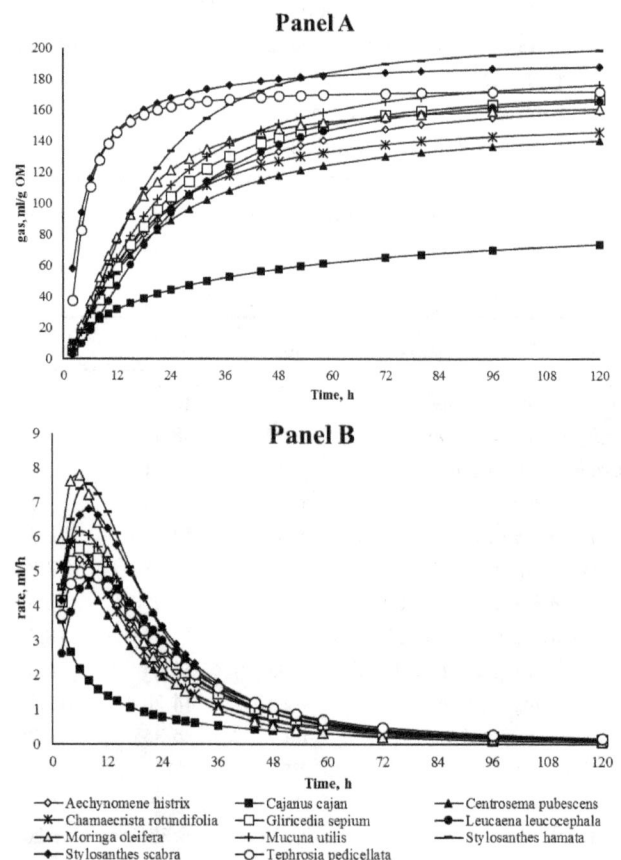

Figure 2. *In vitro* cumulative gas production (Panel A) and fermentation rate (Panel B) over time for legume.

fermentation process was similar for most of the samples (starts within 6 h of incubation, reaches the maximum around 14 h and finishes gradually at 120 h), except for *Vetiveria* that showed a very slow process characterized by a fermentation rate profile quite flat (R_{max}: 2.31 mL/h and B: 40.0 h; p<0.01). With regard to legume forages, the fermentation profiles were more diversified: *Cajanus* showed the slowest fermentation process: the gas production curve quite flat and the fermentation rate curve similar to a hyperbole. As an opposite trend, the fermentation for *Tephrosia pedicellata* and *Stylosanthes scabra* started quickly and finished quickly (Figure 2, Panel A). *Stylosanthes hamata* and *Moringa oleifera* showed a similar shape (Figure 2, Panel B) characterized by very high maximum fermentation rate (7.58 and 7.84 mL/h, respectively). In terms of t_{max}, *Cetrosema pubescens* showed the lowest value, whereas *Stilosanthes* spp. the highest (3.31 vs 7.71 h, respectively; p<0.01). The remaining substrates (*Aeschynomene*, *Chamaecrista*, *Gliricedia*, *Mucuna*, *Thephrosia*, and *Leuchena*) showed a similar profile in terms of time (t_{max} mean value: 6.01 h) and rate

(R_{max} mean value: 5.88 mL/h). Table 7 pictures the significance of correlation between some chemical parameters and *in vitro* fermentation data for grass and legume forages.

DISCUSSION

The chemical composition and *in vitro* degradability of forages appears quite diversified. Indeed, although the values obtained fall, in part, within those found in the literature (Nasrullah et al., 2003; Calabrò et al., 2007; Adjolohoun et al., 2008; Babatoundé et al., 2011), it is important to mention that not only the variety, but also climatic conditions, sampling site, soil and management conditions and vegetative stage at harvest significantly affect the plant nutrient accumulation (Adjolohoun et al., 2013).

Grass vs legume forages

As reported by Calabrò et al. (2007), compared to legumes, tropical grass forages showed a lower energy level

Table 7. Significance of correlation between some chemical parameters and *in vitro* fermentation data for grass and legume forage

	dOM %	Yield[1] mL/g	A[1] mL/g	B[1] h	t_{max}[1] h	R_{max}[1] mL/h	tVFA Mm/g
Grass forage							
CP	0.67	−0.85	−0.12	−0.69	0.01	0.39	0.29
	*	**	NS	*	NS	NS	NS
EE	0.79	−0.66	0.28	−0.84	−0.04	0.82	0.60
	**	*	NS	**	NS	**	NS
NDF	−0.76	0.65	−0.32	0.90	0.05	−0.76	−0.51
	**	*	NS	**	NS	**	NS
ADL	0.43	−0.56	0.06	−0.56	0.21	0.35	0.21
	NS	NS	NS	NS	NS	NS	NS
Ash	0.59	−0.92	−0.25	−0.53	0.18	0.25	0.18
	*	**	NS	NS	NS	NS	NS
NE$_l$	0.64	−0.87	−0.17	−0.64	−0.01	0.35	0.26
	*	**	NS	*	NS	NS	NS
Legume forage							
CP	0.46	−0.87	−0.30	−0.23	−0.20	0.15	−0.29
	NS	**	NS	NS	NS	NS	NS
EE	0.03	−0.60	−0.47	0.14	0.001	0.40	−0.29
	NS	*	NS	NS	NS	NS	NS
NDF	−0.62	0.79	0.07	0.26	−0.12	−0.27	0.14
	*	**	NS	NS	NS	NS	NS
ADL	−0.90	0.40	−0.46	0.68	−0.14	−0.48	0.065
	**	NS	NS	*	NS	NS	NS
Ash	0.77	−0.40	0.36	−0.65	−0.15	0.53	−0.20
	**	NS	NS	*	NS	NS	NS
NE$_l$	−0.59	−0.82	−0.16	−0.36	−0.14	0.32	−0.30
	*	**	NS	NS	NS	NS	NS

tVFA, total volatile fatty acids; dOM, organic matter degradability; CP, crude protein; NS, not significant; EE, ether extract; NDF, neutral detergent fiber; ADL, acid detergent lignin; NE$_l$, net energy for lactation.
[1] Yield = cumulative volume of gas related to degraded OM; A = potential gas production; B = time at which A/2 was formed; t_{max} = time at which maximum rate was reached; R_{max} = maximum fermentation rate.
*, ** p<0.05, p<0.01, respectively.

(8.07 vs 10.57 MJ/kg DM), crude protein (16.10 vs 19.91% DM) and a higher cell wall content (NDF: 63.8 vs 40.45% DM), respectively. However, in general, as reported by Calabrò et al. (2007) on Niger forage, the OM degradability is quite low in legume forage compared to grass (mean values: 56.5% vs 68.4%, respectively), probably due to the high lignin content (mean values: 15.21% vs 10.77% DM, respectively) especially in some samples (*Cajanus, Chamaecrista, Mucuna, Stylosanthes scabra*), associated with the anti-nutritional factors (ANFs) (Adjolohoun et al., 2008). Because of the higher content of carbohydrates in grass than in legume, the total VFA produced is more elevated in the first. The proportion of branched chain fatty acids is higher in legume compared to grass and reflects the protein level in the incubated samples, because these acids derive from the degradation of some amino acids (valine, proline, isoleucine, leucine) (Calabrò et al., 2012). Regarding the kinetics, the greatest part of the fermentation process occurred within the first 60 h for grass and 48 h for legume, to complete slowly in 96 h. In particular, at similar fermentation rates (5.18 and 6.10 mL/h) grass showed a slower fermentation process (t_{max}: 14.43 vs 6.08 h), for grass and legume respectively, probably due to the higher fermentable carbohydrates content. These results are influenced by the different nature of the substrates used; in particular, all the *Graminae* were annual or perennial herbaceous species, whereas *Leguminosae* are shrub, tree or herb.

Grass forage

In the grass forages, the chemical composition is important; in particular, crude protein content was much higher than the critical level of 60 to 80 g/kg DM, below which forage intake is depressed. The *in vitro* degradability is slightly lower than the range of tropical pasture (50% to 80%) reported by Adjolohoun et al. (2008). There are many studies reporting data on different varieties of *Brachiaria, Panicum, Pennisetum* and *Vetiveria*; in particular, Nasrullah et al. (2003) studied the *in vitro* dry matter digestibility (IVDMD) using the Goering and Van Soest (1970) method for these forages grown as natural pasture, never fertilized, without management intervention and collected in different seasons. The values of chemical composition (CP, EE, ADL) and IVDMD were lower compared to our data, indicating that the water availability and fertilization can improve their nutritive value. On average, the nutritive value of *B. ruziziensis* is higher compared to the other selected grasses, but the risks of photosensitization and goiter must be emphasized when *B. ruziziensis* is fed alone to ruminants (Hare et al., 1997). *P. maximum* (local) shows typical legume characteristics with high protein content (25.9% DM), on the other hand *P. maximum* var. C1, despite the low crude protein content (10.1% DM), appears more

interesting than the two other *Panicum* spp., due to its high productivity among the perennial grass forages in Southeast of Benin (Adjolohoun et al., 2013). *P. purpureum*, one of the most productive grass crops in the world (10 to 30 t/ha DM per year), in our trial appears as the best forage in terms of nutritive value (NE_l: 16.1 MJ/kg DM) and crude protein content (27.3% DM), even if it is reported that DM yield rapidly declines if fertility is not maintained (Adjolohoun et al., 2008). The fermentation characteristics of *V. zizanoïdes* are in agreement with the chemical composition, and point out a very slow kinetics and a low OM degradability and VFA production, whose value is similar to that one reported by Nasrullah et al. (2003).

Regarding *Andropogon* spp., *Cynodon* and *Hyparrhenia* few studies were found; in our trial they showed similar chemical characteristics (except for the high lignin content of *Cynodon* and *Hyparrhenia*) and the *in vitro* data indicates a low nutritive value. *Echinochloa stagnina*, is a perennial semi-aquatic tropical grass; during dry season, it is of outmost importance for maintenance of cattle (Dicko et al., 2003), moreover, it is very productive and highly palatable. In our study, it showed interesting *in vitro* data in terms of OM degradability, gas and VFA production, and fermentation rate, probably due to the high sugar content.

Legume forage

As a whole, the woody legumes (*Cajanus, Gliricedia, Leucaena, Moringa*) seem to be richer in protein and energy and lower in structural carbohydrates than herbaceous plants (*Aeschynomene, Centrosema, Chamaecrista, Stylosanthes*); these characteristics do not always agree with their *in vitro* data, probably because of the presence of other complex compounds (i.e. tannins). Except for *Moringa oleifera*, our samples presented lower crude protein and OM degradability values compared to the values reported by Babatoundé et al. (2011) who studied the chemical composition and the *in vitro* OM degradability, using pepsin and cellulose technique, in legume forages hand-plucked in South Benin during the short dry season. The legume forages showed diversified fermentation kinetics profiles.

M. oleifera, which sample was mainly represented by leaves, was the forage with the highest energy content, due to the high protein and low fiber level; similar data were reported by Babatoundé et al. (2011). It also showed the highest OM degradability, VFA production and fermentation rate, associated with a moderate gas production; this result may in part be explained by the high protein and lipid content (Melesse et al., 2013). In general, low gas production would indicate low degradability in the rumen, but feedstuffs with high protein produce less gas during fermentation. Moreover their extent of degradation is high, because protein fermentation produces ammonia, which influences the carbonate buffer equilibrium neutralizing H+ ions from VFA without release of carbon dioxide. The low gas volume for *M. oleifera* can be also explained by a high fat content, which contributes to limit fermentation Akinfemi et al. (2009).

On the contrary, the two varieties of *Stylosanthes* spp. showed the lowest energy content: they were similar in terms of chemical composition and fermentation parameters, even with the higher lignin level and consequent lower NE$_l$ value in *S. scabra*. The high total VFA level was due to the high acetic production, as consequence of the high NDF content.

The low OM degradability, VFA production and very slow fermentation kinetics of *C. cajan* is due to the highest lignin content registered among all the legume samples; this component, bound to tannins or phenols, is found by many authors (Norton, 1994; Adjolohoun et al., 2008; Babatoundé et al., 2011).

G. sepium and *L. leucocephala* showed similar fermentation kinetics, with intermediate values in terms of OM degradability and gas and VFA production, associated with a slow process. This result might be due to the presence of high level of tannins in both plants (Shayo and Udén, 1999). Several authors (Norton, 1994; Babatoundé et al., 2011) reported high phenolic and tannin levels in some African woody legumes. *In vivo* studies evidence that the presence of these ANFs affects the forage nutritive value reducing intake and digestibility (Dzowela and Hove, 1995). In browse species with considerable protein content ANFs interfere with the protein utilization. A low rumen degradation of protein, decreases rumen ammonia concentration; a minimum concentration of ammonia (70 mg N/L) is required from the microbial population in the rumen, lower levels are associated with lower microbial activity and, consequently, lower digestion (Norton, 1994). However, the significance of ANFs becomes more evident when woody foliage is the only feed consumed. The utilization of these species in feeding strategy may be limited, but the interpretation of the protein nutritional value in these species requires information on the nature and actions of tannins, also consideration that they increase aminoacids amount bypassing rumen (Babatoundé et al., 2011).

The other three herbaceous plants (*Aeschynomene, Centrosema, Chamaecrista*) showed similar chemical composition (i.e. crude protein, structural carbohydrates and energy content) as well as *in vitro* parameters (i.e. OM degradability, gas production and fermentation rate), but a different VFA profile; in particular, in *Centrosema* these values were quite low. Among the less-known plants, there are *Mucuna utilis* and *Tephrosia pedicellata*; both are more investigated as seeds for ruminants feed, but can also be used as forage (i.e. pasture and hay). Both forages presented quite low nutritive value in terms of protein, lignin, energy,

and the fermentation characteristics appear intermediate compared to the other studied legumes.

In vitro gas production

The *in vitro* gas production method appears to be a suitable technique for the evaluation of the nutritive value of forages in developing countries (Calabrò et al., 2007; Babatoundé et al., 2011) where financial resources are limited. According to Calabrò et al. (2007), this method gives an assessment of the degradability of both soluble and insoluble fractions of forages and useful information about the fermentation kinetics, and final products (i.e. VFA). In this study, pH value at the end of the incubation, both for grass and legume forages, was adequate for cellulolytic activity, indicating the efficiency of the buffer in the *in vitro* system. In general, the final gas production recorded (OMCV) was similar to the potential value estimated by the adjusted model (A) indicating that the incubation time (120 h) is adequate to complete the fermentation process of these kind of substrates. The ratio of substrate truly degraded to gas volume (partitioning factor, PF) in our trial ranges for all substrates from 3.07 to 4.94 mg/mL, and falls within the range for conventional feed roughages (PF: 2.74 to 4.65 mg/mL) reported by Getachew et al. (1998). The escape of tannins from the feed during fermentation, contributes to the DM loss but does not influence the gas produced; so our result indicates that the potential tannin content of the tested forages did not affect the *in vitro* OM fermentation. The VFA produced during the incubation gives information about the energy released by carbohydrates during fermentation and directly available to the animal. All samples showed moderate levels of total VFA and the proportion among the singular acids, reflects that one produced in the rumen with mixed forage/concentrate ratio. As expected, both in grass and legume, tVFA data reflect the trend of gas production (OMCV) and OM degradability (coefficient of correlation 0.80 and 0.89, p<0.01, respectively) indicating that all the degraded OM fermenting gives energy.

According to Bulgden et al. (2001), the *in vitro* fermentation characteristics follow the trend of their chemical composition. The analysis of correlations between some IVGPT parameters and some chemical data confirmed the influence of certain features on the fermentation process. In particular, for grass forages OM degradability and gas production (Yield) are significantly (p<0.05) correlated with some chemical characteristics (protein, EE, energy and ash); as expected, the structural carbohydrates negatively influenced OM degradability and fermentation rate (correlation coefficient: −0.76 p<0.01) (Calabrò et al., 2007). In legume forages, less significant correlations were found: in particular, kinetic parameters are not affected by chemical data, whereas *in vitro* some parameters where negatively correlated: OM degradability with cell wall (NDF: -0.62, $p<0.05$; ADL: -0.90, $p<0.01$), tVFA with ADL (-0.76; $p<0.01$) and gas production (Yield) resulted correlated with crude protein (-0.87; $p<0.01$), EE (-0.60; $p<0.05$) and ash -0.40; $p<0.05$). These nutrients can negatively interfere with the microbial activity, as reported by Calabrò et al. (2007).

CONCLUSION

Our evaluation of cultivable grasses and legumes in Benin revealed remarkable characteristics for their chemical composition, nutritive value and Dom. Several authors worked on these species, but mainly in natural than artificial pasture. Some data obtained in this investigation confirm that already present in the literature (*Brachiaria, Panicum, Pennisetum, Cajanus, Gliricedia, Leucaena, Moringa*); other results are extremely interesting because they are related to less studied plants (*Andropogon, Cynodon, Echinocloa, Hyparrhenia, Vetiveria, Aechynomene, Mucuna, Stylosanthes, Tephrosia*).

The *in vitro* method utilized was helpful to obtain the nutritive value and describe the fermentation kinetics of most common cultivated legume and grass forages from Benin. However, the complexity of some constituents present in tropical plants (i.e. lignin and secondary compounds), influence the *in vitro* fermentation. The information reported could help to better supplement the animal diet based on poor quality forage and assist the small ruminant breeds in increasing their productivity using cultivated forage in South Benin and in other regions of West Africa.

ACKNOWLEDGMENTS

The authors want to thank Dr. Maria Ferrara for her technical support in chromatographic analysis at the Department of Veterinary Medicine and Animal Production, University of Napoli Federico II, Italy. The study was supported by "Benin Buffalo Breeding Project" (BBB-Project) 2014, financed by University Federico II, Napoli (Italy); Responsible Professor Bianca Gasparrini.

REFERENCES

Adjolohoun, S., A. Buldgen, C. Adandedjan, V. Decruyenaere, and P. Dardenne. 2008. Yield and nutritive value of herbaceous and browse forage legumes in the Borgou region of Benin. Trop. Grasslands 42:104-111.

Adjolohoun S., M. Dahouda, C. Adandedjan, S. Toleba, V. Kindomihou, and B. Sinsin. 2013. Evaluation of biomass production and nutritive value of nine *Panicum maximum* ecotypes in Central region of Benin. Afr. J. Agric. Res. 8:1661-1668.

Agence pour la sécurité de la navigation aérienne en Afrique et à Madagàscar (ASECNA). 2013. Rapport annuel. Dakar, Sénégal.

Akinfemi, A., A. O. Adesanya, and V. E. Aya. 2009. Use of an *in vitro* gas production technique to evaluate some Nigerian feedstuff. American-Eurasian J. Sci. Res. 4:240-245.

AOAC. 2000. Official Methods of Analysis. 2 vol., 18th edn. Association of Official Analytical Chemists, Arlington, VA, USA.

Babatoundé, S., M. Oumorou, V. Tchabi, T. Lecomte, M. Houinato, and C. Adandedjan. 2010. Voluntary feed intake in of dietary preferences Djallonké sheep fed on tropical forage grasses and legumes grown in Benin. Int. J. Biol. Chem. Sci. 4:1030-1043.

Babatoundé, S., M. Oumorou, I. Alkoiret, S. Vidjannagni, and G. A. Mensah. 2011. Relative frequencies, chemical composition and *in vitro* organic matter digestibility of forage consumed by sheep in humid tropic of West Africa. J. Agric. Sci. Technol. A. 1:39-47.

Bauer, E., B. A. Williams, C. Voigt, R. Mosenthin, and M. W. A. Verstegen. 2001. Microbial activities of faeces from unweaned and adult pigs, in relation to selected fermentable carbohydrates. Anim. Sci. 73:313-322.

Buldgen A., B. Michiels, S. Adjolohoun, S. Babatounde, and C. C. Adandedjan. 2001. Research note: Production and nutritive value of grasses cultivated in the coastal area of Benin. Trop. Grass. 35:43-47.

Calabrò, S., M. I. Cutrignelli, G. Piccolo, F. Bovera, F. Zicarelli, M. P. Gazaneo, and F. Infascelli. 2005a. *In vitro* fermentation kinetics of fresh and dried silages. Anim. Feed. Sci. Technol. 123-124:129-137.

Calabrò, S., M. I. Cutrignelli, F. Bovera, G. Piccolo, and F. Infascelli. 2005b. *In vitro* fermentation kinetics of carbohydrate fractions of fresh forage, silage and hay of Avena sativa. J. Sci. Food Agric. 85:1838-1844.

Calabrò, S., S. D'Urso, M. Banoin, V. Piccolo, and F. Infascelli. 2007. Nutritional characteristics of forages from Niger. Ital. J. Anim. Sci. 6 (suppl. 1):272-274.

Calabrò, S., A.C. Carciofi, N. Musco, R. Tudisco, M. O. S. Gomes, and M. I. Cutrignelli. 2012. Fermentation characteristics of several carbohydrate sources for dog diets using the *in vitro* gas production technique. Ital. J. Anim. Sci. 12:e4.

Calabrò, S., M. I. Cutrignelli, V. Lo Presti, R. Tudisco, V. Chiofalo, M. Grossi, F. Infascelli, and B. Chiofalo. 2015. Characterization and effect of year of harvest on the nutritional properties of three varieties of white lupine (*Lupinus albus* L.). J. Sci. Food Agric. http://dx.doi.org/10.1002/jsfa.7049.

Commission Regulation (EC). 2004. No 882/2004 of the European Parliament and Council on "The official controls performed to ensure the verification of compliance with feed and food law, animal health and animal welfare rules". OJL 165, 1-141, 30.04.04.

Dicko, M. B., B. Diarra, S. Samassekou, and A. Ballo. 2003. Inventaire et caractérisation des zones humides au Mali. UICN (World conservation union) and GEPIS/SAWEG Groupe d'experts sur les plaines d'inondation sahéliennes (Sahelian wethlands experts group), Comité local du Mali.

Dzowela, B. H. and L. Hove. 1995. Effect of drying method on chemical composition and *in vitro* digestibility of multi-purpose tree and shrub fodders. Trop. Grasslands 29:263-269.

FAO. 2006. Livestock's long shadow: Environmental issues and option. Roma, Italy.

Getachew G., M. Blümmel, H. P. S. Makkar, and K. Becker. 1998. *In vitro* gas measuring techniques for assessment of nutritional quality of feeds: a review. Anim. Feed Sci. Technol. 72:261-281.

Groot, J. C. J., J. W. Cone, B. A. William, F. M. A. Debersaque. And E. A. Lantiga. 1996. Multiphasic analysis of gas production kinetics for *in vitro* fermentation of ruminant feedstuff. Anim. Feed Sci. Technol. 64:77-89.

Hare, M. D. and C. Phaikew. 1997. Forage seed production in northeast Thailand. In: Forage Seed Production. 2: Tropical and Subtropical Species (Eds. D. S. Loch and J. E. Ferguson). CABI Publishing, Wallingford, UK. 435-443.

INSAE. 2012. Statistiques/Statistiques économiques/Produit intérieur brut. Consulté le 13 Janvier 2014 sur http://www.insae-bj.org/produit-interieur.html.

Koura, B. I., L. H. Dossa, B. Kassa, and M. Houinato. 2015. Adaptation of periurban cattle production systems to environmental changes: Feeding strategies of herdsmen in Southern Benin. Agroecol. Sust. Food. 39:83-98.

Melesse, A., H. Steingass, J. Boguhn, and M. Rodehutscord. 2013. *In vitro* fermentation characteristics and effective utilisable crude protein in leaves and green pods of *Moringa stenopetala* and *Moringa oleifera* cultivated at low and mid-altitudes. J. Anim. Physiol. Anim. Nutr. 97:537-546.

Menke, K. H. and H. Steingass. 1988. Estimation of the energetic feed value obtained from chemical analysis and *in vitro* gas production using rumen fluid. Anim. Res. Dev. 28:7-55.

Nasrullah, M. Niimi, R. Akashi, and O. Kawamura. 2003. Nutritive evaluation of forage plants grown in South Sulawesi, Indonesia. Asian Australas. J. Anim. Sci. 16:693-701.

Norton, B.W. 1994. The nutritive value of tree legumes. In: Forage Tree Lgumes in Tropical Agriculture (Ed. R. C. Gutteridge). CAB International, Wallingford, UK. pp. 177-191.

SAS/STAT. 2000. User's Guide SAS Institute Inc. Version 8.2, Vol. 2. (4th ed.). Cary, NC, USA.

Shayo, C. M. and P. Udén. 1999. Nutritional uniformity of crude protein fractions in some tropical browse plants estimated by two *in vitro* methods. Anim. Feed Sci. Technol. 78:141-151.

Van Soest, P. J., J. B. Robertson, and B. A. Lewis. 1991. Methods for dietary fiber, neutral detergent fiber, and nonstarch polysaccharides in relation to animal nutrition. J. Dairy Sci. 74:3583-3597.

Zicarelli, F., S. Calabrò, M. I. Cutrignelli, F. Infascelli, R. Tudisco, F. Bovera, and V. Piccolo. 2011. *In vitro* fermentation characteristics of diets with different forage/concentrate ratios: comparison of rumen and faecal inocula. J. Sci. Food Agric. 91:1213-1221.

2

Effects of *Candida norvegensis* Live Cells on *In vitro* Oat Straw Rumen Fermentation

Oscar Ruiz, Yamicela Castillo[1,]*, Claudio Arzola, Eduviges Burrola, Jaime Salinas[2],
Agustín Corral, Michael E. Hume[3], Manuel Murillo[4], and Mateo Itza[1]

College of Animal Science and Ecology, Autonomous University of Chihuahua, Chihuahua, Chih. 31000, Mexico

ABSTRACT: This study evaluated the effect of *Candida norvegensis* (*C. norvegensis*) viable yeast culture on *in vitro* ruminal fermentation of oat straw. Ruminal fluid was mixed with buffer solution (1:2) and anaerobically incubated with or without yeast at 39°C for 0, 4, 8, 16, and 24 h. A fully randomized design was used. There was a decrease in lactic acid (quadratic, $p = 0.01$), pH, (quadratic, $p = 0.02$), and yeasts counts (linear, $p<0.01$) across fermentation times. However, *in vitro* dry matter disappearance (IVDMD) and ammonia-N increased across fermentation times (quadratic; $p<0.01$ and $p<0.02$, respectively). Addition of yeast cells caused a decrease in pH values compared over all fermentation times ($p<0.01$), and lactic acid decreased at 12 h ($p = 0.05$). Meanwhile, yeast counts increased ($p = 0.01$) at 12 h. *C. norvegensis* increased ammonia-N at 4, 8, 12, and 24 h ($p<0.01$), and IVDMD of oat straw increased at 8, 12, and 24 h ($p<0.01$) of fermentation. Yeast cells increased acetate ($p<0.01$), propionate ($p<0.03$), and butyrate ($p<0.03$) at 8 h, while valeriate and isovaleriate increased at 8, 12, and 24 h ($p<0.01$). The yeast did not affect cellulolytic bacteria ($p = 0.05$), but cellulolytic fungi increased at 4 and 8 h ($p<0.01$), whereas production of methane decreased ($p<0.01$) at 8 h. It is concluded that addition of *C. norvegensis* to *in vitro* oat straw fermentation increased ruminal fermentation parameters as well as microbial growth with reduction of methane production. Additionally, yeast inoculum also improved IVDMD. (**Key Words:** Rumen, Fermentation, Yeast, Oat Straw, Methane)

INTRODUCTION

Agricultural by-products, such as cereal straw from oats, wheat, and corn, constitute a great potential source of ruminant feed energy. Straws have low nutritional value, because of their low nitrogen and high indigestible fiber content. In recent years, yeast-based additives, primarily *Saccharomyces cerevisiae* (*S. cerevisiae*), have been used to increase rumen feed utilization efficiency (Williams et al., 1991; Miller-Webster et al., 2002; Lila et al., 2004; Doležal et al., 2011; Chaucheyras-Durand et al., 2012). The beneficial effects associated with *S. cerevisiae* in animal studies include a greater dry matter (DM) and neutral detergent fiber digestibility. (Plata et al., 1994), as well as a higher feed utilization and milk production (Moallem et al., 2009). *In vitro* studies have also shown that yeast cultures favourably alter microbial fermentation (Marrero et al., 2013; Ye et al., 2014) and stimulate DM and cellulose digestion (Miller-Webster et al., 2002; Lila et al., 2004; Tang et al., 2008). In the same way, Marrero et al. (2015) showed that inclusion of two strains of yeast (Levazot 15 and Levica 25) in the *in vitro* fermentation of oat straw the accumulated gas production had a twofold increase as a result of yeast effect compared to control. Similar findings were reported by Marrero et al. (2014) when a yeast culture

* Corresponding Author: Yamicela Castillo. Tel: +52-636-110-9227, E-mail: ycastillo75@yahoo.com
[1] Department of Veterinary Medicine, Multidisciplinary Division, Autonomous University of Juarez City, Nuevo Casas Grandes, Chih. 31803, México.
[2] College of Veterinary Medicine and Animal Science, Autonomous University of Tamaulipas, Cd. Victoria, Tamps. 87000, México.
[3] Agricultural Research Service, Southern Plains Research Center, Food and Feed Safety Research Unit, United States Department of Agriculture, College Station, TX 77843, USA.
[4] College of Veterinary Medicine and Animal Science, Juarez University of Durango State, Durango, Dgo. 34000, Mexico.

(Levica 27) was included in the fermentation of corn stover. Aldo et al. (2006) also reported significant increases in the *in vitro* forage degradabilty when rice bran was treated with *Candida utilis* attributing this to the stimulation of rumen microbes by yeast. However, studies examining yeast cultures other than *S. cerevisiae* as rumen feed additives are scarce (Shin et al., 2002; Oeztuerk et al., 2005; Ando et al., 2006). In a previous study, Castillo (2009) isolated and identified the yeast *Candida norvegensis* and demonstrated the favourable effect of a non-*Saccaromyces* yeast addition on some fermentative parameters, such as gas production. The aim of the current experiment was to investigate the inoculation of a yeast (*Candida norvegensis*) and the effects on the *in vitro* ruminal fermentation of oat straw.

MATERIALS AND METHODS

Additives and Substrates

The yeast strain Levazoot 15 (*Candida norvegensis*) from the UACH yeast collection was used, with record number of the Gen Bank: JQ519367.1 GI: 386785959, was obtained in a previous study (Castillo, 2009) when the yeast was selected, isolated and identified from the rumen environment of dairy cows. Yeast strain samples from a culture of the *Candida norvegensis* strain were plated on malt extract broth and incubated for 20 h at 30°C±2°C. The probiotic was prepared by placing 30 mL of this inoculum into 1,200 mL of malt extract broth at identical conditions and incubation time, and containing 2.5×10^8 live organisms/mL. Oat straw was used as substrate on a DM basis for *in vitro* incubation. The oat straw had been ground in a Wiley mill (Thomas-Wiley Model 4 Thomas Scientific, Swedesboro, NJ, USA) to pass through a 1 mm screen. The chemical composition of the cow's diet and the oat straw are shown in Table 1.

In vitro batch fermentation

Two cannulated lactating Holstein dairy cows (550± 25.5 kg BW) were fed twice daily with 4.0 kg of a mix grain and 4 kg of corn silage (DM basis) (Table 1), and used as donor animals for ruminal liquor. The ruminal liquor was drawn before feeding (0600) from each cow with a vacuum pump and deposited in a 2-L hermetically sealed insulated flask that had been brought to the proper temperature and flushed with CO_2, and immediately taken to the laboratory. The mixed sample was strained through several layers of surgical gauze into a 1-L Erlenmeyer flask. The buffer solution had the following composition in 1.0 L: 9.8 g of $NaHCO_3$, 7 g of $Na_2HPO_4.7H_2O$, 0.57 g of KCl, 0.47 g of $CaCl_2$, 0.12 g of $MgSO_4.7H_2O$, 0.917 g of urea, and 0.917 g of glucose. The entire procedure was conducted under a CO_2 atmosphere to ensure anaerobic conditions and efforts were made to keep the temperature at 39°C. For each

Table 1. Chemical composition of feedstuffs

Item	Corn silage	Concentrate[1]	Oat straw[2]
Percentage of daily ration	69.5	30.5	-
Chemical composition[3]			
OM	94.7	96.4	88.7
CP	8.3	19.1	5.3
EE	2.9	2.6	2.3
ADF	31.0	9.5	45.8
NDF	75.4	20.2	70.3

OM, organic matter; CP, crude protein; EE, ether extract; ADF, acid detergent fiber; NDF, neutral detergent fiber.
[1] Contained (DM basis) 51.0% corn, 23.5% wheat bran, 10% cottonseed meal, 8.49% corn gluten meal, 2.0% sugarcane molasses, 1.5% soybean meal, 1.0% bypass fat, 0.8% $CaCo_3$, 0.5% urea, 0.5% animal fat, 0.2% NaCl, 0.5% trace mineral and vitamin premix. Trace mineral and vitamin premix contained: Mg, 0.003 Co, 0.001% Se, 0.140% Zn, 0.092% Mn, 0.052% Cu, 0.140% Fe, 2,500 IU of vitamin A/g, and 50 IU vitamin D/g.
[2] Substrate for *in vitro* fermentation.
[3] Percentage dry mass basis.

sample, 50 mL of filtered rumen fluid and 100 mL of buffer solution were mixed in a 1:2 ratio. After mixing, 120 mL of diluted ruminal fluid was transferred under anaerobic conditions to 250 mL serum bottles containing 1.5 g of the substrate (oat straw) on a DM basis. Forty bottles were anaerobically sealed with butyl rubber stoppers and capped with aluminium. One set of 20 bottles, the control, was incubated at 39°C for 24 h with shaking at 30 rpm (New Brunswick Model Innova 4000, Nijmegen, Netherlands) and each bottle contained only the filtered rumen liquor and buffer solution. The remaining bottles (20) contained the same medium and was inoculated with 30 mL of the yeast probiotic and were incubated under identical conditions as the control group. Five sample times were set, 0, 4, 8, 12, and 24 h, (except for methane at 36 h) from the start of fermentation, with four repetitions for each time point. Consequently, the fermentation was conducted in a total of 40 bottles.

Microbiological determination

At every incubation time, 1 mL samples were collected to determine colony forming units (Log_{10} colony-forming unit [CFU])/mL) of cellulolytic bacteria, cellulolytic fungi, and yeast using the Hungate (1969) culture roll-tube technique under strict anaerobic conditions. The cellulolytic bacteria were grown in culture media as described by Caldwell and Bryant (1966) with the modification described by Elías (1971). Dilutions of 10^3, 10^4, and 10^5 were used to cultivate the bacteria and the fungi. The same dilutions were used for the fungi, and the culture media used was as specified by Joblin (1981). For the yeasts, malt extract agar was utilized with 0.01 g/L of chloramphenicol and three dilutions were used, 10^4, 10^5, and 10^6. The number of CFU was determined by visual inspection of colonies on the tube rolls under a magnifying glass.

Analyses

At the end of incubation methane gas was measured according to Theodorou et al. (1994), and total gas was measured by inserting a needle connected to a pressure transducer (Festo Co. Chihuahua, Chih. Mexico) where amount of gas was recorded and collected in a previously sealed tube using a 50-mL syringe. Tubes were retained and methane was measured by gas chromatography (Pye Unicam LTD, Cambridge, UK) using a flame ionization detector. The carrier gas was nitrogen, temperature of furnace was 52°C and detector temperature was 230°C. Silica gel columns were utilized. Methane concentration produced in each fermentation was calculated according to the general equation of gases (N = PV/RT). Atmospheric pressure was 10.1×10^4 Pa and room temperature was 299.5°K. The pH was immediately determined after bottles were uncapped with a portable pH meter (Hannah Instruments, Model HI 9017, Arvore-Vila do Conde, Portugal). Lactic acid (LA) concentrations (μg/mL) were determined by a colorimetric method. as described by Taylor (1996). For analysis of ammonia nitrogen (mMol) and volatile fatty acids (VFA) (mMol), 1 mL of 25% meta-phosphoric acid (wt/vol) was added to 5 mL of fermentation fluid and was stored at –20°C. The NH_3-N was determined by the colorimetric method of Broderick and Kang (1980) using a spectrophotometer (Hach DR500, Loveland, CO, USA). After thawing, 1 mL of fermentation fluid was centrifuged ($10,000\times g$ for 10 min) and VFA levels (mMol) were determined by gas chromatography (Perkin Elmer Model Clarus 800, Perkin Elmer Inc., Waltham, MA, USA) using a column (30 m×0.32 mm i.d.) and a flame ionization detector (column temperature 180°C, detector temperature 320°C, and injector temperature 240°C). The carrier gas was hydrogen with a flow rate of 1.40 mL/min. For in vitro DM disappearance (IVDMD) was determined according to Capetillo et al. (2002). the bottles' contents were transferred into test tubes (20 cc) previously dried at 105°C and centrifuged at $1,000\times g$ for 5 min and the pellets were dried at 55°C for 48 h. The IVDMD was calculated by subtracting the dry residue weight (pellet) from the original weight of oat straw divided by the original sample weight, and the values multiplied by 100 to derive the percentage of IVDMD.

Statistical analysis

An analysis of variance was performed with the PROC MIXED procedure of SAS. (2002) (SAS Inst., Inc., Cary, NC, USA). The adjusted model included the fixed effects of yeast strain treatment and the control group, fermentation time, and the interaction between treatment and the time effect.

The model as fitted was as follows:

$$Y_{ijk} = \mu+T_i+M_j+L\times M_{(ij)}+\varepsilon_{ijk,}$$

where Y_{ijk} = the dependent variable; μ = overall mean; T_i = treatment effect (i = 1,2); M_j = Time effect (j = 1,2,3,4,5); $L\times M_{(ij)}$ = Interaction effect, and ε_{ijk} = random residual error.

The identity of the nested balloon flask (experimental unit) was used in the treatment as a random effect. A trend analysis was performed on the variables across fermentation times in each treatment group, using fermentation time as an indicator variable. Mean values are reported with standard errors, and p-values are declared statistically different when p<0.05 or as indicated. Comparison of means were made using the predicted difference procedure of SAS (2002).

RESULTS AND DISCUSSION

pH value and lactic acid content

As shown in Table 2, the pH and LA concentrations in the medium decreased quadratically over time (p<0.01 and p<0.02, respectively). Treatment with yeast reduced pH relative to controls at all fermentation times (p<0.01). Williams et al. (1991) reported similar results; they observed decreased pH when S. cerevisiae (10 g/d) was added to oat hay diet of young bulls. and by Lynch and Martin (2002), who studied the in vitro effect of S. cerevisiae on fermentation of Bermuda hay and alfalfa hay. Lila et al. (2004), Oeztuerk et al. (2005), Lattimer et al. (2007), Longuski et al. (2009), and Inal et al. (2010), reported in vitro studies where pH remained unchanged. LA concentration in the yeast treatment was lower than the control group at 12 h of fermentation (p<0.05). At 24 h, the reduction was not statistically significant, although there was an appreciable numerical difference. These results confirmed findings by Williams et al. (1991), Erasmus et al. (1992), and Lila et al. (2004) who reported that the presence of yeast significantly reduced rumen LA concentrations and this reduction could be attributed to yeast cells stimulating the activity of Selenomona ruminantum, which consumes LA and thus contributes to pH stabilization.

Ammonia nitrogen concentration

The concentrations of ammonia-N (Table 3) increased quadratically with time of fermentation (p<0.01). Yeast treatment reduced ammonia-N concentrations in the medium at 4, 8, 12, and 24 h. These results agreed with those of Erasmus et al. (1992) and Moallem et al. (2009) who found significant ammonia-N concentration decreases when yeast was added to dairy cow diets. Chaucheyras-Durand and Fonty (2001) also reported that ammonia-N

Table 2. Least squares means (±SE) of pH and lactic acid during *in vitro* oat straw rumen fermentation

Variable	Time (h)	Control	Yeast	SEM	L	Q	Treatment effect
		Treatment			Trend		
pH	0	7.56[a]	7.36[b]	0.045	***	**	***
	4	7.55[a]	7.16[b]	0.045			***
	8	7.37[a]	7.02[b]	0.045			***
	12	7.28[a]	6.91[b]	0.045			***
	24	7.08[a]	6.68[b]	0.045			***
Lactic acid (μg/mL)	0	20.36[a]	19.08[a]	1.11	***	**	NS
	4	18.19[a]	19.65[a]	1.11			NS
	8	17.42[a]	16.05[a]	1.11			NS
	12	17.18[a]	14.01[b]	1.11			*
	24	16.58[a]	14.68[a]	1.11			NS

SEM, standard error of the mean; L, linear trend across fermentation times; Q, quadratic trend across fermentation times; NS, non significant.
Different letters in the same row indicate significant differences (* $p<0.05$; ** $p<0.01$; *** $p<0.001$) between treatments.
Means (n = 40) in rows with the same superscript are not significantly different (ns) due to treatment effect.

concentrations decreased when yeast was added to alfalfa hay and feeds for growing sheep. This ammonia-N reduction may be due to higher ammonia intake by microbial cells, perhaps as a direct result of stimulation of rumen microbial activity (Williams and Newbold, 1990).

Percentage of *in vitro* dry matter digestibility

As shown in Table 3, oat straw IVDMD increased quadratically with greater fermentation time ($p<0.02$). Yeast treatment only increased IVDMD relative to controls at 24 h ($p<0.01$), in agreement with *in vitro* results reported by Miller-Webster et al. (2002) and Lila et al. (2004). Williams et al. (1991) reported that stimulation of cellulose degradation by yeast cultures is associated with a decrease in the lag phase, which also causes an initial increase in digestion rate. These authors also attributed the improvement in DM digestibility to pH stabilization. Ando et al. (2004) as well, reported an increment in the total

degradability of forage when they added dried beer yeast attributing it to the activation of rumen microbes and partially due to the addition of nitrogen sources. Newbold et al. (1995) reported that some yeast cultures increased total and cellulolytic bacterial counts in the rumen, thus increasing fibre digestion. Another explanation may be that yeast cells in the rumen produce ethanol (Kung et al., 1997), which is converted by rumen microorganisms into valeric and isovaleric acids. The increases in these acids that were observed with the yeast treatment in this study are consistent with this latter possibility (Table 4). Valeric, capric, isobutyric, and isovaleric acids were found to stimulate rumen cellulolysis in studies involving fractional distillation of the rumen fluid (Elias, 1983). Likewise, in our study, higher concentrations of valeric and isovaleric acids were associated with increased rumen cellulolysis and *in vitro* disappearance of DM. The molar concentrations of acetic ($p<0.01$), butyric ($p<0.01$), propionic ($p<0.01$),

Table 3. Least squares means (±SE) of ammonium nitrogen and *in vitro* dry matter disappearance during *in vitro* oat straw rumen fermentation

Variable	Time (h)	Control	Yeast	SEM	L	Q	Treatment effect
		Treatment			Trend		
NH₃-N	0	3.27[a]	2.76[a]	0.25	***	**	NS
	4	7.45[a]	5.33[b]	0.25			***
	8	7.04[a]	5.51[b]	0.25			***
	12	7.72[a]	5.46[b]	0.25			***
	24	9.44[a]	6.46[b]	0.25			***
IVDMD	0	9.76[a]	12.63[a]	2.74	***	**	NS
	4	10.96[a]	14.59[a]	2.74			NS
	8	17.64[a]	21.27[a]	2.74			NS
	12	25.00[a]	30.33[a]	2.74			NS
	24	46.64[a]	56.71[b]	2.74			**

SEM, standard error of the mean; L, linear trend across fermentation times; Q, quadratic trend across fermentation times;
NH₃-N, ammonium nitrogen (mMol); NS, non significant; IVDMD, *in vitro* dry matter disappearance (%).
Different letters in the same row indicate significant differences (** $p<0.01$; *** $p<0.001$) between treatments.
Means (n = 40) in rows with the same superscript are not significantly different (ns) due to treatment effect.

Table 4. Least squares mean (±SE) of concentrations of acetic, propionic, butyric, valeric and isovaleric acids during *in vitro* oat straw rumen fermentation

Volatile fatty acid (mMol)	Treatment			SEM	Trend		Treatment effect
	Time (h)	Control	Yeast		L	Q	
Acetic acid	0	24.40[a]	20.00[a]	1.97	***	*	NS
	4	24.56[a]	26.82[a]	1.97			NS
	8	29.50[a]	37.42[b]	1.97			***
	12	32.36[a]	40.89[b]	1.97			***
	24	45.04[a]	48.84[a]	1.97			NS
Propionic acid	0	8.55[a]	6.72[b]	0.58	***	NS	*
	4	8.22[a]	9.09[a]	0.58			NS
	8	9.99[a]	11.90[b]	0.58			*
	12	11.50[a]	13.46[b]	0.58			*
	24	17.26[a]	17.79[a]	0.58			NS
Butyric acid	0	5.81[a]	4.48[b]	0.42	***	**	*
	4	5.27[a]	6.23[a]	0.42			NS
	8	6.73[a]	8.12[b]	0.42			*
	12	7.61[a]	8.50[a]	0.42			NS
	24	9.73[a]	9.68[a]	0.42			NS
Valeric acid	0	0.50[a]	0.45[a]	0.06	***	NS	NS
	4	0.46[a]	0.63[a]	0.06			NS
	8	0.60[a]	0.99[b]	0.06			***
	12	0.67[a]	1.31[b]	0.06			***
	24	0.96[a]	1.77[b]	0.06			***
Isovaleric acid	0	0.44[a]	0.47[a]	0.08	*	**	NS
	4	0.42[a]	0.46[a]	0.08			NS
	8	0.48[a]	0.78[b]	0.08			**
	12	0.58[a]	1.02[b]	0.08			***
	24	0.99[a]	1.69[b]	0.08			***

SEM, standard error of the mean; L, linear trend across fermentation times; Q, quadratic trend across fermentation times; NS, non significant.
Different letters in the same row indicate significant differences (* $p<0.05$; ** $p<0.01$; *** $p<0.001$) between treatments.
Means (n = 40) in rows with the same superscript are not significantly different (ns) due to treatment effect.

valeric ($p<0.01$), and isovaleric acids ($p<0.02$) increased linearly over fermentation times (Table 4).

Acetic acid content

In the present study (Table 4), acetic acid concentrations were higher in the yeast treatment groups than in the controls at 8 h ($p<0.01$) and 12 h ($p<0.01$). Propionic acid was lower when yeast was added at 0 h ($p<0.03$), but higher at 8 h ($p<0.03$) and 12 h ($p<0.02$). Meanwhile, butyric acid concentration was also lower at 0 h and higher at 8 h with yeast treatment. Valeric and isovaleric acid concentrations with yeast treatment were higher than that in control solutions at 8, 12, and 24 h ($p<0.01$). Oeztuerk et al. (2005) studied the effects of live yeast on *in vitro* rumen fermentation of a hay and grain diet and found that acetic, butyric, valeric, and isovaleric acids increased, while only propionic acid did not increase significantly. Similar results were found by Erasmus et al. (2005); Miller-Webster et al. (2002). Diaz et al. (2011); Křižova et al. (2011); Kowalik et al. (2012). The stimulating effect of the strain under study on production of VFAs may be related to the chemical composition of the yeast cell wall and other cell components. Yeast cell walls, which account for ~20% of the yeast cell weight, are primarily made up of β-1.3 and β-1.6 glucans, and chitin (Moukadiri et al., 1997). These structures, which form appropriate substrates for microbial rumen fermentation, independent of the state of the yeast (Oeztuerk et al., 2005).

Microorganisms counts

Table 5 shows the resulting CFU counts in viable yeast, cellulolytic bacteria, and cellulolytic fungi in the rumen. The quantity of viable yeast inoculated into the rumen decreased linearly with fermentation time ($p<0.01$, Table 5). Similar results were reported by Arambel and Rung-Syin (1987) in studies examining the growth of *S. cerevisiae* in a ruminal environment. These authors indicate that yeasts are incapable of sustaining a productive population within a ruminal environment, because of inhibiting factors for yeast growth, such as non-optimal temperature. The optimal temperature range for yeast growth is 28°C to 30°C, with survival remaining possible up to 37°C through formation

Table 5. Least squares mean (±SE) of yeast and microorganism counts during *in vitro* oat straw rumen fermentation

| Variable | Treatment | | | SEM | Trend | | Treatment effect |
	Time (h)	Control	Yeast		L	Q	
Yeast	0	5.42^a	7.30^b	0.10	***	***	***
	4	5.73^a	6.65^b	0.10			***
	8	5.69^a	6.25^b	0.10			***
	12	5.50^a	5.91^b	0.10			**
	24	5.45^a	5.67^a	0.10			NS
Cellulolytic fungi	0	5.88^a	5.78^a	0.19	NS	NS	NS
	4	5.26^a	6.09^b	0.19			**
	8	5.20^a	6.15^b	0.19			***
Cellulolytic bacteria	0	6.26^a	6.29^a	0.11	NS	NS	NS
	4	6.36^a	6.58^a	0.11			NS
	8	6.23^a	6.50^a	0.11			NS

SEM, standard error of the mean; L, linear trend across fermentation times; q, quadratic trend across fermentation times; NS, non significant.
Different letters in the same row indicate significant differences (**p<0.01; *** p<0.001) between treatments.
Means (Log_{10} CFU/mL, n = 40) in rows with the same superscript are not significantly different (ns) due to treatment effect.

of ascospores (Dengis et al., 1995). At 39°C, the typical temperature in the ruminal environment, growth and viability are reduced. (Mendoza, 1993). Williams et al. (1990) postulated that yeast do not establish themselves permanently in the rumen. In the current study, the inoculated rumen materials maintained a higher yeast population than the control for up to 12 h (p<0.01), and was not significantly different until 24 h post-inoculation. After 12 h, the yeast cells entered a lethal no-growth phase, which corroborates the hypothesis of Williams et al. (1990). Also it was observed that, after 12 h, the differences in yeast populations between the inoculated and control samples were smaller, reaching a nadir at 24 h. In the present study, rumen material was tested for resident yeast soon after collection, and it was demonstrated that a pool of yeast populations existed in the rumen. Fermentation and treatment time did not affect cellulolytic bacterial counts (p<0.05) (Table 5). The effects of live yeast on cellulolytic bacterial counts are very diverse. Similar to the present findings, Erasmus et al. (1992) found no effect of *S. cerevisiae* on the cellulolytic bacterial populations of dairy cows fed an energy-rich diet. Conversely, Newbold et al. (1996) and Lila et al. (2004) reported a positive effect of adding yeast cultures to *in vitro* fermentation of hay and concentrate mixes. Counts of cellulolytic fungi (Table 5) did not exhibit any trends across the fermentation times. However, treatment with yeast cultures positively affected cellulolytic fungal populations in the rumen after 4 h (p< 0.01) and 8 h (p<0.01) of fermentation. Chaucheyras et al. (1995) also report that yeast stimulate the growth of the ruminal fungus *Neocallimastix frontalis*. This stimulation effect on ruminal microorganisms, particularly on those that break down cellulose, such as cellulolytic fungi, may be explained by specific mechanisms of action in the rumen. When yeast are plasmolized, over time they supply growth factors such as peptides, amino acids, B-complex vitamins, and other components that favour bacterial and fungal growth (Elías, 1971). Also, live yeast help to eliminate a small amount of oxygen (~1%), that enters the rumen when the animal ingests feed, through aerobic respiration, and this process facilitates growth of more stringent anaerobic microorganisms, such as cellulolytic bacteria and fungi (Newbold et al., 1996).

Methane production

As shown in Table 6, methane production did not exhibit any trends across fermentation times. A reduction in methane production was observed only at 8 h by yeast

Table 6. Least squares mean (±SE) of methane production during *in vitro* oat straw rumen fermentation

| Item | Treatment | | | SEM | Trend[2] | | Treatment effect |
	Time (h)	Control	Yeast		L	Q	
Methane	4	45.00^a	28.78^a	11.53	NS	NS	NS
(mL)	8	66.00^a	15.06^b	11.53			**
	12	33.05^a	15.91^a	11.53			NS
	24	29.93^a	13.12^a	11.53			NS
	36	34.50^a	24.94^a	11.53			NS

SEM, standard error of the mean; L, linear trend across fermentation times; q, quadratic trend across fermentation times; NS, non significant.
Different letters in the same row indicate significant differences (* p<0.05; ** p<0.01; *** p<0.001) between treatments.
Means (n = 40) in rows with the same superscript are not significantly different (ns) due to treatment effect.

treatment (p<0.01). At other time points, there were noticeable numerical decreases, although not statistically significant. Numerous studies have demonstrated that methane production is affected by addition of yeast. Mutsvangwa et al. (1992) observed a marked decrease in methane production with the addition of *S. cerevisiae* in intensive fattening of bulls. Lynch and Martin (2002) and Lila et al. (2004) also found reduced methane production when they studied the behaviour of yeasts added to *in vitro* fermentation of alfalfa hay and hay-concentrate mix. This drop in methane production may be due to the yeast stimulating utilization of metabolic hydrogen by acetogenic bacteria in the generation of acetic acid. (Chaucheyras et al., 1995). Based on the above results, addition of the yeast *Candida norvegensis* to *in vitro* oat straw fermentation positively influenced ruminal fermentation parameters as well as microbial growth with reduction of methane production. Additionally, yeast inoculum also improved IVDMD.

REFERENCES

Ando, S., Y. Nishiguchi, K. Hayasaka, H. Iefuji, and J. Takahashi. 2006. Effects of *Candida utilis* treatment on the nutrient value of rice bran and the effect of *Candida utilis* on the degradation of forages *in vitro*. Asian Australas. J. Anim. Sci. 19:806-810.

Ando, S., R. I. Khan, J. Takahasi, Y. Gamo, R. Morikawa, Y. Nishiguchi, and K. Hayasaka. 2004. Manipulation of rumen fermentation by yeast: The effects of dried beer yeast on the *in vitro* degradability of forages and methane production. Asian Australas. J. Anim. Sci. 17:68-72.

Arambel, M. J. and T. Rung-Syin. 1987. Evaluation of *Saccharomyces cerevisiae* growth in the rumen ecosystem. Memories 19th Biennial Conference on Rumen Function, Chicago, IL, USA. pp. 17-19.

Broderick, G. A. and J. H. Kang. 1980. Automated simultaneous determination of ammonia and total amino acids in ruminal fluid and *in vitro* media. J. Dairy. Sci. 63:64-75.

Caldwell, D. R. and M. P. Bryant. 1966. Medium without fluid for non-selective enumeration and isolation of rumen bacteria. Appl. Microbiol. 14:794-801.

Capetillo, L. C. M., P. E. Herrera, and C. C. C. Sandoval. 2002. Chemical composition of boherhavia erecta L, digestibility and gas production *in vitro*. Arch. Zootec. 51:461-464.

Castillo, Y. 2009. *In vitro* Fermentation to Obtain the Yeast *Candida norvegensis* in Mixes of Alfalfa with Fermented Apple Waste and Effects on the Microbial Activity. PhD Thesis. Facultad de Zootecnia y Ecología. Universidad Autónoma de Chihuahua. Chihuahua, México.

Chaucheyras-Durand, F., G. Fonty, G. Bertin, and P. Gouet. 1995. Effects of live *Saccharomyces cerevisiae* cells on zoospore germination, growth, and cellulolytic activity of the rumen anaerobic fungus, *Neocallimastix frontalis* MCH3. Curr. Microbiol. 31:201-205.

Chaucheyras-Durand, F. and G. Fonty. 2001. Establishment of cellulolytic bacteria and development of fermentative activities in the rumen of gnotobiotically-reared lambs receiving the microbial additive *Saccharomyces cerevisiae* CNCM I-1077. Reprod. Nutr. Dev. 41:57-68.

Chaucheyras-Durand, F., E. Chevaux, C. Martin, E. Forano. 2012. Use of yeast probiotics in ruminants: Effects and mechanisms of action on rumen pH, fibre degradation, and microbiota according to the diet. Chapter 7. In: Probiotic in Animals. Edited by Everlon Cid Rigobelo. INTECH, http://dx.doi.org/10.5772/50192. pp.119-152.

Dengis, P. D., L. R. Nelissen, and P. G. Rouxhet. 1995. Mechanisms of yeast flocculation comparison of top- and bottom-fermenting strains. Appl. Environ. Microbiol. 61:718-728.

Diaz, A., C. Saro, M. L. Tejido, A. Sosa, M. E. Martinez, J. Galindo, M. D. Carro, and M. J. Ranilla. 2011. Effects of a yeast enzymatic hydrolyzate on *in vitro* ruminal fermentation. In: Challenging strategies to promote the sheep and goat sector in the current global context (Eds. M. J. Ranilla, M. D. Carro, H. Ben Salem, and P. Morand-Fehr). Zaragoza: CIHEAM, CSIC, Universidad de León, FAO. pp. 181-186. http://om.ciheam.org/article.php?IDPDF=801554 Accessed June 2, 2014.

Doležal, P., J. Dvořáček, J. Doležal, J. Čermáková, L. Zeman, and K. Szwedziak. 2011. Effect of feeding yeast culture on ruminal fermentation and blood indicators of Holstein dairy cows. Acta Vet. Brno. 80:139-145.

Elias, A. 1971. The Rumen Bacteria of Animals Fed on a High Molasses-urea Diet. Ph. D Thesis. University of Aberdeen, Aberdeen, UK.

Elias, A. 1983. Digestion of grasslands and tropical forages. In: The grasslands in Cuba, vol. 2. Ed. EDICA. La Habana, Cuba. pp.187-246.

Erasmus, L. J., P. M. Botha, and A. Kistner. 1992. Effect of yeast culture supplement on production, rumen fermentation, and duodenal nitrogen flow in dairy cows. J. Dairy Sci. 75:3056-3065.

Erasmus, L. J., P. H. Robinson, A. Ahmadi, R. Hinders, and J. E. Garrett. 2005. Influence of prepartum and postpartum supplementation of a yeast culture and monensin, or both, on ruminal fermentation and performance of multiparous dairy cows. Anim. Feed. Sci. Technol. 122:219-239.

Hungate, R. E. 1969. A roll tube method for cultivation in microbiology (Eds. J. B. Morris and D. B. Ribbons). Academic Press Inc., New York, NY, USA. 117 p.

Inal, F., E. Gürbüz, B. Coşkun, M. S. Malataş, O. B. Citil, E. S. Polat, E. Seker, and C. Ozcan. 2010. The Effects of live yeast culture (*Saccharomyces cerevisiae*) on rumen fermentation and nutrient degradability in yearling lambs. Kafkas Univ. Vet. Fak. 16:799-804.

Joblin, K. N. 1981. Isolation, enumeration, and maintenance of rumen anaerobic fungi in roll tubes. Appl. Environ. Microbiol. 42:1119-1122.

Kowalik, B., J. Skomiał, J. J. Pająk, M. Taciak, M. Majewska, and G. Bełżecki. 2012. Population of ciliates, rumen fermentation indicators and biochemical parameters of blood serum in heifers fed diets supplemented with yeast (*Saccharomyces cerevisiae*) preparation. Anim. Sci. Pap. Rep. 30: 329-338.

Křižova, L., M. Richter, J. Třinacty, J. Řiha, and D. Kumprechtova. 2011. The effect of feeding live yeast cultures on ruminal pH and redox potential in dry cows as continuously measured by a new wireless device. Czech J. Anim. Sci. 56:37- 45.

Kung, L. Jr., E. M. Kreck, R. S. Tung, A. O. Hession, A. C. Sheperd, M. A. Cohen, H. E. Swain, and J. A. Leedle. 1997. Effects of a live yeast culture and enzymes on *in vitro* ruminal fermentation and milk production of dairy cows. J. Dairy Sci. 80:2045-2051.

Lattimer, J. M., S. R. Cooper, D.W. Freeman, and D. L. Lalman. 2007. Effect of yeast culture on *in vitro* fermentation of a high-concentrate or high-fiber diet using equine fecal inoculums in a Daisy II incubator. J. Anim. Sci. 85:2484-2491.

Lila, Z. A., N. Mohammed, T. Yasui, Y. Kurokawa, S. Kanda, and H. Itabashi. 2004. Effects of a twin strain of *Saccharomyces cerevisiae* live cells on mixed ruminal microorganism fermentation *in vitro*. J. Anim. Sci. 82:1847-1854.

Longuski, R. A., Y. Ying, and M. S. Allen. 2009. Yeast culture supplementation prevented milk fat depression by a short-term dietary challenge with fermentable starch. J. Dairy Sci. 92: 160-167.

Lynch, H. A. and S. A. Martin. 2002. Effects of *Saccharomyces cerevisiae* culture and *Saccharomyces cerevisiae* live cells on *in vitro* mixed ruminal microorganism fermentation. J. Dairy Sci. 85:2603-2608.

Marrero Y., M. E. Burrola-Barraza, Y. Castillo, L. C. Basso, C. A. Rosa, O. Ruiz, and E. González-Rodríguez. 2013. Identification of *Levica* yeasts as a potential ruminal microbial additive. Czech J. Anim. Sci. 58:460-469.

Marrero, Y., O. Ruiz, A. Corrales, O. Jay, J. Galindo, Y. Castillo, and N. Madera. 2014. *In vitro* gas production of fibrous substrates with the inclusion of yeast. Cuban J. Agric. Sci. 48: 119-123.

Marrero, Y., Y. Castillo, O. Ruiz, E. Burrola, and C. Angulo. 2015. Feeding of yeast (*Candida* spp.) improves *in vitro* ruminal fermentation of fibrous substrates. J. Integr. Agric. 14:514-519.

Mendoza, M. G. D. and R. Ricalde-Velasco. 1993. Alimentación de ganado bovino con dietas altas en grano. Universidad Autónoma Metropolitana. Cap. 9. Uso de aditivos alimenticios. p. 97.

Miller-Webster, T., W. H. Hoover, M. Holt, and J. E. Nocek. 2002. Influence of yeast culture on ruminal microbial metabolism in continuous culture. J. Dairy Sci. 85:2009-2014.

Moallem, U., H. Lehrer, L. Livshitz, M. Zachut, and S. Yakoby. 2009. The effects of live yeast supplementation to dairy cows during the hot season on production feed efficiency, and digestibility. J. Dairy Sci. 92:343-351.

Moukadiri, I., J. Armero, A. Abad, R. Sentandreu, and J. Zueco. 1997. Identification of a mannoprotein present in the inner layer of the cell wall of *Saccharomyces cerevisiae*. J. Bacteriol. 179:2154-2162.

Mutsvangwa, T., I. E. Edwards, J. H. Topps, and G. F. M. Paterson. 1992. The effect of dietary inclusion of yeast culture (Yea-Sacc) on patterns of rumen fermentation, food intake and growth of intensively fed bulls. Anim. Prod. 55:35-40.

Newbold, C. J., R. J. Wallace, X. B. Chen, and F. M. McIntosh. 1995. Different strains of *Saccharomyces cerevisiae* differ in their effects on ruminal bacterial numbers *in vitro* and in sheep. J. Anim. Sci. 73:1811-1818.

Newbold, C. J., R. J. Wallace, and F. M. McIntosh. 1996. Mode of action of the yeast *Saccharomyces cerevisiae* as a feed additive for ruminants. Br. J. Nutr. 76:249-261.

Oeztuerk, H., B. Schroeder, M. Beyerbach, and G. Breves. 2005. Influence of living and autoclaved yeasts of *Saccharomyces boulardii* on *in vitro* ruminal microbial metabolism. J. Dairy. Sci. 88:2594-2600.

Plata, P. F., M. G. D. Mendoza, J. R. Barcena-Gama, and M. S. Gonzalez. 1994. Effect of a yeast culture (*Saccharomyces cerevisiae*) on neutral detergent fiber digestion in steers fed oat straw based diets. Anim. Feed. Sci. Technol. 49:203-210.

SAS. Institute. 2002. SAS. User's Guide. SAS Institute Inc. Cary, NC, USA.

Shin, H. T., Y. Beom Lim, J. Ho Koh, J. Yun Kim, S. Young Baig, and J. Heung Lee. 2002. Growth of *Issatchenkia orientalis* in aerobic batch and fed-batch cultures. J. Microbiol. 40:82-85.

Tang, S. X., G. O. Tayo, Z. L. Tan, Z. H. Sun, L. X. Shen, C. S. Zhou, W. J. Xiao, G. P. Ren, X. F. Han, and S. B. Shen. 2008. Effects of yeast culture and fibrolytic enzyme supplementation on *in vitro* fermentation characteristics of low-quality cereal straws. J. Anim. Sci. 86:1164-1172.

Taylor, K. A. C. C. 1996. A simple colorimetric assay for muramic acid and lactic acid. Appl. Biochem. Biotechnol. 56:49-58.

Theodorou, M. K., B. A. Williams, M. S. Dhanoa, A. B. McAllan, and J. France. 1994. A simple gas production method using a pressure transducer to determine the fermentation kinetics of ruminant feeds. Anim. Feed. Sci. Technol. 48:185-197.

Williams, P. E., A. Tait, G. M. Innes, and C. J. Newbold. 1991. Effects of the inclusion of yeast culture (*Saccharomyces cerevisiae* plus growth medium) in the diet of dairy cows on milk yield and forage degradation and fermentation patterns in the rumen of steers. J. Anim. Sci. 69:3016-3026.

Williams, J. G. K., A. R. Kubelik, K. J. Livak, J. A. Rafalski, and S. V. Tingey. 1990. DNA-polymorphism amplified by arbitrary primers is useful as genetic markers. Nucl. Acids Res. 18:6531-6535.

Williams, P. E. V. and C. J. Newbold. 1990. Rumen probiosis: the effects of novel microorganisms on rumen fermentation and rumen productivity. In: (Eds. W. Haresing, and D J. A. Cole), Recent Advances in Animal Nutrition. Butterworths, London, England. p. 211.

Ye, G., Y. Zhu, J. Liu, X. Chen, and K. Huang. 2014. Preparation of glycerol-enriched yeast culture and its effect on blood metabolites and ruminal fermentation in goats. PLOS ONE 9(4):e94410.

3

Effects of Cellulase Supplementation on Nutrient Digestibility, Energy Utilization and Methane Emission by Boer Crossbred Goats

Lizhi Wang and Bai Xue*

Institute of Animal Nutrition, Sichuan Agricultural University, Yaan 625014, Sichuan, China

ABSTRACT: This study examined the effect of supplementing exogenous cellulase on nutrient and energy utilization. Twelve desexed Boer crossbred goats were used in a replicated 3×3 Latin square design with 23-d periods. Dietary treatments were basal diet (control, no cellulase), basal diet plus 2 g unitary cellulase/kg of total mixed ration dry matter (DM), and basal diet plus 2 g compound cellulase/kg of total mixed ration DM. Three stages of feeding trials were used corresponding to the three treatments, each comprised 23 d, with the first 14 d as the preliminary period and the following 9 d as formal trial period for metabolism trial. Total collection of feces and urine were conducted from the 4th d of the formal trial, and gas exchange measures were determined in indirect respiratory chambers in the last 3 d of the formal trial. Results showed that cellulase addition had no effect (p>0.05) on nutrient digestibility. Dietary supplementation of cellulase did not affect (p>0.05) N intake and retention in goats. Gross energy (GE) intake, fecal energy and urinary energy excretion, heat production were not affected (p>0.05) by the cellulase supplementation. Total methane emission (g/d), CH_4 emission as a proportion of live weight or feed intake (DM, organic matter [OM], digestible DM or digestible OM), or CH_4 energy output (CH_4-E) as a proportion of energy intake (GE, digestible energy, or metabolizable energy), were similar (p>0.05) among treatments. There was a significant (p<0.001) relationship between CH_4 and live weight (y = 0.645x+0.2, R^2 = 0.54), CH_4 and DM intake (y = 16.7x+1.4, R^2 = 0.51), CH_4 and OM intake (y = 18.8x+1.3, R^2 = 0.51) and CH_4-E and GE intake. Results from this study revealed that dietary supplementation of cellulase may have no effect on nutrient digestibility, nitrogen retention, energy metabolism, and methane emission in goat. (**Key Words:** Cellulase, Digestibility, Energy, Goat, Nutrients, Methane)

INTRODUCTION

Cellulase is believed to be effective in improving nutrient digestibility of ruminants, especially rumen degradability of fiber. Ballard et al. (2003) reported that dietary addition of compound cellulase (CC) increased the apparent digestibility of dry matter (DM), organic matter (OM), and neutral detergent fiber (NDF) of dairy cows. Bilik and Łopuszańska-Rusek (2010) reported that the addition of exogenous cellulase may increase total volatile fatty acid (VFA) of rumen fluid of dairy cows without disturbing the VFA profile. Rumen VFA is the energy source for ruminants, and therefore it could be deduced that exogenous cellulase might affect energy metabolism of ruminants. There have been considerable research on the

energy metabolism of different species of ruminants, but no research on the effect of exogenous cellulase supplementation on energy metabolism of ruminants has been reported. As the evidence for nutrient digestibility being elevated by exogenous cellulase accumulated (McAllister et al., 2000; Tang et al., 2013; Vijay Bhasker et al., 2013), this study hypothesized that energy utilization could also be affected by the dietary addition of exogenous cellulase. Therefore, the objective of this study was to determine the effects of exogenous cellulase on nutrient utilization by Boer crossbred goats.

MATERIALS AND METHODS

All animals in this study were cared for according to the standards set by the Chinese guidelines for animal welfare and the experimental protocol was approved by the Animal Care and Use Committee of the Chinese Academy of

---

* Corresponding Author: Bai Xue.
E-mail: Xuebai2000@yahoo.com

Sciences.

Animals and diets

Twelve castrated Boer crossbred goats (Boer×Jianchang Black), average body weight (BW) 20±1.51 kg and aged around 10 months, were used in a four replicate 3×3 Latin square design. These goats were purchased from farmers around Chengdu. The sexual maturity of this species is at 7 to 8 months of age, and the growing age until 12 months of age. They were housed individually in metabolism cages designed to separately collect feces and urine. Before the experiment, goats were dewormed for internal parasites with albendazole (Yakang Animal pharmacy, Weifang, Shandong, China) at 12 mg/kg BW. There were three experimental periods corresponding to three treatments. Each stage comprised 21-d periods, with the first 14 d as the adaptation period and the following 9 d as the measurement period including 6 d in crates and 3 in chambers.

From d 1 to d 6 of the formal trial, goats were kept in metabolism cages for total collection of feces and urine. On d 7 of the formal trial, the goats were transferred into respiratory chambers for adaptation; on d 8 and d 9, the goats were subject to gas exchange measurements in respiratory chambers for measuring oxygen consumption, carbon dioxide output, and methane output. Total feces and urine were collected daily both in chambers and in metabolism cages.

The experimental diet comprised of 30% concentrate, 35% alfalfa and 35% Chinese wildrye (Table 1), and was formulated to provide adequate digestible energy (DE) and

Table 1. The ingredient and chemical composition of the diet (DM basis)

Items	Content (%)
Ingredients	
Alfalfa	35
Wildrye	35
Corn	11
Soybean meal	5
Wheat bran	13
Premix[1]	0.5
NaCl	0.5
Total	100
Nutrient levels[2]	
Gross energy (MJ/kg)	17.0
Dry matter (%)	88.8
Crude protein (%)	11.9
Lipid	2.0
Ash	10.7
Neutral detergent fiber (%)	38.9
Acid detergent fiber (%)	28.9

[1] Premix provides: Fe 30 ppm; Cu 10 ppm; Zn 50 ppm; Mn 60 ppm; Vit. A 2,937 IU; Vit. D 343 IU; Vit. E 30 IU.
[2] Neutral detergent fibers are analyzed values.

crude protein (CP) for a 20-kg goat gaining 100 g/d of body weight according to NRC (1981). Dietary treatments were basal diet (control, no cellulase), basal diet plus 2 g unitary cellulase (UC)/kg of ration DM, and basal diet plus 2 g CC/kg of ration DM. The enzymes (UC and CC) used in this study were commercial preparations of fungal extracts (Youter Biotechology Shanghai Co., Ltd., Shanghai, China). The UC additive contained 10,000 IU/g of cellulase (endoglucanase, EC 3.2.1.4) activity, while CC additive contained 7,000 IU/g of cellulase (endoglucanase, EC 3.2.1.4) and 5,000 IU/g of xylanase (1, 4-β-xylanase, EC 3.2.1.8) activity. Cellulase activity was determined at 39°C and pH 6.0 using carboxymethyl cellulose (catalog no. C-5678; Sigma Chemical Co., St. Louis, MO, USA) as substrates and xylanase activity was determined at 50°C and pH 5.5 using birchwood xylan (catalog no. X-0502; Sigma Chemical Co., USA) as substrates. The enzyme additive (i.e., 2 g of the enzyme additive per kg of ration DM for UC and CC treatment) was first diluted in 40 mL of water, and the dilution was added to the concentrate at the time of mixing, usually within 1 h before feeding. An equal amount of water (40 mL/kg of concentrate DM) was added to the control diet. Experimental diets were fed to goats for *ad libitum* intake twice daily with equal amounts at 1,000 and 1,600 h, and fresh water was available freely at all times.

Measurements

Live weight was recorded for 3 consecutive days in the morning before feeding at the beginning and end of each period. Feed intake for individual goats was recorded daily during the 9-d formal trial in each period. Experimental diets, dietary ingredients, and refusals for individual goats were sampled at the beginning and end of formal trial of each period and composited by period. Periodical refusals, experimental diets, and ingredients were stored at –20°C until analyzed. Feces were collected daily in wire-screen baskets placed under the floor of metabolism crates, and urine was collected through a funnel into plastic buckets. Urine and feces were weighed daily. Urine was acidified with 10% HCl (vol/vol, 5% of urine), and feces were preserved with 10% formaldehyde (vol/vol, 3% of feces) and thoroughly mixed. Urine and feces were sub-sampled daily after acidification and preservation, and stored frozen at –20°C for later analysis.

Feed offered, refusals, and fecal samples were dried in a forced-air oven at 65°C for 48 h, then ground to pass through a 1-mm screen, and were analyzed for DM, gross energy (GE) (AOAC, 1990), and NDF (filter bag technique; China Agricultural University). Urine samples were assayed for N and GE concentrations.

For calorimetry measures, goats were moved into indirect open-circuit calorimetry chambers. The concentration of O_2, CO_2 and CH_4 were analyzed using gas chromatography. Before the gas exchange measurements, validity and accuracy of expired CO_2 and inspired O_2 flows were checked with ethane combustion with the same flow rates as used during measurements. Before each test, analyzers were calibrated with reference gases: 99.99% O_2, 99.99% CO_2, and 99.99% CH_4. Temperature and humidity in the chambers was maintained at 23°C and 50% to 55%, respectively.

Calculations

Metabolizable energy intake (MEI) was calculated by the difference between GE intake (GEI) and energy losses in feces (FE), urine (UE), and CH_4 (CH_4E)

$$MEI = GEI - FE - UE - CH_4 E$$

Heat production (HP) was estimated based on the Brouwer (1965) equation:

$$HP (Kj) = 16.1735 \times O_2 (L) + 5.0208 \times CO_2 (L) - 5.9873 \times UN (g)$$

CH_4 contains 39.5388 kJ/L energy as reported by Brouwer (1965). Retained energy (RE) was determined as the difference between MEI and HP.

Statistical analyses

Data were analyzed using the MIXED procedure (SAS Institute Inc., Cary, NC, USA; Littell et al., 1996). Goat was the experimental unit for all variables. The full general linear model (GLM) model included the fixed effects of square, period nested within square, treatment (control, UC and CC enzyme), sampling time (day), and the interaction of treatment and sampling time. For all data, if the interaction of treatment by sampling time was not significant (p>0.05), the fixed effect of sampling time and its related interaction were removed from the full GLM model. Sampling time was a repeated effect in the model. Goat nested within square was used in the random statement. Differences were declared significant at p≤0.05, and a tendency to significance was declared at 0.05<p≤0.10. Means were separated using Duncan's multiple contrasts.

RESULTS AND DISCUSSION

Nutrient digestibility

There was no significant difference (p>0.05) in live weight, live weight gain and DM intake among treatments. Digestibility of DM, OM, CP, crude fat, NDF and acid detergent fiber (ADF) were not affected (p>0.05) by cellulase treatments (Table 2).

There are controversial reports on total tract digestibility of DM and OM or both, following cellulase treatments. Some researchers observed a positive effect of treatment with exogenous cellulase on DM or organic digestibility (Lewis et al., 1996; Rode et al., 1999) while some other studies observed no effect (Lewis et al., 1999; McAllister et al., 2000; Yang et al., 2000). However, present results found that cellulase addition had no effect on DM and OM digestibility. Some researchers found that NDF and ADF digestibility increased following cellulase treatments (Lewis et al., 1999; McAllister et al., 2000), but these results were not observed in this study. The contradictory reports on the effect of cellulase on nutrient digestibility may be due to the digestion characteristics of the rations. Yang et al. (2000) noted that exogenous cellulase improved digestibility of low quality rations, but no effect was evident with high quality rations. This viewpoint still needs confirmation, because O'Connor et al. (2007) found that nutrient

Table 2. The effects of enzyme treatments on apparent nutrient digestibility and N metabolism in growing goats

Item	Control	Cellulase	Cellulase/xylanase	SE	p-value
Live weight (kg)	22.8	22.7	22.9	0.45	0.967
DM intake (kg)	0.86	0.84	0.87	0.017	0.538
DM digestibility	62.30	65.94	63.61	0.75	0.129
OM digestibility	63.92	67.28	65.08	0.71	0.151
EE digestibility	51.26	52.19	50.70	2.01	0.957
NDF digestibility	49.54	53.36	50.55	0.89	0.187
ADF digestibility	47.15	56.86	54.68	2.02	0.157
N intake (g/d)	16.37	16.01	16.51	0.012	0.698
Fecal N (g/d)	6.05	5.54	6.10	0.007	0.097
N digestibility (%)	62.99	65.34	63.13	0.46	0.063
Urine N (g/d)	4.12	5.29	4.66	0.217	0.075
Urine N/N intake (%)	25.67	33.14	26.38	0.015	0.135
Digestible N (g/d)	10.32	10.48	10.39	0.011	0.647
NR (g/d)	5.36	5.94	6.28	0.163	0.369
NR/N intake (%)	32.74	37.10	38.04	1.463	0.371
NR/digestible N (%)	53.46	56.82	57.78	2.319	0.205

SE, standard error; DM, dry matter; OM, organic matter; EE, ether extract; NDF, neutral detergent fiber; ADF, acid detergent fiber; NR, N retention.

Table 3. Effects of dietary addition of exogenous cellulase on enteric methane emission in growing goats

Items	Control	Cellulase	Cellulase/xylanase	SE	p-value
CH_4 (L/d)	21.84	21.84	21.70	0.47	0.976
CH_4/live weight (L/kg)	0.96	0.96	0.95	0.022	0.746
CH_4/DM intake (L/kg)	25.40	26.00	24.94	0.39	0.414
CH_4/OM intake (L/kg)	28.56	29.26	28.14	0.44	0.414
CH_4/DDMI (L/kg)	40.76	39.45	39.22	0.72	0.563
CH_4/ DOMI (L/kg)	44.70	43.48	43.23	0.77	0.634
CH_4-E/GE (%)	6.0	6.1	5.9	0.13	0.435
CH_4-E/DE (%)	9.3	9.0	9.0	0.23	0.589
CH_4-E/ME (%)	10.6	10.3	10.4	0.29	0.725

SE, standard error; DM, dry matter; OM, organic matter; DDMI, digestible dry matter intake; DOMI, digestible organic matter intake; GE, gross energy; ME, metabolizable energy.

digestibility even decreased when cellulase was added to high quality forage diets of horses. Therefore, the effect and mechanisms of exogenous cellulase on nutrient digestion are still incompletely understood. Yang et al. (1999) speculated that improved digestibility caused by exogenous cellulase was due to improved microbial colonization. Wang et al. (2001) and Morgavi et al. (2004) found associative effects between exogenous cellulase and cellulase from rumen micro-organisms. Therefore, exogenous cellulase may improve nutrient digestibility by improving colonization and by stimulating the endogenous enzyme activity within the rumen. But such improved colonization and stimulated endogenous enzyme activity may be related to forage type or ration type.

Nitrogen balance

There was no difference (p>0.05) on N intake, fecal N and urinary N among the treatments. N retention (NR) and N retention rate (NR/NI) were not affected (p>0.05) by the addition of cellulase (Table 2).

Awawdeh and Obeidat (2011) found no effect on NR of lamb after the addition of cellulase at 20 g/d. Although it has been documented that ruminal microbial protein synthesis could be stimulated by the addition of cellulase (Hristov et al., 1998; Rode et al., 1999), however, this result may not be sufficient to explain the subsequent effect on NR. Gado et al. (2009) added cellulase at the rate of 40 g/d in cow diets (30% concentrate), and found no effect on NR and utilization, also microbial protein synthesis was increased. In the present study, the supplementation of cellulase had no effect on NR in goats which is in agreement with the report of Gado et al. (2009) with cows, but in conflict with McAllister et al. (2000) in lambs.

Effect of exogenous cellulase on energy metabolism

No significant (p>0.05) difference was evident on GE intake, fecal energy and urinary energy excretion, CH_4 energy, heat production and RE among treatments (Table 4). The addition of UC and CC could significantly increase the DE/GEI and ME/GEI, but did not affect HP/GEI (Table 4).

Few studies concerning the effect of cellulase on energy utilization of livestock have been reported. O'Connor et al. (2007) reported that cellulase supplementation did not improve GE digestibility of horses. In fact, horses consuming the control ration digested a greater (p<0.05) GE than those consuming the ration with cellulase

Table 4. Effects of dietary addition of exogenous cellulase on energy metabolism of growing goats

Item	Control	Cellulase	Cellulase/xylanase	SE	p-value
GE intake (MJ/d)	14.57	14.23	14.69	0.270	0.473
Fecal energy (MJ/d)	5.21	4.58	5.08	0.194	0.072
DE intake (MJ/D)	9.37	9.65	9.61	0.252	0.694
UE (MJ/d)	0.28	0.32	0.32	0.023	0.420
CH_4 (MJ/d)	0.87	0.87	0.86	0.026	0.976
ME intake (MJ/d)	8.24	8.46	8.42	0.223	0.746
HP (MJ/d)	4.21	4.28	4.25	0.065	0.889
CH_4/GE	0.059	0.060	0.058	0.007	0.735
CH_4/DE	0.092	0.089	0.089	0.003	0.854
ME (MJ/d)	8.30	8.54	8.49	0.183	0.864
RE (MJ/d)	4.09	4.26	4.24	0.198	0.942
DE/GE (%)	64.25	67.88	65.54	0.007	0.127
ME/GE (%)	56.57	59.61	57.53	0.007	0.202
HP/ME (%)	29.03	30.09	29.02	0.006	0.699

SE, standard error; GE, gross energy; DE, digestible energy; UE, urine; ME, metabolizable energy; HP, heat production; RE, retained energy.

Table 5. The relationships between methane emission and live weight, feed intake or energy intake in growing goats

Item	Equations[a]	SE	R^2
CH_4 (g/d) =	$0.674_{(0.108)}$ Live weight (kg)+$0.22_{(2.47)}$	1.15	0.54
	$16.8_{(2.82)}$ DM intake (kg/d)+$1.3_{(2.41)}$	1.18	0.51
	$18.8_{(3.16)}$ OM intake (kg/d)+$1.3_{(2.41)}$	1.18	0.51
	$18.5_{(3.36)}$ DDM intake (kg/d)+$5.5_{(1.84)}$	1.22	0.47
	$21.0_{(3.75)}$ DOM intake (kg/d)+$5.2_{(1.88)}$	1.21	0.48
CH_4-E (MJ/d) =	$0.055_{(0.0092)}$ GE intake (MJ/d)+$0.073_{(0.1340)}$	0.0655	0.51
	$0.058_{(0.0109)}$ DE intake (MJ/d)+$0.315_{(0.1050)}$	0.0691	0.45
	$0.057_{(0.0127)}$ ME intake (MJ/d)+$0.393_{(0.1070)}$	0.0742	0.37

SE, standard error; DM, dry matter; OM, organic matter; DDM, digestible DM; DOM, digestible OM; GE, gross energy; DE, digestible energy; ME, metabolizable energy.
[a] The data in brackets are SE values.

supplementation. Results from the present study showed that GE intake, fecal energy and urinary energy excretion, methane energy and heat production were not affected by the addition of cellulase.

Effect on methane emission

There were no significant differences among dietary treatments on total CH_4 emissions (g/d), CH_4 emission as a proportion of live weight or feed intake (DM, OM, digestible DM or digestible OM), or CH_4 energy output (CH_4-E) as a proportion of energy intake (GE, DE, or ME) (Table 3). There was a significant (p<0.001) relationship between CH_4 and live weight (y = 0.645x+0.2, R^2 = 0.54), CH_4 and DM intake (y = 16.7x+1.4, R^2 = 0.51), CH_4 and OM intake (y = 18.8x+1.3, R^2 = 0.51) and CH_4-E and GE intake (Table 5 and Figure 1).

Methane emission from ruminants is receiving an increased level of public attention. Different varieties of additives were tested for the mitigation of methane emission. Except for plant extracts, ionophores such as monensin or lasalocid (Guan et al., 2006), supplemental fat (Jordan et al., 2006), halogenated analogues such as bromochloromethane (Abecia et al., 2012) were studied for the possible usage in the mitigation of methane emission from ruminants. But till now, very few studies concerning the effect of cellulase on methane emission have been reported (Tang et al., 2013). Eun and Beauchemin (2007) reported that properly formulated cellulase can lower the acetate-to-propionate ratio, which is thought to be the primary mechanism of methane abatement. Apparently this hypothesis was not supported by our study. Chung et al. (2012) even found negative results, e.g. enteric methane of dairy cow increased linearly with an increasing level of cellulase supplement.

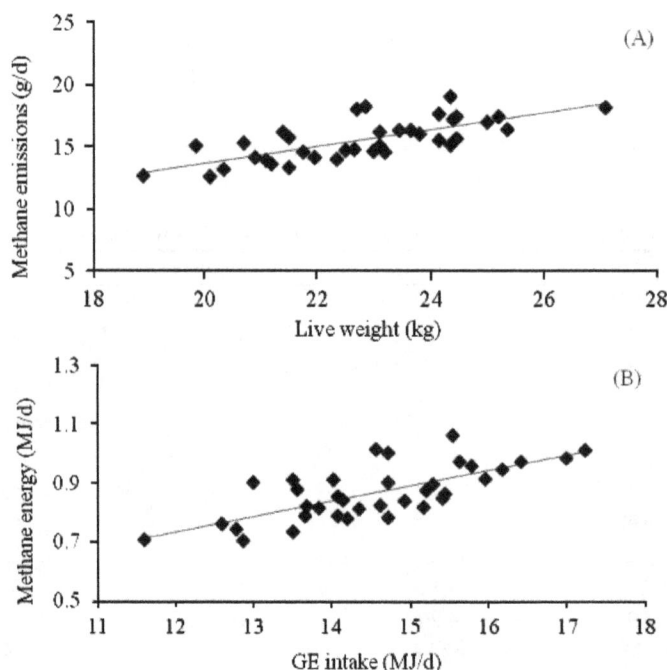

Figure 1. The linear relationships between methane emission and live weight (A), and between methane energy put and gross energy (GE) intake (B) in growing goats.

The CH$_4$-E/GE intake ranged from 0.059 to 0.061, which is within recommendations of IPCC (2006) for lambs (0.045) and mature sheep (0.065) for development of Tier 2 emission inventories. There is no recommendation of CH$_4$-E/GE intake for goats in IPCC (2006).

CONCLUSION

On the basis of above results, it may be concluded that cellulase addition had no effect on dry matter intake, nutrients digestibility and N utilization in goats. GE intake, fecal energy and urinary energy excretion, methane emission, heat production were also not affected by the cellulase supplementation.

IMPLICATIONS

The results from this study implicated that exogenous cellulase has no effect on nutrients digestion and N utilization in goats fed on high roughage diet irrespective of sources used. Supplementation of cellulase may not be an effective additive for methane mitigation.

REFERENCES

Abecia, L., P. G. Toral, A. I. Martín-García, G. Martínez, N. W. Tomkins, E. Molina-Alcaide, C. J. Newbold, and D. R. Yanez-Ruiz. 2012. Effect of bromochloromethane on methane emission, rumen fermentation pattern, milk yield, and fatty acid profile in lactating dairy goats. J. Dairy Sci. 95:2027-2036.

Awawdeh, M. S. and B. S. Obeidat. 2011. Effect of supplemental exogenous enzymes on performance of finishing Awassi lambs fed olive cake-containing diets. Livest. Sci. 138:20-24.

Ballard C. S., M. P. Carter, K. W. C. Tach, C. J. Sniffen, T. Sato, K. Uchida, A. Teo, U. D. Nhan, and T. H. Meng. 2003. Feeding fibrolytic enzymes to enhance DM and nutrient digestion and milk production by dairy cows. J. Dairy Sci. 86(Suppl. 1):150 (Abstr.).

Beauchemin, K. A., M. Kreuzer, F. O'Mara, and T. A. McAllister. 2008. Nutritional management for enteric methane abatement: A review. Aust. J. Exp. Agric. 48:21-27.

Bhasker, T. V., D. Nagalakshmi, and D. S. Rao. 2013. Development of appropriate fibrolytic enzyme combination for maize stover and its effect on rumen fermentation in sheep. Asian Australas. J. Anim. Sci. 26:945-951.

Bilik, K. and M. Łopuszańska-Rusek. 2010. Effect of adding fibrolytic enzymes to dairy cow rations on digestive activity in the rumen. Ann. Anim. Sci. 10:127-137.

Brouwer E. 1965. Report on subcommittee on constants and factors. Proc. 3rd EAAP Symp on Energy Metabolism. pp 441-443. Troon Publ., 11, Academic Press, London, UK.

Chung, Y. H., M. Zhou, L. Holtshausen, T. W. Alexander, T. A. ruminal fermentation, rumen microbial populations, and enteric methane emissions. J. Dairy Sci. 95:1419-1427.

Eun, J. S. and K. A. Beauchemin. 2007. Assessment of the efficacy of varying experimental exogenous fibrolytic enzymes using in vitro fermentation characteristics. Anim. Feed Sci. Technol. 132:298-315.

Gado, H. M., A. Z. M. Salem, P. H. Robinson, and M. Hassan. 2009. Influence of exogenous enzymes on nutrient digestibility, extent of ruminal fermentation as well as milk production and composition in dairy cows. Anim. Feed Sci. Technol. 154:36-46.

Guan, H., K. M. Wittenberg, K. H. Ominski, and D. O. Krause. 2006. Efficacy of ionophores in cattle diets for mitigation of enteric methane. J. Anim. Sci. 84:1896-1906.

Hristov, A. N., T. A. McAllister, and K. J. Cheng. 1998. Effect of dietary or abomasal supplementation of exogenous polysaccharide-degrading enzymes on rumen fermentation and nutrient digestibility. J. Anim. Sci. 76:3146-3156.

IPCC, "2006 IPCC Guidelines for National Greenhouse Gas Inventories" http://www.ipcc-nggip.iges.or.jp/public/2006gl/ Accessed on October 10, 2014.

Jordan, E., D. K. Lovett, F. J. Monahan, J. Callan, B. Flynn, and F. P. O'Mara. 2006. Effect of refined coconut oil or copra meal on methane output and on intake and performance of beef heifers. J. Anim. Sci. 84:162-170.

Lewis, G. E., W. K. Sanchez, C. W. Hunt, M. A. Guy, G. T. Pritchard, B. I. Swanson, and R. J. Treacher. 1999. Effect of direct-fed fibrolytic enzymes on the lactational performance of dairy cows. J. Dairy Sci. 82:611-617.

Lewis, G. E., C. W. Hunt, W. K. Sanchez, R. Treacher, G. Pritchard, and P. Feng. 1996. Effect of direct-fed fibrolytic enzymes on the digestive characteristics of a forage-based diet fed to beef steers. J. Anim. Sci. 74: 3020-3028.

Littell R. C., G. A. Milliken, W. W. Stroup, and R. D. Wolfinger. 1996. SAS System for Mixed Models. SAS Institute, Inc. Cary, NC, USA.

McAllister, T. A., K. Stanford, H. D. Bae, R. J. Treacher, A. N. Hristov, J. Baah, J. A. Shelford, and K. J. Cheng. 2000. Effect of a surfactant and exogenous enzymes on digestibility of feed and on growth performance and carcass traits of lambs. Can. J. Anim. Sci. 80:35-44.

Morgavi, D. P., K. A. Beauchemin, V. L. Nsereko, L. M. Rode, T. A. McAllister, and Y. Wang. 2004. Trichoderma enzymes promote Fibrobacter succinogenes S85 adhesion to, and degradation of, complex substrates but not pure cellulose. J. Sci. Food Agric. 84:1083-1090.

O'Connor-Robison, C. I., B. D. Nielsen, and R. Morris. 2007. Cellulase supplementation does not improve the digestibility of a high-forage diet in horses. J. Equine Vet. Sci. 27:535-538.

Rode, L. M., W. Z. Yang, and K. A. Beauchemin. 1999. Fibrolytic enzyme supplements for dairy cows in early lactation. J. Dairy Sci. 82:2121-2126.

Tang, S. X., Y. Zou, M. Wang, A. Z. M. Salem, N. E. Odongo, C. S. Zhou, X. F. Han, Z. L. Tan, M. Zhang, and Y. F. Fu. 2013. Effects of exogenous cellulase source on in vitro fermentation McAllister, L. L. Guan, M. Oba, and K. A. Beauchemin. 2012. A fibrolytic enzyme additive for lactating Holstein cow diets:

characteristics and methane production of crop straws and grasses. Anim. Nutr. Feed Technol. 13:489-505.

Wang, Y., T. A. McAllister, L. M. Rode, K. A. Beauchemin, D. P. Morgavi, V. L. Nsereko, A. D. Iwaasa, and W. Yang. 2001. Effects of an exogenous enzyme preparation on microbial protein synthesis, enzyme activity and attachment to feed in the Rumen Simulation Technique (Rusitec). Br. J. Nutr. 85:325-332.

Yang, W. Z., K. A. Beauchemin, and L. M. Rode. 1999. Effects of an enzyme feed additive on extent of digestion and milk production of lactating dairy cows. J. Dairy Sci. 82:391-403.

Yang, W. Z., K. A. Beauchemin, and L. M. Rode. 2000. A comparison of methods of adding fibrolytic enzymes to lactating cow diets. J. Dairy Sci. 83:2512-2520.

4

Rumen Degradability and Small Intestinal Digestibility of the Amino Acids in Four Protein Supplements

Y. Wang*, L. Jin, Q. N. Wen[2], N. K. Kopparapu, J. Liu, X. L. Liu, and Y. G. Zhang[1]*

College of Food and Biological Engineering, Qiqihar University, Qiqihar 161006, China

ABSTRACT: The supplementation of livestock feed with animal protein is a present cause for public concern, and plant protein shortages have become increasingly prominent in China. This conflict may be resolved by fully utilizing currently available sources of plant protein. We estimated the rumen degradability and the small intestinal digestibility of the amino acids (AA) in rapeseed meal (RSM), soybean meal (SBM), sunflower seed meal (SFM) and sesame meal (SSM) using the mobile nylon bag method to determine the absorbable AA content of these protein supplements as a guide towards dietary formulations for the dairy industry. Overall, this study aimed to utilize protein supplements effectively to guide dietary formulations to increase milk yield and save plant protein resources. To this end, we studied four cows with a permanent rumen fistula and duodenal T-shape fistula in a 4×4 Latin square experimental design. The results showed that the total small intestine absorbable amino acids and small intestine absorbable essential amino acids were higher in the SBM (26.34% and 13.11% dry matter [DM], respectively) than in the SFM (13.97% and 6.89% DM, respectively). The small intestine absorbable Lys contents of the SFM, SSM, RSM and SBM were 0.86%, 0.88%, 1.43%, and 2.12% (DM basis), respectively, and the absorbable Met contents of these meals were 0.28%, 1.03%, 0.52%, and 0.47% (DM basis), respectively. Among the examined food sources, the milk protein score of the SBM (0.181) was highest followed by those of the RSM (0.136), SSM (0.108) and SFM (0.106). The absorbable amino acid contents of the protein supplements accurately reflected protein availability, which is an important indicator of the balance of feed formulation. Therefore, a database detailing the absorbable AA should be established. (**Key Words:** Amino Acid, Protein, Small Intestinal Digestibility, Rumen Degradability, Absorbable Amino Acids)

INTRODUCTION

A proper balance of the amino acids (AA) in the diet of dairy cattle is desirable because it may increase the level of milk protein (Paz et al., 2014). Many researchers report that increases in dietary protein levels do not necessarily improve the milk production of dairy cows (Chiou et al., 1995), which may be due to imbalances in the AA profile of

* Corresponding Authors: Y. Wang.
E-mail: wangyan0468@126.com / Y. G.
Zhang., E-mail: zhangyonggen@sina.com
[1] Animal Science and Technology Institute, Northeast Agriculture University, Harbin 150030, China.
[2] Agricultural Machinery Research Institute of Liaoning, Shenyang 110036, China.

the rumen undegraded protein (RUP) or dietary protein degradation in the rumen that exceeds the capability of microbial proteins synthesized by rumen microorganisms (Chiou et al., 1995; Van Straalen et al., 1997; Abu-Ghazealeh et al., 2001). For the diet of dairy cows, profile AA and digestibility of RUP are estimated in a similar value as the initial protein supplements via the dairy NRC (2001) model. However, these estimates may be inaccurate, and studies have shown that the AA profiles of the RUP of protein concentrates differ from those in the original feed (Erasmus et al., 1994; Mjoun et al., 2010; Maxin et al., 2013). To satisfy the AA demands of highly lactating cows, a sufficient amount of RUP must complement the AA supplied by the microbial crude protein (CP) and endogenous sources (NRC, 2001). Therefore, the rumen degradability (RD) of the feed AA and small intestinal digestibility (SID) of rumen undegraded AA of the feed need to be determined (Borucki-Castro et al., 2007). However, little information is available on the SID of individual AA in common feeds (Von Keyserlingk and Mathison, 1989; Mjoun et al., 2010), which is not easily determined. The reason was that the mobile nylon bag method is a most widely accepted method to determine SIDs of AAs of protein supplements, but it requires the animal to be fitted with ruminal, duodenal and sometimes ileal fistula, which is expensive and may cause harm to the animals. Consequently, data of SIDs of AAs of protein supplements determined by mobile nylon bag (MNB) method are scarce. Domestic data on the SIDs of AAs are small, although there are several foreign reports for the RDs and SIDs of AAs of the protein supplements. However, protein supplements produced in China are different from those produced in other countries, due to China's climatic conditions, light conditions and variety etc., and methods of handicraft workshops for extraction of oil in China are various. Therefore, these foreign databases applied to evaluate the nutritional values of protein supplements are impractical, and it is necessary to evaluate the RD and SID of AAs systematically in China. The AA composition of the RUP fraction and the SID of RUP are necessary to determine the absorbable protein content accurately (Harstad and Prestløkken, 2001). The absorbable AA content of protein supplements accurately reflects protein availability and is an important indicator to guide feed diet formulation.

The supplementation of livestock feed with animal protein is a present cause for public concern, and plant protein shortages have become increasingly prominent in China. This conflict may be resolved by fully utilizing currently available sources of plant protein, which has motivated the study of plant protein supplements, including different processing distillers dried grains with soluble (DDGS) (Li et al., 2012). The effect of substitution of soybean meal (SBM) with cotton seed meal, high-protein DDG, or wheat DDG on milk production and composition has been widely studied, and these studies have shown that feeding these protein supplements may be as effective as feeding SBM to dairy cows (Christen et al., 2010; Oba et al., 2010; Abdelqader and Oba, 2012). This study was conducted to determine the RD and SID of the CP as well as the AA contents of rapeseed meal (RSM), SBM, sunflower seed meal (SFM) and sesame meal (SSM) using the MNB method. Overall, this study aimed to determine and compare the absorbable AA contents of these protein supplements to guide diet formulation to increase milk yield and save plant protein resources. The results of this study may serve as a reference to establish a database of absorbable AA content.

MATERIALS AND METHODS

Material

The SBM, RSM, SFM, and SSM used in this experiment were collected from six different locations in China (SBM and SFM from the Northeast, RSM from the Southwest, and SSM from the North). All feed samples were ground to pass through a 2-mm screen. The SBM was hot peeled at 45°C to 50°C, and the bran and nuts were then mechanically separated. The RSM, SFM, and SSM were extracted using mechanical methods.

Animals and feeding

Four lactating Holstein cows fitted with a rumen fistula and T-shaped duodenal fistula were allocated in a 4×4 Latin square experiment design to study the RD and SID. The protocols for the ruminal fistula surgery and the small intestinal fistula surgery in this experiment were approved by both the Animal Science and Technology College of Northeast Agricultural University and the Animal Care and Use Committee. The animals were housed in tie stalls and had free access to water, and fed *ad libitum* twice per day (08:00 am and 17:00 pm). The diet was formulated according to the NRC (2001) (Table 1), and the ration consisted of roughage and concentrate at a ratio of 60:40. Table 2 shows the chemical composition of the basal diet. The cows were adapted to the diet for 1 week prior to the study. The rumen degradation and small intestine digestion experiments were all divided into four periods.

In situ rumen incubation and *in situ* intestinal incubation of feeds

Feed samples (2.5 g) were placed in nitrogen-free polyester bags (10×20 cm; 47-μm pore size) according to the guidelines set forth by the NRC (2001); the ratio of the sample size to the surface area of nylon bags was 12.3 mg/cm^2 (Maiga et al., 1996). Forty-eight replicates of feed samples (3 replicates per cow and period) were incubated in the rumen for 16 h. The nylon bags in the rumen were

Table 1. Ingredients of the basal diet (dry matter basis)

Ingredient	% Dry matter	kg/d
Soybean meal	7	1.47
Cottonseed Meal	4	0.84
Cracked maize	21	4.41
Wheat Bran	6	1.26
Corn silage	60	12.6
Premix[1]	2	0.42
Total	100	21

[1] One kilogram of premix contains the following: 400 g limestone, 100 g Calcium perphosphate, 200 g salt, MgO 90 g, Vit A 320,000 IU, Vit D 75,000 IU, Vit E 165 mg/kg, Fe, 1,500 mg, Cu 685 mg, Zn 2,500 mg, Mn 1,500 mg, Se 80 mg, I 30 mg, Co 25 mg, and rice husk powder was used as carrier.

Table 2. Chemical composition of the basal diet (DM basis)

Nutrients	% DM
CP (%)	12.62
NDF (%)	36.18
ADF (%)	21.25
Ash (%)	5.87
pe NDF (%)	29.02
SP	4.84
Ca	0.53
P	0.60
ME (MJ/kg)	10.93
NEL (MJ/kg)	7.04
ADICP	0.76
NDICP	1.52
NPN	3.73

DM, dry matter; CP, crude protein; NDF, neutral detergent fiber; ADF, acid detergent fiber; Ash, crude ash; pe NDF, physically effective NDF; SP, soluble protein; Ca, calcium; P, phosphorus; ME, metabolic energy; NEL, lactation net energy; ADICP, acid detergent insoluble crude protein; NDICP, neutral detergent insoluble crude protein; NPN, non-protein nitrogen.

attached to a polyester rope and removed from the rumen at the same time to be immediately washed. The washes were repeated until the rinsing water ran clear. The purine derivatives of microorganisms were determined by rinsing the residue in a neutral detergent solution. The samples were then dried in an oven at 55°C until a constant weight was achieved, and the sample residues of four replicates (one bags per cow and period) collected from the mobile nylon bag of the same feed were ground through a 0.5-mm sieve to analyze CP contents and AA composition. The sample residues of the remaining eight replicates (two bags per cow and period) were ground and passed through a 2 mm sieve and transferred into eight mobile nylon bags (3.5×5.5 cm, R510 Ankom products; Ankom, Fairport, NY, USA) with pore size of 50±15 um to estimate the SID. These nylon bags were inserted into the small intestine of the cows via the T-shaped fistula and were collected from the feces according to the technique proposed by Hveplund et al. (1992).

Chemical analyses

The dry matter (DM), ether extract (EE), CP contents, crude fiber analyzed according to the AOAC (1990) procedures. Phosphorus (P) concentrations of the samples were assayed photometrically, and calcium (Ca) concentrations were determined with anatomic absorption spectrophotometer (model 5100 PC, Perkin-Elmer, Norwalk, CT, USA). The neutral detergent fiber (NDF) and acid detergent fiber (ADF) were analyzed according to the methods of Van Soest et al. (1991) using the Ankom system (Ankom 220 fiber analyzer; Ankom, USA) and heat-stable α-amylase without sodium sulfate. Physically effective

NDF (pe NDF) was calculated by the method of Sova et al. (2014). Net energy for lactation and metabolic energy were calculated based on NRC (2001) equations. The nitrogen fractions, defined according to the Cornell Net Carbohydrate and Protein System (CNCPS), were determined using the methods described by Licitra et al. (1996). The AA analysis was performed after the samples were hydrolyzed in 6 M hydrochloric acid for 24 h at 100°C. Single AA was analyzed with a Hitachi L8800 analyzer (Hitachi Co., Tokyo, Japan), with the exception of Met and Trp, whose contents were measured after hydrolysis in formic acid for 24 h (Hagen et al., 1989).

Calculations and statistical analysis

The percentage of AAs absorbed by the rumen after a 16 h incubation was calculated based on the difference in the AA content between the feed and rumen residues (Table 5). Similarly, the percentage of AAs absorbed by the small intestinal tract was calculated based on the difference between the AA content in the rumen residue after 16 h of incubation and the AA content of the feces (Table 6). The total absorbable AA content was estimated by calculating the sum of 60% of the AA that degraded in the rumen and undegraded AA in the small intestine. The milk protein score (MPS) was obtained by determining the ratio of the absorbable AA profile to the AA composition of milk protein.

The data were analyzed using a Latin design with the MIXED procedure of SAS (SAS, 2012), and differences among treatments were assessed using LSMEANS with the PDIFF in SAS (2012). Differences were considered significantly at p<0.05. The effect of the feed was considered to be fixed, whereas that of the cows was considered to be random. The following model was adopted: $Y_{ij} = u+F_i+C_j+E_{ij}$, where Y_{ij} is the value of the variable studied for the ith feed and the jth cow, u is the overall mean, F_i is the fixed effect of the ith feed (i = 1-4), C_j is the random effect of the jth cow (j = 1-4), and E_{ij} is random error.

RESULTS

Composition of feed

The chemical composition of the protein supplements is presented in Table 3. The ammonia nitrogen (AAN) content varied from 4.94% DM (SFM) to 7.96% DM (SBM). The contribution of N (nitrogen) from AAN to the total N content ranged from 69.19% (SFM) to 86.26% (RSM). The content of CP was highest in the SBM (49.73%) and lowest in the SFM (30.87%). The other two feed sources contained intermediate CP values. The soluble protein concentration of the SBM was significantly higher than those of the SFM and RSM (p<0.05). The concentrations of NDF of the SBM,

Table 3. Chemical composition of the SBM, SFM, SSM and RSM (DM basis)

Item (%)	SBM	SFM	SSM	RSM	SEM
CP	49.73[a]	30.87[d]	44.24[b]	39.33[c]	2.10
SP	12.99[a]	10.69[b]	11.53[ab]	10.2[b]	0.40
NDICP	0.67[d]	2.04[c]	2.56[b]	6.33[a]	0.63
ADICP	0.46[c]	1.10[b]	1.10[b]	2.55[a]	0.24
NPN	5.95[c]	9.46[b]	9.57[b]	6.64[a]	0.50
NDF	15.94[d]	49.94[a]	30.31[b]	29.31[c]	3.65
ADF	6.97[d]	29.34[a]	13.97[c]	19.33[b]	2.46
EE	1.77[c]	2.76[b]	7.55[a]	2.70[b]	0.68

SBM, soybean meal; SFM, sunflower meal; SSM, sesame meal; RSM, rapeseed meal; DM, dry matter; SEM, standard error; CP, crude protein; SP, soluble protein; NDICP, neutral detergent insoluble crude protein; ADICP, acid detergent insoluble crude protein; NPN, non-protein nitrogen; NDF, neutral detergent fiber; ADF, acid detergent fiber; EE, ether extract.
[a-d] Different lowercase superscripts indicate significant differences (p<0.05) in the same row.

RSM, SSM, and SFM were 15.94%, 29.31%, 30.31%, and 49.94% (DM basis), respectively, and the concentrations of ADF in these sources was 6.97%, 19.33%, 13.97%, and 29.34% (DM basis), respectively. The EE concentrations of the SBM, RSM, SFM, and SSM were 1.77%, 2.70%, 2.76%, and 7.55% (DM basis), respectively. The neutral detergent insoluble crude protein (NDICP) values of the SBM, SFM, SSM, and RSM were 0.67%, 2.04%, 2.56% and 6.33% (DM basis), respectively, the acid detergent insoluble crude protein (ADICP) values of these sources were 0.46%, 1.10%, 1.10%, and 2.55% (DM basis), respectively.

Table 4 shows that the AA compositions of the samples differed among the protein supplements. Specifically, the total amino acid (TAA) content was highest in the SBM but lowest in the SFM. Moreover, the Lys concentration was significantly higher in the SBM than in the RSM, SSM, and SFM (p<0.05); overall, the Lys concentration of the SFM was the lowest among the feeds (p<0.05). The SBM contained the most Met, whereas the SFM contained the least Met; the Met concentrations of the other two sources were intermediate.

Rumen degradability

The RDs of the feeds after 16 h of incubation are shown in Table 5. Overall, the RD contents of TAA of the RSM were highest, whereas those of the SFM were lowest (p< 0.05). For each protein source, the RD of EAA followed a pattern similar to that of the TAA. Of all examined feed sources, the RD of Met was highest in the RSM (p<0.05) and lowest in the SBM (p<0.05). The RD values of Tyr exceeded 75% in the SSM, RSM, SFM, and SBM, but the RD values of Leu and Ile were lower than 55% in these samples, as low as 13.23% in the SFM. The RDs of individual AAs also differed for the same food source. Overall, ruminal degradation modifies the AA profile from

Table 4. Amino acid composition of the SBM, SFM, SSM and RSM (DM basis)

Item (%)	SBM	SFM	SSM	RSM	SEM
TAA	42.17[a]	21.36[d]	38.01[b]	33.92[c]	2.35
CP	49.73[a]	30.87[d]	44.24[b]	39.33[c]	2.10
AAN	7.96[a]	4.94[d]	7.08[b]	6.29[c]	0.34
Lys	3.17[a]	1.02[d]	1.64[c]	2.41[b]	0.24
Asp	4.39[a]	1.75[b]	0.78[c]	2.16[b]	0.41
Thr	1.64[b]	1.02[d]	2.30[a]	1.43[c]	0.14
Ser	2.05[a]	1.08[c]	1.43[b]	1.99[a]	0.12
Glu	7.30[a]	3.86[c]	3.64[c]	5.18[b]	0.44
Gly	2.21[b]	1.65[c]	5.26[a]	2.26[b]	0.43
Ala	2.36[b]	1.16[d]	4.86[a]	1.70[c]	0.43
Cys	0.35[c]	0.47[c]	2.84[a]	0.97[b]	0.30
Val	2[b]	1.11[c]	0.84[d]	2.16[a]	0.17
Met	0.65[c]	0.45[d]	1.57[a]	1.00[b]	0.13
Ile	1.90[a]	0.85[c]	1.42[b]	1.73[b]	0.12
Leu	3.18[a]	1.54[d]	2.52[b]	2[c]	0.19
Tyr	1.7[b]	0.63[b]	2.53[a]	1.39[c]	0.21
Phe	2.86[a]	1.14[c]	1.11[c]	1.81[b]	0.22
His	1.16[a]	0.70[b]	1.00[b]	1.27[a]	0.07
Arg	3.25[a]	1.89[c]	3.33[a]	2.46[b]	0.18
Pro	2.00[a]	1.04[b]	0.94[b]	2.00[a]	0.15
EAA	19.81[a]	9.72[d]	15.73[c]	16.27[b]	1.09
NEAA	22.36[a]	11.64[c]	22.28[a]	17.65[b]	1.33
BCAA	7.08[a]	3.50[d]	4.78[c]	5.89[b]	0.40

SBM, soybean meal; SFM, sunflower meal; SSM, sesame meal; RSM, rapeseed meal; DM, dry matter; SEM, standard error; TAA, total amino acid; CP, crude protein; AAN, ammonia nitrogen; EAA, essential amino acid, NEAA, nonessential amino acid; BCAA, branched chain amino acid.
[a-d] Different lowercase superscripts indicate significant differences (p<0.05) in the same row.

feedstuffs.

Small intestinal digestibility

The SIDs of AAs of the rumen residues are presented in Table 6. The SID of TAA varied from 83.92% (RSM) to 89.76% (SBM). The SIDs of CP, TAA, EAA, branched chain amino acid, Arg, Ser, and Thr were higher than 80% for all feed sources. The SID of RUP was the highest for the SBM, followed by those of the RSM and SSM, and lowest for the SFM. The SIDs of Lys for the SSM, RSM, SBM, and SFM were 73.92%, 90.93%, 91.56%, and 94.89% (DM basis), respectively, and the SIDs of Met in these food sources were 98.58%, 72.42%, 94.78%, and 79.35% (DM basis), respectively. For all feed samples, the SIDs of Arg and the Thr were higher than 90% and 85%, respectively.

Small intestine absorbable amino acids

The small intestine absorbable AA contents of all feed samples are presented in Table 7. The levels of intestinal absorbable dietary proteins were higher in the SBM and SSM than in the SFM and RSM. The TAA was highest in

Table 5. Rumen degradability of CP and amino acid of SBM, SFM, SSM and RSM (16 h)

Item (%)	SBM	SFM	SSM	RSM	SEM
CP	68.29[a]	55.02[b]	46.83b[c]	56.05[b]	2.59
Lys	56.93[b]	23.56[c]	78.46[a]	73.32[a]	6.52
Asp	59.30[a]	58.56[a]	40.51[b]	49.99[a]	2.58
Thr	53.14[a]	55.82[a]	62.38[a]	57.45[a]	1.28
Ser	62.87[a]	75.14[a]	30.45[c]	70.94[a]	5.33
Glu	73.96[a]	55.07[b]	26.61[c]	52.33[b]	5.10
Gly	83.42[a]	43.56[c]	79.21[b]	86.35[a]	5.23
Ala	34.89[b]	33.08[b]	64.97[a]	65.51[a]	4.94
Cys	77.64[d]	85.50[c]	97.11[a]	92.13[b]	2.25
Val	67.11[ab]	48.94[c]	76.89[a]	58.57[bc]	3.39
Met	47.40[d]	53.15[c]	65.22[b]	84.58[a]	4.3
Ile	52.66[a]	13.23[c]	34.39[b]	51.36[a]	4.89
Leu	42.74[b]	35.20[c]	34.99[c]	49.10[a]	2.12
Tyr	76.32[b]	83.84[a]	76.02[b]	82.98[a]	1.15
Phe	66.47[a]	57.73[a]	29.42[b]	65.88[a]	4.84
His	57.62[c]	67.15[b]	56.64[c]	87.92[a]	3.96
Arg	60.40[b]	72.04[a]	43.72[c]	66.55[a]	3.30
Pro	80.79[b]	91.87[a]	58.22[d]	66.55[c]	3.93
TAA	62.67[b]	54.82[d]	58.94[c]	66.28[a]	1.30
EAA	56.76[b]	48.81[d]	51.64[c]	66.45[a]	2.05
NEAA	67.88[a]	59.84[d]	64.09[bc]	66.12[ab]	0.96
BCAA	52.38[b]	34.21[d]	42.21[c]	58.79[a]	2.95

CP, crude protein; SBM, soybean meal; SFM ,sunflower meal; SSM, sesame meal; RSM, rapeseed meal; SEM, standard error; TAA, total amino acid; EAA, essential amino acid; NEAA, nonessential amino acid; BCAA, branched chain amino acid.
[a-d] Different lowercase superscripts indicate significant differences (p<0.05) in the same row.

Table 6. Small intestinal digestibility of CP and amino acids of SBM, SFM, SSM and RSM

Item (%)	SBM	SFM	SSM	RSM	SEM
CP	98.13[a]	82.60[d]	84.95[c]	88.47[b]	1.80
Lys	91.56[b]	94.89[a]	73.92[d]	90.93[b]	2.47
Asp	86.07[b]	90.68[a]	77.96[c]	85.73[b]	1.42
Thr	89.04[a]	91.12[a]	90.16[a]	86.09[b]	0.63
Ser	86.19[bc]	84.18[c]	91.47[a]	88.03[ab]	1.04
Glu	79.12[c]	94.74[a]	94.17[a]	89.03[b]	1.91
Gly	52.81[a]	36.01[a]	58.97[a]	21.85[c]	4.48
Ala	93.95[bc]	95.03[ab]	96.74[a]	74.24[d]	2.77
Cys	73.47[b]	14.51[d]	67.30[c]	98.74[a]	9.27
Val	85.99[bc]	89.15[a]	53.15[d]	87.58[ab]	4.51
Met	94.78[b]	79.35[c]	98.58[a]	72.42[d]	3.28
Ile	89.67[b]	94.34[a]	94.36[a]	90.24[b]	0.71
Leu	90.88[b]	94.44[b]	95.17[a]	80.56[c]	1.78
Tyr	84.90[b]	73.35[c]	94.23[a]	73.78[c]	2.78
Phe	88.54[bc]	91.13[ab]	93.04[a]	85.40[d]	0.98
His	88.95[b]	92.24[a]	92.84[a]	74.23[c]	2.29
Arg	90.07[c]	94.97[b]	96.73[a]	91.14[c]	0.85
Pro	66.54[b]	72.31[b]	68.03[b]	78.37[a]	1.58
TAA	86.72[b]	86.49[b]	89.76[a]	83.92[c]	0.64
EAA	89.99[b]	92.67[a]	92.74[a]	86.68[c]	0.75
NEAA	82.86[b]	79.92[c]	86.94[a]	81.34[b]	0.82
BCAA	89.61[c]	93.17[a]	91.95[b]	86.16[d]	0.82

CP, crude protein; SBM, soybean meal; SFM, sunflower meal; SSM, sesame meal; RSM, rapeseed meal; SEM, standard error; TAA, total amino acid; EAA, essential amino acid; NEAA, nonessential amino acid; BCAA, branched chain amino acid.
[a-d] Different lowercase superscripts indicate significant differences (p<0.05) in the same row.

the SBM (26.33%) and lowest in the RSM (13.98%). The small intestine absorbable essential amino acid (EAA) followed the same pattern as that of the small intestine absorbable TAA. The content of small intestine absorbable Lys varied from 0.86% (SFM) to 2.12% (SBM), whereas the content of absorbable Met ranged from 0.28% (SFM) to 1.03% (SSM).

Milk protein score

The AA content relative to milk protein level is ranked in Table 8, and the following respective first and second limiting AAs of the protein supplements were determined: Met and Val for the SBM, Val and Lys for the SSM, Lys and Met for the SFM, and Leu and Lys for the RSM. The MPS of the SBM (0.181) was highest, followed by those of the RSM (0.136), SSM (0.108) and SFM (0.106).

DISCUSSION

Of all the examined protein sources, the CP of SBM was highest, which corroborated previously reported data (Robision et al., 2008; FOBI Network, 2011). The NDF and

ADF concentrations of the SBM were within the expected ranges reported in the literature (Borucki-Castro et al., 2007; Mjoun et al., 2010; Li et al., 2012). The NDICP and ADICP concentrations of the RSM were lower than the values reported in Maxin et al. (2013), which may be due to low NDF and ADF concentrations of the RSM in this study. The NDICP and ADICP values of the protein supplements are considered to be associated with NDF and ADF concentrations, and high NDICP proportions may reflect higher slowly degradable protein fraction in the rumen. Conversely, high ADICP values may result from heat treatment (ADICP being an indicator of heat-damaged proteins), which may lead to low rumen protein digestibility (Mustafa et al., 2000). In addition, the processing method (Extraction method of oil) may affect the nutrient content of the protein supplements.

Generally, most AA concentrations of the feedstuffs were within the ranges reported by the NRC (2001). The AA profiles (except the Lys and Met levels) of SBM and RSM were within the expected ranges reported in the literature (NRC, 2001; Oba et al., 2010; Heendeniya et al., 2012). The Lys concentrations of the SBM and the RSM

Table 7. Small intestinal absorbable AA contents of SBM, SFM, SSM and RSM (DM basis)

Item (%)	SBM	SFM	SSM	RSM	SEM
CP	26.33^a	13.98^c	24.74^a	20.53^b	2.28
Lys	2.12^a	0.86^c	0.88^c	1.43^b	0.16
Asp	2.77^a	1.15^c	0.51^d	1.44^b	0.25
Thr	1.10^b	0.69^d	1.47^a	0.92^c	0.09
Ser	1.27^a	0.62^b	1.12^a	1.19^a	0.08
Glu	4.10^a	2.66^a	2.98^c	3.50^b	0.17
Gly	1.08^b	0.68^c	2.64^a	1.00^b	0.23
Ala	1.84^b	0.92^c	3.16^a	0.97^c	0.27
Cys	0.17^c	0.20^c	1.37^a	0.48^b	0.09
Val	1.21^b	0.77^c	0.41^d	1.39^a	0.12
Met	0.47^b	0.28^c	1.03^a	0.52^b	0.08
Ile	1.29^a	0.75^b	1.11^a	1.19^a	0.06
Leu	2.30^a	1.20^d	1.98^b	1.32^c	0.14
Tyr	0.96^b	0.33^d	1.49^a	0.73^c	0.13
Phe	1.76^a	0.76^c	0.88^c	1.10^b	0.12
His	0.76^a	0.44^c	0.67^b	0.65^b	0.04
Arg	2.10^b	1.15^d	2.51^a	1.54^c	0.16
Pro	1.03^b	0.52^c	0.53^c	1.16^a	0.09
TAA	26.34^a	13.97^d	24.76^b	20.52^c	1.44
EAA	13.11^a	6.89^d	10.95^b	10.06^c	0.67
NEAA	13.23^b	7.08^d	13.81^a	10.47^c	0.80
BCAA	4.80^a	2.72^d	3.51^c	3.90^b	0.23

AA, amino acid; SBM, soybean meal; SFM, sunflower meal; SSM, sesame meal; RSM, rapeseed meal; DM, dry matter; SEM, standard error; CP, crude protein; TAA, total amino acid; EAA, essential amino acid; NEAA, nonessential amino acid; BCAA, branched chain amino acid.
$^{a-d}$ Different lowercase superscripts indicate significant differences (p<0.05) in the same row.

Table 8. Essential AA milk protein score of SBM, SFM, SSM and RSM

	Milk	SBM	SFM	SSM	RSM
Lys	8.1	0.261(6)[1]	0.106(1)	0.108(2)	0.177(2)
Thr	4.6	0.240(5)	0.150(6)	0.319(7)	0.200(4)
Val	6.6	0.183(2)	0.116(3)	0.063(1)	0.211(6)
Met	2.6	0.181(1)	0.109(2)	0.396(8)	0.199(3)
Ile	5.9	0.218(3)	0.128(5)	0.188(4)	0.201(5)
Leu	9.7	0.237(4)	0.124(4)	0.204(5)	0.136(1)
Phe	4.9	0.358(8)	0.155(7)	0.180(3)	0.225(7)
His	2.7	0.281(7)	0.162(8)	0.249(6)	0.241(8)
Arg	3.6	0.584(9)	0.321(9)	0.697(9)	0.427(9)
MPS		0.181	0.106	0.108	0.136

AA, amino acid; SBM soybean meal, SFM sunflower meal, SSM sesame meal, RSM rapeseed meal; SEM, standard error; MPS, milk protein score.
[1] Limiting AA, 1st to 9th.

reported values due to differences in the maturity, variety, source, processing, and fertilization for the same type of feed (Piepenbrink and Schingoethe, 1998; Taghizadeh et al., 2005). The Glu contents of all samples were high. Therefore, the Arg contents of all samples were high because Glu is the synthetic precursor of Arg.

The RDs of most AAs differed among protein supplements, which corroborated data reported by Maxin et al. (2013). These differences indicated that rumen fermentation altered the AA profile of RUP compared with the original feed, which agreed with previous reports (NRC, 2001; Taghizadeh, et al., 2005). The tyrosine in the feed samples in this study was strongly degraded in the rumen, and the RDs of all AAs, except for that of Tyr, primarily depended on the feed, including the chemical and physical properties of the feed, which agreed with previous findings (Crooker et al., 1987; Erasmus et al., 1994). Sniffen et al. (1992) observed that Met was not easily degraded in the rumen, but the RD of Met in the RSM exceeded 80% in this study, whereas those of the other protein supplements were low. This finding was consistent with results reported by Crooker et al. (1987), who indicated that the RD of Met depends on the feed source. In this study, the RDs of Thr, Cys, His and Pro were highest among the RDs of AAs in the four protein sources, which was consistent with a report by Paz et al. (2014). His is a very reactive AA and sensitive to degradation, which corroborates a study by Gerrard (2002). In this study, branched-chain AAs (BCA), particularly the Leu in the SBM, SSM, and SFM, appeared to be resistant to rumen microbial degradation. This finding was similar to results reported by Crooker et al. (1987), who indicated that BCAs are consistently less degradable than non-branched AAs. Many previous reports obtained a similar conclusion, stating that the Leu content of the SBM was lowest (Borucki-Castro et al., 2007; Mjoun et al., 2010; Maxin et al., 2013). Of the essential AAs in the protein

were lower than the values reported in NRC (2001) (3.17% vs 6.20% for SBM; 2.41% vs 5.62% for RSM), but the average Lys concentration (3.17%) of the SBM was much higher than that (average 2.24%) reported by others (Mjoun et al., 2010; Li et al., 2012). The Met concentrations of the SBM and the RSM reported in NRC (2001) were both lower than those measured in this study (0.65% vs 1.44% for SBM; 1% vs 1.87% for RSM). These differences may be due to the different processing methods and different sources for the same feed. Lys and Met are reportedly co-limiting AAs for growth and milk synthesis in dairy cattle (Socha et al., 2005). Specifically, the ratios of Lys and Met are known to affect milk protein synthesis; the contents of Lys and Met should be maximized, and their ratio should be as close to three as possible (NRC, 2001). Although the SBM contained high levels of Lys, the levels of sulfur-containing AAs (Met and Cys) were low, which agreed with the results reported by Taghizadeh et al. (2005). In the RSM, the Lys content was high and the Met content was low, which was consistent with data reported by Piepenbrink and Schingoethe (1998). The differences in the AA contents of these feed samples in this study may differ from previously supplements, the RDs of Ile and Leu were lowest, which may be due to their low solubility (Maxin et al., 2013). Only small and inconsistent differences were observed in the RDs of TAA and CP in RSM, which agrees with findings reported by Piepenbrink and Schingoethe (1998). The RDs of Cys, His, Gly, Met and Tyr were highest among the AAs in RSM, whereas those of Asp and Leu were lowest, which was similar to findings reported by Maxin et al. (2013) for RSM. The differences in the RDs of individual AAs among protein supplements may be related to the physical properties (such as the solubility), AA composition of the feed, characteristics of rumen digestive enzymes and amounts of protozoa and bacteria present in the rumen (Messman et al., 1992; Van Straalen et al., 1997). The variation in the AA profile of the original feed emphasizes the importance of determining the AA profile of the undegradable protein and indicates that the feed AA profile should be adjusted to account for differential rumen degradation (Gonzalez et al., 2001).

The SIDs for the majority of AAs significantly differed from that of TAA in this study, but most of these differences were small, which was consistent with Borucki-Castro et al. (2007). In this study, the SID of the RUP was the highest for the SBM, followed by those of RSM and SSM; the observations for SBM and RSM were consistent with published values (Borucki-Castro et al., 2007). Information about the SID for CP and AA is scarce for SSM and SFM. The SID of RUP can vary widely depending on the feedstuff and specific AA (Hvelplund et al, 1992; O'Mara et al., 1997). The SID of Lys in the SBM was higher than 90%, which was consistent with findings by Borucki-Castro et al. (2007) and Boucher et al. (2009a). Hastad and Prestløkken (2001) reported that the SID of the RUP in RSM was 94.6%, which was higher than the value determined in the present study (88.47%). Furthermore, the SIDs of most AAs in SBM in the present study were lower than values reported by Mjoun et al. (2010), which may be due to differences in heat processing temperatures for the feed. In addition, when using the mobile bag technique, bags should be recovered at the terminal ileum. However, bags were commonly recovered from feces for practical reasons, which may be responsible for the differences between the results of this study and those of other studies. The SID of Arg exceeded 90% for all protein supplements, which may be due to the action of trypsin, which can hydrolyze the bonds between Lys and Arg. Arg participates in the ornithine cycle, which allows it to not only provide energy but also urea being detoxified (Van Straalen et al., 1997).

For all feedstuffs, intestinal incubation considerably affected the AA profiles. Compared with the RSM and SFM, the SBM and SSM were better sources of intestinal absorbable dietary protein. For all feed samples, Pro, Gly, Ala, and Cys were minimally absorbed in the small intestine, but Glu, Leu, Tyr, Phe, and Arg were easily absorbed in the small intestine, which was consistent with the SIDs of AAs for Horse beans and White kidney beans obtained by Cros et al. (1992). The profiles of small intestinal absorbable AAs differed by feedstuff, which emphasized the importance of models that account for this factor to improve the accuracy of estimating the dietary supply of AA. As price fluctuates, small intestinal absorbable protein and AA may be used as a tool to aid in the selection of feedstuffs of differing protein quality.

The MPS remained low for the SFM and SSM. Thus, animal feed supplemented with SFM and SSM requires further supplementation with other AA or combination with other feeds to ensure a complete AA profile. The combination of these protein supplements likely improved the AA profile by supplementing the feed with AAs that were deficient. The efficiency of microbial proteins synthesized by the rumen absorbed in the small intestine was almost 60%, and the efficiency of RUP absorbed in the small intestine almost as high as 80%. Therefore, increasing the content of undegraded rumen AA may improve the utilized AA efficiency. To increase milk production, diets should be formulated to specifically include absorbable AAs, as opposed to AAs in general.

Before rumen residues were placed into the small intestine, they were incubated with pepsin-HCl to imitate abomasum digestion, and the resultant SIDs were found to be similar to those observed by Voigt et al. (1985) and Van Straalen et al. (1993). The mobile nylon bags with the rumen residues were recovered from the feces, and the digestibility of AAs was then determined by calculating the sum of the AAs that had been absorbed by the small and large intestine. Many studies reported that large intestine fermentation exerts limited effects on total intestinal absorption (Voigt et al., 1985; Van Straalen et al., 1997). Eramus et al. (1994) concluded that microorganisms from the large intestine affected between 0.9% and 8.6% of the SID of the feed, and these values were lower for concentrates. Kohn and Allen (1992) also found that the feed samples containing higher levels of protein were less contaminated by microbes. Therefore, the large intestine digestibility was not determined in our experiments, and the data were not corrected for microbial contamination by the large intestine.

Across feedstuffs, the rumen incubation time (16 h) used in this study was recommended by Erasmus et al. (1994) and Boucher et al. (2009) to simulate the retention time. Therefore, the RD was generally higher in our study than in other studies, which examined rumen incubation times of 12 h (De Boer et al., 1987; Von keyseling and Mathison, 1989). The rumen incubation time affects the

SID of feed samples and total absorption of protein. These results confirmed previous findings: longer incubation times may correlate with increased degradation in the rumen (De Boer et al., 1987; Von Keyseling and Mathison, 1989).

CONCLUSIONS

The digested proportions of AAs differed by feed sample and AA. The absorbable AA content is an important index to adjust the balance of AAs in the feed of dairy cows to increase milk yield and save plant protein resources. The absorbable AA data should be incorporated in a large integral database.

ACKNOWLEDGMENTS

The authors thank Natural Science Foundation of Heilongjiang Province (C2015049), FARA project of the Agriculture Ministry in China (CARS-37), the Major Application Technology Research and Development Program in Heilongjiang Province (2013G0880), National Natural Science Foundation (C200101) and the Youth Fund of Qiqihar University (2014k-M26) for financial support.

REFERENCES

AOAC. 1990. Association of Official Analytical Chemists, Official Methods of Analysis. 15th Edition. Washington, DC, USA.

Abdelqader, M. M. and M. Oba. 2012. Lactation performance of dairy cows fed increasing concentrations of wheat dried distillers grains with solubles. J. Dairy Sci. 95:3894-3904.

Abu-Ghazealeh, A. A., D. J. Schingoethe, and A. R. Hippen. 2001. Blood amino acids and milk composition from cows fed soybean meal, fish meal, or both. J. Dairy Sci. 84:1174-1181.

Boucher, S. E., S. Calsamiglia, C. M. Parsons, M. D. Stern, M. Ruiz Moreno, M. Vázquez-Añón, and C. G. Schwab. 2009a. In vitro digestibility of individual amino acids in rumen-undegraded protein: The modified three-step procedure and the immobilized digestive enzyme assay. J. Dairy Sci. 92:3939-3950.

Borucki-Castro, S. I., L. E. Phillip, H. Lapierre., P. W. Jardon and, and R. Berthiaume. 2007. Ruminal degradability and intestinal digestibility of protein and amino acids in treated soybean meal products. J. Dairy Sci. 90:810-812.

Chiou, P. W. S., K. Chen, K. Kuo, J. Hsu, and B. Yu. 1995. Studies on the protein degradabilities of feedstuffs in Taiwan. Anim. Feed Sci. Technol. 55:215-226.

Christen, K. A., D. J. Schingoethe, K. F. Kalscheur, A. R. Hippen, K. K. Karges, and M. L. Gibson. 2010. Response of lactating dairy cows to high protein distillers grains or 3 other protein supplements. J. Dairy Sci. 93:2095-2104.

Crooker, B. A., J. H. Clark, R. D. Shanks, and G. C. Fahey Jr.. 1987. Effects of ruminal exposure on the amino acid profile of feeds. Can. J. Anim. Sci. 67:1143-1148.

Cros, P., M. Vernay, C. Bayourthe, and R. Moncou-Ion. 1992. Influence of extrusion on ruminal and intestinal disappearance of amino acids in whole horsebean. Can. J. Anim. Sci. 72: 359-366.

De Boer, G., J. J. Murphy, and J. J. Kennedy. 1987. Mobile nylon bag for estimating intestinal availability of rumen undegradable protein. J. Dairy Sci. 70:977-982.

Erasmus, L. J., P. M. Botha, and C. W. Cruywagen. 1994. Amino acid profile and intestinal digestibility in dairy cows of rumen-undegradable protein from various feedstuffs. J. Dairy Sci. 77:541-551.

FOBI Network. 2011. Wheat DDGS Feed Guide, 1st ed. Feed Opportunities for Biofuels Industries. Canadian International Grains Institute, Winnipeg, Canada.

Gerrard, J. A. 2002. Protein–protein crosslinking in food: methods, consequences, applications. Trends Food Sci. Technol. 13:391-399.

Gonzalez, J., C. Centeno, F. Lamrani, and C. A. Rodrigez. 2001. In situ rumen degradation of amino acids from different feeds corrected for microbial contamination. Anim. Res. 50:253-264.

Hagen, S. R., B. Frost, and J. Augustin. 1989. Precolumn phenylisothiocyanate derivatization and liquid chromatography of amino acids in food. J. Assoc. Off. Anal. Chem. 72:912-916.

Harstad, O. M. and E. Prestløkken. 2001. Rumen degradability and intestinal indigestibility of individual amino acids in corn gluten meal, canola meal and fish meal determined in situ. Anim. Feed Sci. Technol. 94:127-135.

Heendeniya, R. G., D. A. Christensen, D. D. Maenz, J. J. McKinnon, and P. Yu. 2012. Protein fractionation byproduct from canola meal for dairy cattle. J. Dairy Sci. 95:4488-4500.

Hvelplund, T., M. R. Weisbjerg, and L. S. Andersen. 1992. Estimation of the true digestibility of rumen undegraded dietary protein in the small intestine of ruminants by the mobile bag technique. Acta Agric. Scand. Sec. A. Anim. Sci. 42:34-39.

Kohn, R. A. and M. S. Allen. 1992. Storage of fresh and ensiled forages by freezing affects fiber and crude protein fractions. J. Sci. Food Agric. 58:215-220.

Li, C., J. Q. Li, W. Z. Yang, and K. A. Beauchemin. 2012. Ruminal and intestinal amino acid digestion of distiller's grain vary with grain source and milling process. Anim. Feed Sci. Technol. 175:121-130.

Licitra, G., T. M. Hernandez, and P. J. Van Soest. 1996. Standardization of procedures for nitrogen fractionation of ruminant feeds. Anim. Feed Sci. Technol. 57:347-358.

Mjoun, K., K. F. Kalscheur, A. R. Hippen, and D. J. Schingoethe. 2010. Ruminal degradability and intestinal digestibility of protein and amino acids in soybean and corn distillers grains products. J. Dairy Sci. 93:4144-4154.

Maxin, G., D. R. Ouellet, and H. Lapierre. 2013. Ruminal

degradability of dry matter, crude protein, and amino acids in soybean meal, canola meal, corn, and wheat dried distillers grains. J. Dairy Sci. 96:5151-5160.

Maiga, H. A., D. J. Schingoethe, and J. E. Henson. 1996. Ruminal degradation, amino acid composition, and intestinal digestibility of the residual components of five protein supplements. J. Dairy Sci. 79:1647-1653.

Messman, M. A., W. P. Weiss, and D. O. Erickson. 1992. Effects of nitrogen fertilization and maturity of bromegrass on nitrogen and amino acid utilization by cows. J. Anim. Sci. 70: 566-575.

Mustafa, A. F., D. A. Christensen, J. J. McKinnon, and R. Newkirk. 2000. Effects of stage of processing of canola seed on chemical composition and in vitro protein degradability of canola meal and intermediate products. Can. J. Anim. Sci. 80: 211-214.

National Research Council (NRC). 2001. Nutrient Requirement of Dairy Cattle, 7th revised Edition. National Academy Press. Washington, DC, USA

Oba, M., G. B. Penner, T. D. Whyte, and K. Wierenga. 2010. Effects of feeding triticale dried distillers grains plus solubles as a nitrogen source on productivity of lactating dairy cows. J. Dairy Sci. 93:2044-2052.

O'Mara, F. P., J. J. Murphy, and M. Rath. 1997. The amino acid composition of protein feedstuffs before and after ruminal incubation and after subsequent passage through the intestines of dairy cows. J. Anim. Sci. 75:1941-1949.

Piepenbrink, M. S. and D. J. Schingoethe. 1998. Ruminal degradation, amino acid composition, and estimated intestinal digestibilities of four protein supplements. J. Dairy Sci. 81: 454-461.

Paz., H. A., T. J. Klopfenstein, D. Hostetler, S. C. Fernando, E. Castillo-Lopez, and P. J. Kononoff. 2014. Ruminal degradation and intestinal digestibility of protein and amino acids in high-protein feedstuffs commonly used in dairy diets. J. Dairy Sci. 97:6485-6498.

Robinson, P. H., K. Karges, and M. L. Gibson. 2008. Nutritional evaluation of four co-product feedstuffs from the motor fuel ethanol distillation industry in the Midwestern USA. Anim. Feed Sci. Technol. 146:345-352.

SAS Institute Inc. 2012. SAS Online Doc 9.3.1. SAS Institute Inc , Cary, NC, USA.

Socha, M. T., D. E. Putnam, B. D. Garthwaite, N. L. Whitehouse, N. A. Kierstead, C. G. Schwab, G. A. Ducharme, and J. C. Robert. 2005. Improving intestinal amino acid supply of pre- and postpartum dairy cows with rumen-protected methionine and lysine. J. Dairy Sci. 88:1113-1126.

Sova, A. D., S. J. LeBlanc, B. W. McBride, and T. J. DeVries. 2014. Accuracy and precision of total mixed rations fed on commercial dairy farms. J. Dairy Sci. 97:562-571.

Sniffen, C., J. D. O'connor, P. J. Van Soest, D. G. Fox, and J. B. Russell. 1992. A net carbohydrate and protein system for evaluating cattle diets: II. Carbohydrate and protein availability. J. Anim. Sci. 70:3562-3577.

Taghizadeh, A., M. Danesh Mesgaran, R. Valizadeh, F. Eftekhar Shahroodi, and K. Stanford. 2005. Digestion of feed amino acids in the rumen and intestine of steers measured using a mobile nylon bag technique. J. Dairy Sci. 88:1807-1814.

Van Straalen, W. M., J. J. Odigaand, and W. Mostert. 1997. Digestion of feed amino acids in the rumen and small intestine of dairy cows measured with nylon-bag techniques. Br. J. Nutr. 77:83-97.

Von Keyserlingk, M. A. G. and G. W. Mathison. 1989. Use of the in situ technique and passage rate constants in predicting voluntary intake and apparent digestibility of forages by steers. Can. J. Anim. Sci. 69:973-987.

Van Soest, P. J., J. B. Robertson, and B. A. Lewis. 1991. Methods for dietary fiber, neutral detergent fiber, and nonstarch polysaccharides in relation to animal nutrition. J. Dairy Sci. 74:3583-3597.

Van Straalen, W. M., F. M. Dooper, A. M. Antoniewicz, I. Kosmala, and A. M. Van Vuuren. 1993. Intestinal digestibility of protein from grass and clover in dairy cows measured with the mobile nylon bag and other methods. J. Dairy Sci. 76: 2970-2981.

Van Straalen, W. M., J. J. Odinga, and W. Mostert. 1997. Digestion of feed amino acids in the rumen and small intestine of dairy cows measured with nylon-bag techniques. Br. J. Nutr. 77:83-97.

Voigt, J., B. Piatkowski, H. Englelman, and E. Rudolph. 1985. Measurement of the postruminal digestibility of crude protein by the bag technique in cows. Arch. Tierernahr. 35: 555-562.

5

Supplementing Vitamin E to the Ration of Beef Cattle Increased the Utilization Efficiency of Dietary Nitrogen

Chen Wei, Shixin Lin, Jinlong Wu[1], Guangyong Zhao*, Tingting Zhang, and Wensi Zheng

State Key Laboratory of Animal Nutrition, College of Animal Science and Technology,
China Agricultural University, Beijing 100193, China

ABSTRACT: The objectives of the trial were to investigate the effects of supplementing vitamin E (VE) on nutrient digestion, nitrogen (N) retention and plasma parameters of beef cattle in feedlot. Four growing Simmental bulls, fed with a total mixed ration composed of corn silage and concentrate mixture as basal ration, were used as the experimental animals. Four levels of VE product, i.e. 0, 150, 300, 600 mg/head/d (equivalent to 0, 75, 150, 300 IU VE/head/d), were supplemented to the basal ration (VE content 38 IU/kg dry matter) in a 4×4 Latin square design as experimental treatments I, II, III and IV, respectively. Each experimental period lasted 15 days, of which the first 12 days were for pretreatment and the last 3 days for sampling. The results showed that supplementing VE did not affect the nutrient digestibility (p>0.05) whereas decreased the urinary N excretion (p<0.01), increased the N retention (p<0.05) and tended to increase the microbial N supply estimated based on the total urinary purine derivatives (p = 0.057). Supplementing VE increased the plasma concentrations of VE, glucose and triglycerol (TG) (p<0.05) and tended to increase the plasma concentration of total protein (p = 0.096) whereas did not affect the plasma antioxidant indices and other parameters (p>0.05). It was concluded that supplementing VE up to 300 IU/head/d did not affect the nutrient digestibility whereas supplementing VE at 150 or 300 IU/head/d increased the N retention and the plasma concentrations of VE and TG (p<0.05) of beef cattle. (**Key Words:** Vitamin E, Digestibility, Nitrogen Retention, Plasma, Beef Cattle)

INTRODUCTION

Vitamin E (VE) plays important roles in animal growth, development and reproduction (Liu et al., 1995; McDowell et al., 1996; Rooke et al., 2004). In the quality of beef and mutton, it was reported that supplementing VE enhanced the anti-oxidation and palatability characteristics. When 500 IU VE/head/d (equivalent to 76 IU/kg dry matter [DM]) was added to the ration of cross bred steers fed with wet distiller's grain-based diets it reduced the lipid oxidation and drip loss (Bloomberg et al., 2011). When 45 mg VE/head/d was supplemented to the barley-based ration during a 75-day fattening period of *Awassi* male lambs it tended to maintain meat redness of mutton (Macit et al.,

2003). In the performance of beef cattle, it was observed that adding up to 500 IU VE/head/d to the finishing ration of cross breed steers (equivalent to 76 IU/kg DM) resulted in a linear increase in carcass-adjusted body weight in average daily gain (Burken et al., 2012) and heifers receiving 570 IU VE/head/d tended to have a higher dressing percentage than receiving 285 IU VE/head/d (Rivera et al., 2002). In *in vitro* rumen fermentation, it was reported that supplementing VE increased the total volatile fatty acid production and improved the growth of rumen microorganisms (Hino et al., 1993; Naziroğlu et al., 2002; Hou et al., 2013). It could be hypothesized that the positive influence of VE on growth and fattening of beef cattle could result from the effects of VE on nutrient digestibility and N metabolism.

In China, the typical rations for beef cattle usually contain more than 50% of roughages including dried corn stover, corn silage, wheat straw, rice straw and other agricultural byproducts which contain a low content of VE.

* Corresponding Author: Guangyong Zhao.
E-mail: zhaogy@cau.edu.cn
[1] DSM China Animal Nutrition Centre, Bazhou, Hebei 065700, China.

In beef production, VE is not supplemented to the rations of beef cattle since it is expensive. It is unclear if supplementing VE to the ration of beef cattle is beneficial to the dietary nutrient digestion and utilization. The objectives of the trial were to study the effects of supplementing VE on nutrient digestion, nitrogen (N) retention and plasma parameters of beef cattle.

MATERIALS AND METHODS

Animals and feeding

Four Simmental bulls, aged at 12 months, with average live weight of 320±15 kg and fitted with permanent rumen fistulas made of polyethylene, were used as the experimental animals. The cattle were housed in separated pens with rubber mattress in a feedlot and fed with a total mixed ration (TMR) (Table 1) prepared everyday and mixed by hand. The TMR included 8.0 kg/d corn silage and 2.0 kg/d concentrate mixture (DM intake 3.73 kg/d), supplying major nutrients to the cattle at the levels of about 1.1 times of maintenance requirements (Feng, 2000). The corn silage was made from a conventional dent hybrid corn species harvested in October at early milk-line period, chopped to 1-2 cm in length and then filled in a bunker silo (length×width×height: 70 m×25 m×3 m). The silage was used after ensiled for 45 days. The corn silage contained DM 24.43%, organic matter (OM) 89.80% DM; ether extract (EE) 1.97% DM, neutral detergent fibre (NDF) 59.18% DM, acid detergent fibre (ADF) 36.73% DM and crude protein (CP) 7.14% DM. The TMR was divided into

Table 1. Ingredients and nutritional composition of experimental ration

Items	Content (% DM)
Ingredients	
Corn	17.82
Wheat bran	1.86
Soybean meal	8.10
Corn gluten meal	18.48
Sodium chloride	0.48
Sodium bicarbonate	0.91
Corn silage	52.35
Total	100.00
Nutritional composition[1]	
OM	91.96
EE	1.93
CP	11.70
NDF	41.35
ADF	22.69
VE (IU/kg DM)	38.00

DM, dry matter; OM, organic matter; EE, ether extract; CP, crude protein; NDF, neutral detergent fiber; ADF, acid detergent fiber; VE, Vitamin E.
[1] Analyzed values.

two equal meals and fed at 7:00 and 17:00, respectively. The cattle had free access to clean drinking water.

Experimental design

Four levels of VE product (all-rac-α-tocopheryl acetate, VE purity 50%) from DSM China Ltd, i.e. 0, 150, 300, 600 mg/head/d (equivalent to 0, 75, 150, 300 IU/d or 0, 20.1, 40.2, 80.4 IU/kg DM, 1 mg all-rac-α-tocopheryl acetate = 1 International Unit), were supplemented to the basal ration (VE content 38 IU/kg DM) in a 4×4 Latin square design as experimental treatments I, II, III, and IV, respectively. Since the VE content of the TMR was 38 IU/kg DM, the total VE supply to the cattle in four treatments was 142, 217, 292, and 442 IU/d, respectively. Each experimental period was 15 days, of which the first 12 days were for pretreatment and the last 3 days for sampling.

Sampling

During the last 3 days of each experimental period, the faeces and the urine were completely collected at 10:00 daily using the method similar to Dermauw et al. (2013). The faeces from each animal were collected using a plastic bucket and 3% of the faeces were sampled. The urine from each animal was collected using a urine collecting apparatus consisted of a rubber funnel connected to a polyvinyl chloride pipe and a plastic barrel. The rubber funnel was harnessed in the position to the penis by flexible hose straps. An aliquot of 1% of the urine was sampled. On the last sampling day at 11:00, 10 mL of blood sample was taken from the jugular vein of each animal using vacutainers containing Na_2-EDTA (Greiner Bio-One GmbH, Frickenhausen, Germany) and were centrifuged at 2,200×g for 15 min to obtain plasma. All the samples were kept at – 20°C for later analysis.

Determinations and chemical analysis

The DM, ash, CP, and EE of the samples were determined according to AOAC (2007). The NDF and ADF of the samples were analyzed using the methods of Van Soest et al. (1991). The OM was calculated by DM minus ash. The VE content of TMR and the VE product was analyzed using high performance liquid chromatography (Agilent 1200 HPLC, Agilent Technologies Inc., Santa Clara, CA, USA) according to China National Standards GB/T 17812-2008 and GB/T 7293-2006, respectively. The urinary purine derivatives (PD) including allantoin and uric acid were analysed using the method of Chen and Gomes (1992) on a spectrophotometer (UV-9100, Beijing *Beifen Ruili* Instrument Co. Ltd, Beijing, China). The VE concentration in plasma was determined using the VE kit (Nanjing *Jiancheng* Bioengeneering Institute, Nanjing, China). The plasma malondialdehyde (MDA), glutathione peroxidase (GSH-PX), superoxide dismutase (SOD), total antioxidation capacity (T-AOC) and catalase (CAT) were analyzed using the kits HY-50116, HY-60005, HY-60001, and HY-50121, respectively (Beijing Sino-UK Institute of Biological Technology, Beijing, China). The plasma total protein (TP), plasma urea nitrogen, glucose (GLU) and triglycerol (TG) were analyzed using the biuret method, the enzyme coupling ratio method, the GLU oxidative method and the glycerophosphate oxidase-peroxisome method using kits, respectively (Beijing Sino-UK Institute of Biological Technology, China). The plasma insulin-like growth factor-1 (IGF-1), growth hormone (GH) and insulin (INS) were analyzed using the radioimmunoassay kits HY-082, HY-10035, and HY-10069, respectively (Beijing Sino-UK Institute of Biological Technology, China). The plasma samples were analyzed twice as duplicates for every plasma parameter to minimize the errors from analytical procedures.

Calculations and statistical analysis

The absorption of microbial purines and the rumen microbial N supply were estimated according to Chen and Gomes (1992):

$$X = (Y - 0.385 \times W^{0.75})/0.85$$

where X refers to the absorption of microbial purines, mmol/d; Y, the excretion of PD in urine, mmol/d; $W^{0.75}$, the metabolic body weight of cattle, kg; 0.385, the endogenous contribution to PD excretion, mmol/kg $W^{0.75}$; 0.85, the recovery rate of absorbed purines in urine.

Rumen microbial N supply (g/d)
$$= X \times 70/(0.83 \times 0.116 \times 1,000) = 0.727 \times X$$

where X refers to the absorption of microbial purines, mmol/d; 70, the N content of purine, mg N/mmol; 0.83, the digestibility of microbial purines; 0.116, the ratio of the purine-N in the total N of mixed rumen microbes.

The data were analyzed using the general linear model procedure of SAS 9.1 (SAS Institute Inc, 2003). The model used for the analysis was: $Y = \mu + VE_i + Period_j + Cattle_k + \varepsilon$, where Y = observation, μ = general mean, VE = effect of V_E (i = 1 to 4), $Period$ = effect of period (j = 1 to 4), and $Cattle$ = effect of cattle (k = 1 to 4), ε = residual error. Differences between treatments were determined by Student-Newman-Keuls multiple-range test and considered to be significant at $p<0.05$, extremely significant at $p<0.01$ and tended to be significant at $0.05<p<0.10$.

RESULTS

The results in Table 2 show that supplementing VE up to 300 IU/head/d did not affect the digestibility of DM, OM, EE, NDF, ADF, and CP of ration ($p>0.05$) while it decreased the urinary N excretion ($p<0.01$). Supplementing VE at 150 or 300 IU/head/d increased the N retention

Table 2. Effects of supplementing vitamin E on nutrient digestibility, N balance, urinary purine derivatives excretion and estimated microbial N supply in beef cattle

Items	Treatments				SEM	p-value
	I	II	III	IV		
Nutrient digestibility (%)						
DM	71.8	70.9	71.6	70.9	1.76	0.164
OM	74.1	73.3	74.1	73.3	1.64	0.162
EE	77.3	75.4	77.5	77.0	2.63	0.284
NDF	58.4	56.6	56.6	55.9	4.47	0.641
ADF	57.3	55.0	54.6	54.5	4.25	0.237
CP	69.8	69.5	69.4	69.1	1.90	0.153
N balance (g/d)						
N intake	69.85	69.86	69.86	69.85	0.01	0.455
Faecal N	21.09	21.58	21.23	21.37	1.29	0.187
Urinary N	38.81[a]	37.17[ab]	36.42[ab]	35.81[b]	1.55	0.003
N retention	9.96[b]	11.11[ab]	12.58[a]	12.68[a]	2.55	0.027
Urinary PD excretion (mmol/d)						
Allantoin	35.05	35.90	42.50	43.77	4.54	0.074
Uric acid	5.28	5.74	5.38	4.84	0.50	0.187
Total urinary PD	40.33	41.65	47.88	48.62	4.43	0.082
Estimated microbial N (g/d)	9.15	10.49	15.97	16.24	3.47	0.057

SEM, standard error of means; DM, dry matter; OM, organic matter; EE, ether extract; NDF, neutral detergent fiber; ADF, acid detergent fiber; CP, crude protein; N, nitrogen; PD, purine derivatives.
n = 4.
[a,b] Values in the same row with different small letter superscripts mean significant difference (p<0.05).

(p<0.05) and tended to increase the urinary PD excretion (p = 0.082) and consequently the estimated rumen microbial N synthesis (p = 0.057) of the cattle.

The results in Table 3 showed that supplementing VE at 150 or 300 IU/head/d increased the plasma concentration of VE and TG (p<0.05) and supplementing VE at 300 IU/head/d increased the plasma concentration of GLU (p<0.05). Supplementing VE up to 300 IU/head/d tended to increase the plasma concentration of TP (p = 0.096) whereas did not affect the plasma antioxidants and hormones including IGF-1, GH and INS of the cattle (p>0.05).

DISCUSSION

Vitamin E requirement of beef cattle

McDowell et al. (1996) suggested that the optimal VE requirement of feedlot cattle (growing and finishing) was 200 to 500 IU/head/d. In the present trial, the VE supply from the basal ration for the cattle was 142 IU/head/d (38 IU/kg DM×3.73 DM intake) which was lower than the amount suggested by McDowell et al. (1996). Therefore, supplementing VE at 75, 150, and 300 IU/head/d in treatments II, III, and IV, respectively was presumed to have met the VE requirement of the cattle in the present trial (Table 2).

Nutrient digestibility

Khodamoradi et al. (2013) reported that supplementing 145 mg VE/head/d (equivalent to 7.5 mg/kg DM) to a TMR of lactating Holstein cows had no effects on the digestibility of DM, OM, CP, and EE and on the milk yield and the contents of milk fat, protein and lactose and the N utilization efficiency. The reason could be that the VE content of the basal ration met the VE requirement of the cows or the supplementation level of 145 mg VE/head/d was not high enough to influence the nutrient digestibility and the performance of the cows. The results in the present trial showed that supplementing VE up to 300 IU VE/head/d (equivalent to 80.4 IU/kg DM) did not affect the nutrient digestibility of beef cattle. Chikunya et al. (2004), however, reported that supplementing 500 IU VE (α-tocopheryl acetate)/kg DM to the ration of sheep with 50 g fatty acids/kg DM using three lipid sources increased the whole tract digestibility of cellulose. The differences found between different trials could be attributed to the VE requirements of different animal species, the VE contents in the basal rations and the VE doses.

N utilization

The rumen environment of adult ruminants is suitable for the colonization and growth of rumen anaerobic microorganisms. However, a limited amount of oxygen may go into the rumen from ingestion, rumination and drinking water. Some oxygen may also diffuse from the blood into the rumen. Stewart and Bryant (1988) reported that the concentration of oxygen present in rumen fluid could be as high as 3 mmol/L. Therefore, supplementing VE would be

Table 3. Effects of supplementing vitamin E on plasma parameters in beef cattle

Items	Treatments				SEM	p-value
	I	II	III	IV		
VE and antioxidants						
VE (µg/mL)	6.99[b]	7.18[b]	7.93[a]	8.20[a]	0.76	0.019
MDA (nM/L)	3.31	3.77	3.40	3.50	0.29	0.236
GSH-PX (U/mL)	863.65	846.40	859.58	861.25	20.86	0.706
SOD (U/mL)	56.15	52.33	54.33	53.71	2.57	0.242
T-AOC (U/L)	8.79	9.05	9.20	9.15	0.53	0.934
CAT (U/mL)	47.88	48.23	47.92	48.21	2.59	0.969
Nutrients						
TP (g/L)	69.51	74.47	71.48	70.80	2.29	0.096
PUN (mM/L)	3.13	2.57	3.16	3.71	0.76	0.202
GLU (mM/L)	3.88[b]	4.23[ab]	3.96[ab]	4.64[a]	0.36	0.039
TG (mM/L)	0.14[b]	0.16[ab]	0.18[a]	0.18[a]	0.01	0.034
Hormones						
IGF-1 (ng/mL)	222.18	198.08	221.85	215.16	35.39	0.854
GH (ng/mL)	3.83	3.48	3.87	4.21	0.74	0.923
INS (µIU/mL)	12.18	12.29	12.12	12.56	1.07	0.957

SEM, standard error of means; MDA, malondialdehyde; GSH-PX, glutathione peroxidase; SOD, superoxide dismutase; T-AOC, total antioxidation capacity; CAT, catalase; TP, total protein; PUN, plasma urea N; GLU, glucose; TG, triglycerol; IGF-1, insulin-like growth factor-1; GH, growth hormone; INS, insulin.

n = 4.

[a,b] Values in the same row with different small letter superscripts mean significant difference (p<0.05).

beneficial to the rumen environment for the colonization and growth of rumen microorganisms.

The results in the present trial showed that supplementing VE at 300 IU/head/d decreased the urinary N excretion and supplementing VE at 150 or 300 IU/head/d increased the N retention of beef cattle. The reasons for the results could be that VE was an antioxidant and maintained the integrity of cell membranes from oxidation (Burton et al., 1990) resulted from oxygen taken in from rumination, drinking water and diffusion through the rumen epithelium from the blood and therefore improved the growth of rumen protozoa and bacteria (Hino et al., 1993; Naziroğlu et al., 2002). The results were in agreement with the increased urinary PD excretion, the estimated microbial N supply and the increasing tendency of the plasma concentration of TP.

Urinary purine derivatives and estimated microbial N supply

The urinary PD excretion was directly associated with the intestinal absorption of purine and could be used for the estimation of rumen microbial N supply to ruminants (Chen and Gomes, 1992). The results in the present trial indicated that supplementing VE to the ration of beef cattle tended to increase the urinary PD excretion and consequently the estimated rumen microbial N supply. The results were in agreement with Chikunya et al. (2004) who reported that supplementing 500 IU VE (α-tocopheryl acetate)/kg DM to the rations of sheep increased the estimated microbial N yield compared to 100 IU/kg DM. The results implied that the rumen microorganisms could possibly require VE which was an antioxidant and VE could protect the integrity of microbial membranes from oxidation (Burton et al., 1990).

Plasma vitamin E and antioxidants

Weiss and Wyatt (2003) reported that supplementing up to 5,500 IU VE/head/d (equivalent to 250 IU VE/kg DM) to the ration of mid-lactation Holstein dairy cows increased the plasma concentration of VE (α-tocopherol). Lindqvist et al. (2011) reported that supplementing 2,400 IU α-tocopheryl acetate/head/d (equivalent to total 150 to 169 IU/kg DM) to the ration of Holstein dairy cows during the transition period tended to increase the plasma concentration of VE (α-tocopherol). Chikunya et al. (2004) reported that supplementing 500 IU VE/kg DM to the ration of sheep with 50 g fatty acids/kg DM increased the plasma concentration of VE (α-tocopherol). The results in the present trial indicated that supplementing VE at 150 or 300 IU/head/d increased the plasma VE concentration of beef cattle. The reason for the results could be that VE was not destroyed by ruminal microbes (Leedle et al., 1993). Therefore, supplementing VE increased the plasma concentration of VE. Liu et al. (2008) reported that

supplementing 5,000 or 10,000 IU DL-α-tocopheryl acetate/head/d (equivalent to 262 or 563 IU/kg DM) to the ration of Holstein dairy cows did not increase the plasma concentration of GSH-PX, but increased GSH-PX when combined with selenium. In the present trial, supplementing VE up to 300 IU/head/d did not affect the plasma concentration of GSH-PX and other antioxidants including SOD, MDA, T-AOC, and CAT. The results of the present trial were in agreement with Liu et al. (2008) and Bourne et al. (2007). Mahmoud et al. (2013), however, reported that injecting 5 mg sodium selenite and 450 mg VE injections (Viteselen 15, Adwia Company, Tanta, Egypt) two times per week for 1 month increased the serum concentration of GSH-PX of *Ossimi* rams (VE content in basal ration 15 mg/kg on fed basis). Similarly, Liu et al. (2008) reported that supplementing 5,000 or 10,000 IU DL-α-tocopheryl acetate/head/d (equivalent to 262 or 563 IU/kg DM) to the ration of Holstein dairy cows combined with selenium increased the plasma concentration of GSH-PX. The results indicated that the antioxidants such as the plasma concentration of GSH-PX could not be increased only if the plasma concentration of VE reached a certain level and the combination of VE and selenium was effective to increase the antioxidants of animals.

CONCLUSION

Under present feeding regimes, supplementing VE up to 300 IU/head/d did not affect the nutrient digestibility when the VE content of the basal ration was 38 IU/kg DM. Supplementing VE at 150 or 300 IU/head/d improved the utilization efficiency of dietary N and increased the plasma concentrations of VE, GLU, and TG of beef cattle.

ACKNOWLEDGMENTS

The authors thank DSM China Ltd. for financial support for the trial and also thank Beijing Dairy Industry Innovation Team for assistance in research.

REFERENCES

AOAC. 2007. Official Methods of Analysis. 18th ed. Association of Official Analytical Chemists, Gaithersburg, MD, USA.

Bloomberg, B. D., G. G. Hilton, K. G. Hanger, C. J. Richards, J. B. Morgan, and D. L. VanOverbeke. 2011. Effects of vitamin E on color stability and palatability of strip loin steaks from cattle fed distillers grains. J. Anim. Sci. 80:3769-3782.

Bourne, N., D. C. Wathes, M. McGowan, and R. Laven. 2007. A comparison of the effects of parenteral and oral administration of supplementary vitamin E on plasma vitamin E concentrations in dairy cows at different stages of lactation. Livest. Sci. 106:57-64.

Burton, G. W. and M. G. Traber. 1990. Vitamin E: antioxidant activity, biokinetics, and bioavailability. Annu. Rev. Nutr. 10:357-382.

Burken, D. B., R. B. Hicks, D. L. VanOverbeke, G. G. Hilton, J. L. Wahrmund, B. P. Holland, C. R. Krehbiel, P. K. Camfield, and C. J. Richards. 2012. Vitamin E supplementation in beef finishing diets containing 35% wet distillers grains with solubles: Feedlot performance and carcass characteristics. J. Anim. Sci. 90:1349-1355.

Chen, X. B. and M. J. Gomes. 1992. Estimation of Microbial Protein Supply to Sheep and Cattle Based on Purine Derivatives: An Overview of Technical Details. Occasional publication of International Feed Resources Unit. Rowett Research Institute, Bucksburn, Aberdeen, UK.

Chikunya, S., G. Demirel, M. Enser, J. D. Wood, R. G. Wilkinson, and L. A. Sinclair. 2004. Biohydrogenation of dietary n-3 PUFA and stability of ingested vitamin E in the rumen, and their effects on microbial activity in sheep. Br. J. Nutr. 91:539-550.

China National Standard, 2006. GB/T 7293-2006. Feed additive —Vitamin E powder. Standards Press of China, Beijing, China.

China National Standard, 2008. GB/T 17812-2008. Determination of vitamin E in feeds—High-performance liquid chromatography. Standards Press of China, Beijing, China.

Dermauw, V., K. Yisehak, E. S. Dierenfeld, G. Du Laing, J. Buyse, B. Wuyts, and G. P. J. Janssens. 2013. Effects of trace element supplementation on apparent nutrient digestibility and utilization in grass-fed zebu (Bos indicus) cattle. Livest. Sci. 155: 255-261.

Feng, Y. L. 2000. The Nutrient Requirements and Feeding Standards of Beef Cattle. China Agricultural University Press, Beijing, China.

Hino, T., N. Andoh, and H. Ohgi. 1993. Effects of β-carotene and α-tocopherol on rumen bacteria in the utilization of long-chain fatty acids and cellulose. J. Dairy Sci. 76:600-605.

Hou, J. C., F. Wang, Y. T. Wang, and F. Liu. 2013. Effects of vitamin E on the concentration of conjugated linoleic acids and accumulation of intermediates of ruminal biohydrogenation in vitro. Small Rumin. Res. 111:63-70.

Khodamoradi, S. H., F. Fatahnia, K. Taherpour, V. Pirani, L. Rashidi, and A. Azarfar. 2013. Effect of monensin and vitamin E on milk production and composition of lactating dairy cows. J. Anim. Physiol. Anim. Nutr. 97:666-674.

Leedle, R. A., J. A. Leedle, and M. D. Butine. 1993. Vitamin E is not degraded by ruminal microorganisms: assessment with ruminal contents from a steer fed a high-concentrate diet. J.

Anim. Sci. 71:3442-3450.

Lindqvist, H., E. Nadeau, K. P. Waller, S. K. Jensen, and B. Johansson. 2011. Effects of RRR-α-tocopheryl acetate supplementation during the transition period on vitamin status in blood and milk of organic dairy cows during lactation. Livest. Sci. 142:155-163.

Liu, Q., M. C. Lanari, and D. M. Schaefer. 1995. A review of dietary vitamin E supplementation for improvement of beef quality. J. Anim. Sci. 73:3131-3140.

Liu, Z. L., D. P. Yang, P. Chen, W. X. Dong, and D. M. Wang. 2008. Supplementation with selenium and vitamin E improves milk fat depression and fatty acid composition in dairy cows fed fat diet. Asian Australas. J. Anim. Sci. 21:838-844.

Macit, M., V. Aksakal, E. Emsen, N. Esenbuğa, and M. İ. Aksu. 2003. Effects of vitamin E supplementation on fattening performance, non-carcass components and retail cut percentages, and meat quality traits of Awassi lambs. Meat Sci. 64:1-6.

Mahmoud, G. B., S. M. Abdel-Raheem, and H. A. Hussein. 2013. Effect of combination of vitamin E and selenium injections on reproductive performance and blood parameters of Ossimi rams. Small Rumin. Res. 113:103-108.

McDowell, L. R., S. N. Williams, N. Hidiroglou, C. A. Njeru, G. M. Hill, L. Ochoa, and N. S. Wilkinson. 1996. Vitamin E supplementation for the ruminant. Anim. Feed Sci. Technol. 60:273-296.

Naziroğlu, M., T. Güler, and A. Yüce. 2002. Effect of vitamin E on ruminal fermentation in vitro. J. Vet. Med. A 49:251-255.

Rivera, J. D., G. C. Duff, M. L. Galyean, D. A. Walker, and G. A. Nunnery. 2002. Effects of supplemental vitamin E on performance, health, and humoral immune response of beef cattle. J. Anim. Sci. 80:933-941.

Rooke, J. A., J. J. Robinson, and J. R. Arthur. 2004. Effects of vitamin E and selenium on the performance and immune status of ewes and lambs. J. Agric. Sci. 142:253-262.

SAS. 2003. Statistical Analysis System, Version 9.1. SAS Institute Inc., Cary, NC, USA.

Stewart, C. S. and M. P. Bryant. 1988. The rumen bacteria. In: The Rumen Microbial Ecosystem (Ed. P. N. Hobson). Elsevier Science Publishers, New York, NY, USA. pp. 15-26.

Van Soest, P. J., J. B. Robertson, and B. A. Lewis. 1991. Methods for dietary fiber, neutral detergent fiber, and nonstarch polysaccharides in relation to animal nutrition. J. Dairy Sci. 74:3583-3597.

Weiss, W. P. and D. J. Wyatt. 2003. Effect of Dietary Fat and Vitamin E on α-Tocopherol in milk from dairy cows. J. Dairy Sci. 86:3582-3591.

6

Effect of Dietary Beta-Glucan on the Performance of Broilers and the Quality of Broiler Breast Meat

Sun Hee Moon[1], Inyoung Lee[2], Xi Feng[1], Hyun Yong Lee[1], Jihee Kim[1], and Dong Uk Ahn[1,3,*]

[1] Department of Animal Science, Iowa State University, Ames, IA 50011, USA

ABSTRACT: A total of 400, one day-old commercial broiler chicks were divided into five diet groups (negative control, positive control group with 55 ppm Zn-bacitracin, 15 ppm β-glucan, 30 ppm β-glucan, and 60 ppm β-glucan) and fed for six weeks. Ten broilers were allotted to each of 40 floor pens. Eight floor pens were randomly assigned to one of the 5 diets. Each diet was fed to the broilers for 6 weeks with free access to water and diet. The survival rate, growth rate, feed efficiency, and feed conversion rate of the broilers were calculated. At the end of the feeding trial, the birds were slaughtered, breast muscles deboned, and quality parameters of the breast meat during storage were determined. The high level of dietary β-glucan (60 ppm) showed better feed conversion ratio and survival rate than the negative control. The survival rate of 60 ppm β-glucan-treated group was the same as that of the antibiotic-treated group, which showed the highest survival rate among the treatments. There was no significant difference in carcass yield, water holding capacity, pH, color, and 2-thiobarbituric acid reactive substances values of chicken breast meat among the 5 treatment groups. Supplementation of 60 ppm β-glucan to broiler diet improved the survival rate and feed conversion rate of broilers to the same level as 55 ppm Zn-bacitracin group. The result indicated that use of β-glucan (60 ppm) can be a potential alternative to antibiotics to improve the survival and performance of broilers. However, dietary β-glucan showed no effects on the quality parameters of chicken breast meat. (**Key Words:** β-Glucan, Broiler, Growth Performance, Physicochemical Properties)

INTRODUCTION

β-1,3-Glucan is a functional polymer consisting of glucose with β-1,3 linkage and can be isolated from various sources, including grains, mushrooms and bacteria. β-1,3-Glucan is known to enhance immunity and bioactivity by promoting secretion of cytokines, activating macrophages, natural killer cells and neutrophils, and have antitumor, antibacterial and antiviral effects (Brown and Gordon, 2005). β-Glucan also functions as an adjuvant for monoclonal antibody immunotherapy because it can induce cellular cytotoxicity by recruiting tumoricidal granulocytes as killer cells. β-1,3-Glucan is known to increase antibody production by activating the B cells, has a

complementing function in mAB-mediated cancer immunology, and activates the secretion of IL-1, IL-2, TNF-α. Therefore, β-glucan can stimulate the cell-mediated immune reactions, which activates the macrophage, NK cell, and cytotoxic T cell (Bohn and BeMiller, 1995; Vetvicka et al., 2007; Chen et al., 2008). As a result, β-1,3-glucan can enhance the resistance against infection of microorganisms and virus by improving i) non-specific immunity, which may protect animals against infection, ii) host defense mechanism, and iii) growth rate and reduce mortality. Thus, β-glucan may be used as a replacement for dietary antibiotics in animal feeds.

Beta Polo, mainly composed of β-1,3-glucan, is a natural feed additive for poultry. Beta-Polo stimulates immune system and improves host defense mechanism, consequently reducing mortality and enhancing growth (Guo et al., 2003). The quality of meat is significantly influenced by the degree of stresses in animals during growing, transportation, pre-slaughter handling and processing. Therefore, dietary β-1,3-glucan can have

* Corresponding Author: Dong Uk Ahn.
E-mail: duahn@iastate.edu
[2] Naturence Co., Ltd., Sejong 339-824, Korea.
[3] Department of Animal Science and Technology, Sunchon National University, Sunchon 540-742, Korea.

significant impact to the quality of meat because it can reduce oxidative stress in birds during growing periods. However, few literature is available in the area.

The objectives of this study were to determine i) the effects of dietary β-glucan on the survival rate, growth rate, feed efficiency, and feed conversion rate of broilers, and ii) the effects of dietary β-glucan on the carcass yield, and color, pH, water holding capacity, composition and storage stability of broiler breast meat.

MATERIALS AND METHODS

Experimental design and diets

The study was approved by the Institutional Animal Care and Use Committee at Iowa State University (Approval # 2-12-7313-G). Four hundred, one-day-old commercial broiler chicks were divided into five dietary groups (eight replications×10 birds each replication) and fed the following diets for six weeks. NC was the negative control group (basal diet, antibiotics-free); PC, the positive control group (55 ppm Zn-bacitracin, approved for broilers and commonly used); 15 BG, 15 ppm β-glucan; 30 BG, 30 ppm β-glucan; and 60 BG, 60 ppm β-glucan. The β-glucan product containing 25% 1,3-β-glucan was obtained from Naturence Co., Ltd. (Sejong, Korea) and used to formulate the BG treatments.

All five diets were prepared on corn-soybean basal diet, which met or exceed the NRC requirements (NRC, 1994) for birds during the trial. The formula and chemical composition of the basal diets are shown in Table 1. Crude protein, metabolizable energy, Ca, P, lysine and methionine levels in the four diets were adjusted to the same levels. Ten broilers were allotted to each of 40 floor pens (experimental units), weighed, and wing banded. Eight floor pens were randomly assigned to one of the five experimental diets with different amounts of 1,3-β-glucan. Each of the dietary treatment was fed to the respective broiler groups for six weeks. Broilers had free access to water and diet. The growth and feed consumption of broilers were measured weekly during the feeding trial. At the end of the feeding trial, survival rate, feed consumption, and feed conversion rate were calculated.

Slaughtering

At the end of the feeding trial, half of the birds (200) were slaughtered in the Meat Lab at Iowa State University following USDA guidelines (Brant et al., 1982) and carcass weight were obtained 24 h after slaughter. Breast muscles were deboned from the carcasses the next day and used to measure color, water holding capacity, cooking loss, ultimate pH, and storage stability.

Table 1. Composition of the basal (control) diet (%)

Items	Starter (1 to 14 d)	Grower (14 to 28 d)	Finisher (14 to 42 d)
Ingredient			
Corn	56.3	60.02	67.87
DDGS	5	5	0
Meat/bone meal	3	3	3
Soybean meal 48	31.3	26.71	24.02
Soy oil	1.12	2.16	2.52
Salt	0.36	0.36	0.28
DL methionine	0.27	0.24	0.19
Threonine	0	0.01	0
Bio-Lys	0.27	0.31	0.22
Limestone	0.69	0.74	0.74
Dicalcium Phos	0.96	0.72	0.57
Choline chloride 60	0.1	0.1	0.1
Vitamin premix[1]	0.63	0.63	0.5
Calculated values			
Crude protein (%)	22.95	21.08	18.99
Poult (ME kcal/kg)	3,000	3,100	3,200
Calcium (%)	0.9	0.85	0.8
Phos (%)	0.74	0.67	0.61
Avail Phos (%)	0.45	0.4	0.35
Fat (%)	4.44	5.52	5.7
Fibre (%)	2.83	2.76	2.51
Met (%)	0.63	0.58	0.49
Cys (%)	0.37	0.34	0.32
Me+Cys (%)	1	0.92	0.81
Lys (%)	1.33	1.21	1.07
His (%)	0.6	0.55	0.5
Tryp (%)	0.25	0.22	0.2
Thr (%)	0.86	0.8	0.71
Arg (%)	1.5	1.35	1.23
Iso (%)	0.95	0.86	0.76
Leu (%)	1.96	1.83	1.69
Phe (%)	1.09	1	0.9
Tyr (%)	0.8	0.73	0.67
Val (%)	1.07	0.98	0.88
Gly (%)	1.04	0.97	0.91
Ser (%)	1.07	0.98	0.88
Phe+Tyr (%)	1.89	1.73	1.57
Phytate P (%)	0.22	0.21	0.2
Na (%)	0.19	0.19	0.15
Cl (%)	0.29	0.29	0.23
K (%)	0.91	0.82	0.75
Linoleic acid (%)	1.69	2.16	2.34
Na+K-Cl	233.52	211.54	192.7

DDGS, distillers dried grains with soluble; ME, metabolizable energy.
[1] Vitamin premix; 0.2 ppm Selenium, 6,608 IU vitamin A, 2,203 ICU vitamin D_3, 14 IU vitamin E, 0.88 mg menadione, 9.35 μg vitamin B_{12}, 33 μg biotin, 358 mg choline, 1.1 mg folic acid, 33 mg niacin, 8.8 mg pantothenic acid, 0.88 mg pyridoxine, 4.4 mg riboflavin, 1.1 mg thiamine)/kg basal diet. The diets were formulated to be iso-caloric based on energy values for feed ingredients published by the National Research Council, and were formulated on a total amino acid basis for methionine, threonine, and lysine.

Growth performance

Body weight and feed intake per cage were recorded, and feed conversion rate was calculated based on feed intake divided by body weight gain throughout the experiment after adjusting mortality.

Physicochemical properties of broiler breast meat

The analyses of the color, pH and water holding capacity in the broiler meat samples were conducted on one of the two breast muscles obtained from each animal by random selection.

Color measurements: Color was measured using a Labscan spectrophotometer (Hunter Associated Labs Inc., Reston, VA, USA) (AMSA, 1991) that had been calibrated against white and black reference tiles covered with the same film as those used for meat samples. Commission Internationale de l'Eclairage (CIE) L* (lightness), a* (redness), and b* (yellowness) values were obtained using illuminant A (light source). Area view and port size were 0.64 and 1.02 cm, respectively. An average value from two random locations of the meat surface was used for statistical analysis.

pH: The pH values of the breast muscle were measured in duplicate with a pH meter. About 10 g of the sample was minced to small pieces and homogenized with 90 mL of distilled water for 60 s using a Polytron homogenizer. The pH values were measured immediately after the homogenization.

Water holding capacity: Water-holding capacity was measured using the centrifugation method of Bertram et al. (2001). Breast samples were cut parallel to the muscle fiber direction, which is about 2.0 cm long and 0.5 cm×0.2 cm in cross-sectional area. The samples were weighed and placed in test tubes with a filter paper (Whatman No. 1) cushion. The tubes were sealed with parafilm and then centrifuged at 400×g at 4°C for 60 min. After centrifugation, the samples were weighed again. Water holding capacity was calculated as the percentage difference in weight before and after centrifugation. Eight replications were conducted for each treatment.

Lipid oxidation (2-thiobarbituric acid reactive substances, TBARS): Lipid oxidation of breast meat was assessed on the basis of malondialdehyde (MDA) formed during the refrigerated storage. Lipid oxidation was determined using a 2-thiobarbituric acid reactive substances (TBARS) method (Ahn et al., 1998). Meat sample (5 g) was placed in a 50-mL test tube and homogenized with 15 mL deionized distilled water for 15 s at high speed (Type PT 10/35; Brinkman Instrument Inc., Westbury, NY, USA). The meat homogenate (1 mL) was transferred to a disposable test tube, and butylated hydroxytoluene (7.2%, 50 μL) and thiobarbituric acid (TBA)/trichloroacetic acid (TCA) solution (2 mL) were added. The sample was mixed using a vortex mixer, and then incubated in a 90°C water bath for 15 min to develop color. After cooling, the samples were centrifuged at 3,000×g for 15 min at 4°C. The absorbance of the resulting upper layer was read at 531 nm against a blank prepared with 1 mL deionized distilled water and 2 mL TBA/TCA solution. The amounts of TBARS were expressed as mg of MDA per kg of meat.

Statistical analysis

Experiments were carried out in eight replications, and the results represent the average values of the replications. Samples were compared by One way analysis of variance followed by Tukey's multiple comparison test (SPSS version 18, SPSS Inc., Chicago, Illinois). Statistical significance was set at p<0.05.

RESULTS AND DISCUSSION

Growth performance

The feed consumption and body weight gain of broiler

Table 2. Effect of β-glucan on the feed consumption and body weight gain of broiler chickens

Items	1 wk	2 wk	3 wk	4 wk	5 wk	6 wk	Total
Feed consumed (kg/pen/week)							
NC	1.33±0.12	3.63±0.29	6.51±0.37	9.00±0.51	10.20±0.46	9.67±1.06	40.34±2.81
PC	1.49±0.11	3.80±0.20	6.78±0.38	8.83±0.58	10.00±0.63	9.65±1.09	40.55±3.00
15 BG	1.44±0.16	3.81±0.18	6.77±0.39	9.15±0.51	10.01±0.39	9.33±0.67	40.51±2.29
30 BG	1.45±0.13	3.83±0.18	6.92±0.22	9.21±0.39	10.17±0.29	9.21±0.52	40.79±1.73
60 BG	1.34±0.11	3.63±0.25	6.32±0.45	8.56±0.68	9.66±0.78	9.25±0.87	38.76±3.15
Body weight gain (kg/pen/week)							
NC	1.50±0.14	2.58±0.31	4.28±0.48	5.38±0.51	5.11±0.53	4.18±0.89	23.03±0.48
PC	1.58±0.07	2.71±0.17	4.58±0.17	5.40±0.49	5.39±0.23	4.36±0.63	24.04±0.29
15 BG	1.49±0.09	2.61±0.14	4.40±0.29	5.38±0.40	4.73±0.78	4.37±0.16	22.98±0.31
30 BG	1.52±0.11	2.55±0.23	4.33±0.28	5.36±0.25	5.11±0.35	4.45±0.81	23.36±0.34
60 BG	1.53±0.08	2.66±0.14	4.54±0.36	5.60±0.41	5.15±0.26	4.26±0.50	23.74±0.29

Values are mean±standard deviation of each treatment group. n = 8.
NC, negative control; PC, positive control (adding Zinc bacitracin); 15 BG, adding 15 ppm β-glucan (6 g/100 kg diet); 30 BG, adding 30 ppm β-glucan (12 g/100 kg diet); 60 BG, adding 60 ppm β-glucan (24 g/100 kg diet).

chicks are shown in Table 2. No differences in weekly feed consumption and body weight gain among dietary treatment groups were found during the 6-week feeding trial (p>0.05). The feed consumption of birds with PC, 15 BG and 30 BG were not different from that of the control (NC). However, high level of beta glucan (60 ppm, 60 BG) treatment showed numerically lower feed consumption than other treatments. In agreement with our results, other studies observed no effects of β-glucan on growth performance (Morales-Lopez et al., 2009; Cox et al., 2010). Hahn et al. (2006) also reported that β-glucan did not show any effects on average daily feed intake and gain to feed ratio (G:F ratio) as the β-glucan level of the diet (0, 0.1, 0.2, 0.3, and 0.4 g/kg) increased in weanling pigs.

Table 3 showed the feed conversion rate of chickens fed with diets containing various concentrations of β-glucan. There was no significant difference in feed conversion rate among the five treatment groups. However, the high-level β-glucan treatment (60 ppm, 60 BG) showed numerically better feed conversion rate than the negative control (Table 3). In fact, 60 ppm β-glucan group showed better feed conversion rate than that of 55 ppm bacitracin group (PC), which is encouraging. Mao et al. (2005) reported that dietary supplement of 1,3-1,6-β-glucans from Chinese herb did not show the improvement of performances. Also, 1,3-1,6,-β-glucan extracted from *Paenibacillus polymyxa* showed no significant improvement in growth performances (Hwang et al., 2008). These findings suggested that β-glucans from various sources were able to cause divergent responses in relation with their structures and sources. All β-glucan treatment groups (15 BG - 60 BG)

Table 3. Effect of β-glucan on the feed conversion rate and survival rate of broiler chickens

Group	Feed conversion rate	Survival rate (%)
NC	1.752±0.170	93.75
PC	1.687±0.101	98.75
15 BG	1.763±0.188	96.25
30 BG	1.746±0.096	96.25
60 BG	1.633±0.218	98.75

Values are mean±standard deviation of each treatment group. n = 8.
NC, negative control; PC, positive control (adding Zinc bacitracin); 15 BG; adding 15 ppm β-glucan; 30 BG, adding 30 ppm β-glucan; 60 BG, adding 60 ppm β-glucan.

showed numerically higher survival rate than the control, and the survival rate of 60 ppm β-glucan-treated group (60 BG) was the same as that of the antibiotic-treated group (PC), which showed the highest survival rate among the treatments (Table 3). Our results suggested that >60 ppm of dietary β-glucan can have a possibility of replacing antibiotics to improve survival rate and promote growth of broilers.

Physicochemical properties of breast meat

Color is the most important perceivable quality in meat products in terms of consumer acceptance. Many factors have been shown to affect poultry meat color, such as bird sex, age, strain, method of processing, exogenous chemicals, cooking method irradiation, and freezing (Froning, 1995). Table 4 shows the color L*-values (lightness), a*-values (redness), and b*-values (yellowness) of chicken breast meat from chickens fed with various concentrations of β-glucan. The lightness of chicken breast meat decreased

Table 4. Dietary effects of β-glucan on the meat color quality in broiler chickens

		0 d	1 d	3 d	7 d
L*	NC	60.87±2.22[ay]	59.98±3.17[ay]	59.75±2.25[ay]	57.49±2.19[ax]
	PC	61.45±1.65[az]	59.27±1.39[ay]	59.07±2.77[ay]	55.61±2.73[ax]
	15 BG	60.76±2.39[az]	58.16±2.08[axy]	57.49±1.35[ax]	59.64±1.87[byz]
	30 BG	59.54±2.30[ax]	58.64±2.19[ax]	57.76±3.36[ax]	57.38±2.66[ax]
	60 BG	61.33±2.54[ay]	60.34±3.90[axy]	57.86±2.57[ax]	57.90±2.81[ax]
a*	NC	7.50±1.70[ax]	7.14±1.52[ax]	7.05±0.97[ax]	8.43±2.01[ax]
	PC	7.93±1.55[ax]	8.27±1.22[axy]	7.48±1.37[abx]	9.29±1.44[ay]
	15 BG	7.83±1.14[ax]	8.44±1.54[ax]	7.54±2.06[abx]	8.07±1.37[ax]
	30 BG	7.73±1.60[ax]	8.31±0.98[ax]	8.86±1.64[bx]	8.44±2.11[ax]
	60 BG	8.72±1.45[ax]	8.28±1.33[ax]	8.54±1.53[bx]	7.89±1.39[ax]
b*	NC	12.14±1.72[ax]	11.67±3.08[ax]	10.75±0.93[ax]	13.74±2.79[ay]
	PC	13.32±2.68[bx]	12.38±1.50[ax]	12.98±2.14[ax]	13.37±1.88[ax]
	15 BG	12.61±1.45[ax]	12.38±2.32[ax]	11.80±3.20[ax]	14.09±3.11[ax]
	30 BG	11.87±1.42[ax]	11.85±2.16[ax]	12.81±3.24[ax]	13.95±3.71[ax]
	60 BG	14.48±1.11[by]	12.17±2.65[ax]	12.59±1.87[ax]	14.34±1.71[ay]

Values are mean±standard deviation of each treatment group. n = 8.
NC, negative control; PC, positive control (adding Zinc bacitracin); 15 BG, adding 15 ppm β-glucan; 30 BG, adding 30 ppm β-glucan; 60 BG, adding 60 ppm β-glucan.
[a-c] Statistically significant differences (p<0.05) between column.
[x-z] Statistically significant differences (p<0.05) between incubation row.

significantly during storage due to pigments oxidation, but no significant difference among 5 treatment groups was found. Redness and yellowness also were not influenced by the dietary treatments and storage even though there were some ups and downs in the values. This result indicated that dietary Zinc bacitracin and β-glucan had no effects on the color values of chicken breast meat. Other research agreed that β-glucan (13.45% Nutrim-10 which contain 10% β-glucan) did not affect the color characteristics in beef patties (Pinero et al., 2008).

Meat pH is known to influence parameters related to meat quality including color, tenderness, flavor and shelf-life. The pH of the chicken breast with various concentrations of β-glucan is shown in Table 5. The pH of chicken breast meat was not significantly different among the dietary treatments groups even though there were some decrease in pH after 1 day of storage in all groups (p>0.05). Water holding capacity (WHC) of the chicken breast from chickens fed with various concentrations of β-glucan is shown in Table 5. The water holding capacity of the chicken breast meat was not significantly different among the treatment groups and during storage (p>0.05). There were very large variations in WHC among the breast muscles even from the same dietary treatment group. No difference in carcass yield among the treatment groups was also detected (data not shown).

Lipid oxidation of the raw chicken breast meat during storage is shown in Table 6. The TBARS values of chicken breast were not differ significantly among the treatment groups. The TBARS values of chicken breast meat during the 7-day storage time differ significantly: 3d-stored samples had the lowest and 7 day-stored samples had the highest values. However, the difference does not have much

Table 5. Dietary effects of β-glucan on the pH values and the water holding capacity in broiler chickens

Items	0 d	1 d	3 d	7 d
pH				
NC	6.30±0.14[bx]	6.22±0.17[ax]	6.20±0.12[ax]	6.29±0.09[ax]
PC	6.12±0.10[ax]	6.24±0.08[ax]	6.27±0.07[ax]	6.18±0.09[ax]
15 BG	6.42±0.09[cy]	6.23±0.10[ax]	6.22±0.09[ax]	6.24±0.11[ax]
30 BG	6.48±0.12[cy]	6.21±0.15[ax]	6.22±0.14[ax]	6.23±0.10[ax]
60 BG	6.12±0.14[ax]	6.18±0.13[ax]	6.20±0.09[ax]	6.21±0.07[ax]
Water holding capacity				
NC	83.66±2.16[ax]	82.91±4.47[ax]	81.76±4.78[ax]	81.51±5.47[ax]
PC	83.18±2.55[ax]	81.90±3.02[ax]	85.06±5.12[ax]	81.18±6.22[ax]
15 BG	84.49±2.43[ay]	83.59±6.51[ay]	85.38±5.35[ay]	77.09±6.09[ax]
30 BG	82.50±3.26[ax]	82.40±4.38[ax]	82.31±5.89[ax]	79.34±5.30[ax]
60 BG	81.27±3.25[ax]	81.12±4.39[ax]	81.02±3.28[ax]	82.86±3.06[ax]

Values are mean±standard deviation of each treatment group. n = 8.
NC, negative control; PC, positive control (adding Zinc bacitracin); 15 BG, adding 15 ppm β-glucan; 30 BG, adding 30 ppm β-glucan; 60 BG, adding 60 ppm β-glucan.
[a-c] Statistically significant differences (p<0.05) between column.
[x-z] Statistically significant differences (p<0.05) between row.

Table 6. Dietary effects of β-glucan on the lipid oxidation in raw broiler chicken breast meat

	0 d	3 d	7 d
NC	0.16±0.07[ay]	0.09±0.01[ax]	0.20±0.03[ay]
PC	0.17±0.04[axy]	0.14±0.16[ax]	0.26±0.03[ay]
15 BG	0.14±0.04[ay]	0.08±0.02[ax]	0.21±0.04[az]
30 BG	0.15±0.04[ax]	0.11±0.01[ax]	0.22±0.06[ay]
60 BG	0.15±0.04[ay]	0.08±0.01[ax]	0.21±0.03[az]

Values are mean±standard deviation of each treatment group. n = 8.
NC, negative control; PC, positive control (adding Zinc bacitracin); 15 BG, adding 15 ppm β-glucan; 30 BG, adding 30 ppm β-glucan; 60 BG, adding 60 ppm β-glucan.
[a-c] Statistically significant differences (p<0.05) between column.
[x-z] Statistically significant differences (p<0.05) between row.

practical meaning to the meat quality at these low values. Dileep et al. (2011) demonstrated that β-glucan had radical scavenging ability while trying to use as hemopoietic stimulant/radioprotectant, and Thondre et al. (2011) reported that the free radical scavenging ability of β-glucan is due to the presence of polyphenol and antioxidant content in the commercial β-glucan sample. All the TBARS values of raw chicken breast meat are very low, indicating that raw chicken breasts are highly resistant to oxidative changes during storage. Although dietary β-glucan showed some effects to broiler performances, but did not show significant effect to the meat quality.

CONCLUSION

Dietary supplementation with β-glucan improved survival rate and feed efficiency. In general, these responses indicated that β-glucan can be a potential alternative to antibiotic growth promoter in order to improve growth performance. However, dietary β-glucan showed no effects on the quality parameters of chicken breast meat.

ACKNOWLEDGMENTS

This study was supported jointly by Naturence Co., Ltd., Korea and the Next-Generation BioGreen 21 Program (No. PJ00964305), Rural Development Administration, Republic of Korea.

REFERENCES

Ahn, D. U., D. G. Olson, C. Jo, X. Chen, C. Wu, and J. I. Lee. 1998. Effect of muscle type, packaging, and irradiation on lipid oxidation, volatile production, and color in raw pork patties. Meat Sci. 49:27-39.

AMSA (American Meat Science Association). 1991. Guidelines for meat color evaluation. In Proceedings of the 44th reciprocal meat conference. National Live Stock and Meat Board, Chicago, IL, USA.

Bertram, H. C., H. J. Andersen, and A. H. Karlsson. 2001. Comparative study of low-field NMR relaxation measurements and two traditional methods in the determination of water holding capacity of pork. Meat Sci. 57:125-132.

Brant, A. W., J. W. Goble, J. A. Hamann, C. J. Wabeck, and R. E. Walters. 1982. Guidelines for establishing and operating broiler processing plants. United States Department of Agriculture, Agricultural Research Service, Agriculture Handbook Number 581.

Brown, G. D. and S. Gordon. 2005. Immune recognition of fungal β-1, 3–1, 6- glucans. Cell. Microbiol. 7:471-479.

Bohn, J. A. and J. N. BeMiller. 1995. (1-3)-β-D-glucans as biological response modifiers: A review of structure-functional activity relationships. Carbohydr. Polym. 28:3-14.

Chen, K. L., B. C. Weng, M. T. Chang, Y. H. Liao, T. T. Chen, and C. Chu. 2008. Direct enhancement of the phagocytic and bactericidal capability of abdominal macrophage of chicks by β-1, 3-1, 6-glucan. Poult. Sci. 87:2242-2249.

Cox, C. M., L. H. Stuard, S. Kim, A. P. McElroy, M. R. Bedford, and R. A. Dalloul. 2010. Performance and immune responses to dietary beta-glucan in broiler chicks. Poult. Sci. 89:1924-1933.

Dipeep, A. O., P. Graham, and B. Mirko. 2011. Effect of different ingredients on color and oxidative characteristics of high pressure processed chicken breast meat with special emphasis on use of β-glucan as a partial salt replacer. Innov. Food Sci. Emerg. Technol. 12:244-254.

Froning, G. W. 1995. Color of poultry meat. Poult. Avian Biol. Rev. 6: 83-93.

Guo, Y., R. A. Ali, and M. A. Qureshi. 2003. The influence of beta-glucan on immune responses in broiler chicks. Immunopharmacol. Immunotoxicol. 25:461-472.

Hahn, T. W., J. D. Lohakara, S. L. Lee, W. K. Moon, and B. J. Chae. 2006. Effects of supplementation of β-glucans on growth performance, nutrient digestibility, and immunity in weanling pigs. J. Anim. Sci. 84:1422-1428.

Hwang, Y. H., B. K. Park, J. H. Lim, M. S. Kim, I. B. Song, S. C. Park, H. K. Jung, J. H. Hong, and H. I. Yun. 2008. Effects of β-glucan from *Paenibacillus polymyxa* and L-theanine on growth performance and immunomodulation in weanling piglets. Asian Australas. J. Anim. Sci. 21:1753-1759.

Mao, X. F., X. S. Piao, C. H. Lai, D. H. Li, J. J. Xing, and B. L. Shi. 2005. Effects of β-glucan obtained from the Chinese herb astragalus membranaceus and lipopolysaccharide challenge on performance, immunological, adrenal, and somatotropic responses of weanling pigs. J. Anim. Sci. 83:2775-2782.

Morales-Lopez, R., E. Auclair, F. Garcia, E. Esteve-Garcia, and J. Brufau. 2009. Use of yeast cell walls; β-1,3/1,6-glucans; and mannoproteins in broiler chicken diets. Poult. Sci. 88: 601-607.

National Research Council (NRC). 1994. Nutrient requirement for poultry. Ninth revision, National Academy Ptress, Washington, DC, USA.

Pinero, M. P., K. Parra, N. Huerta-Leidenz, A. Moreno, M. Ferrer, S. Araujo, and Y. Barboza. 2008. Effect of oat's soluble fibre (β-glucan) as a fat replacer on physical, chemical, microbiological and sensory properties of low-fat beef patties. Meat Sci. 80:675-680.

Thondre, P. S., L. Ryan, and C. J. K. Henry. 2011. Barley β-glucan extracts as rich sources of polyphenols and antioxidants. Food Chem. 126:72-77.

Vetvicka, V., B. Dvorak, J. Vetvickova, J. Richter, P. Krizan, P. Sima, and J. C. Yvin. 2007. Orally administered marine (1→3)-β-d-glucan Phycarine stimulates both humoral and cellular immunity. Int. J. Biol. Macromol. 40:291-298.

7

Bioavailability of Phosphorus in Two Cultivars of Pea for Broiler Chicks

T. A. Woyengo, I. A. Emiola, I. H. Kim[1], and C. M. Nyachoti*

Department of Animal Science, University of Manitoba, Winnipeg, MB R3T 2N2, Canada

[1] Department of Animal Resources and Science, Dankook University, Cheonan 330-714, Korea

ABSTRACT: The aim was to determine the relative bioavailability of phosphorus (P) in peas for 21-day old broiler chickens using slope-ratio assay. One hundred and sixty eight male Ross 308 broiler chicks were divided into 42 groups 4 balanced for body weight and fed 7 diets in a completely randomized design (6 groups/diet) from day 1 to 21 of age. The diets were a corn-soybean meal basal diet, and the corn-soybean meal basal diet to which monosodium phosphate, brown- or yellow-seeded pea was added at the expense of cornstarch to supply 0.5% or 1% total phosphorus. Monosodium phosphate was included as a reference, and hence the estimated bioavailability of P in pea cultivars was relative to that in the monosodium phosphate. Birds and feed were weighed weekly and on d 21 they were killed to obtain tibia. The brown-seeded pea contained 23.4% crude protein, 0.47% P, whereas the yellow-seeded pea contained 24.3% crude protein and 0.38% P. Increasing dietary P supply improved (p<0.05) chick body weight gain and tibia ash and bone density. The estimated relative bioavailability of p values for brown- and yellow-seeded peas obtained using final body weight, average daily gain, tibia ash, and bone mineral density were 31.5% and 36.2%, 35.6% and 37.3%, 23.0% and 5.60%, and 40.3% and 30.3%, respectively. The estimated relative bioavailability of p values for brown- and yellow-seeded peas did not differ within each of the response criteria measured in this study. In conclusion, the relative bioavailability of P in pea did not differ depending on the cultivar (brown- vs yellow-seed). However, the relative bioavailability of P in pea may vary depending on the response criterion used to measure the bioavailability. (**Key Words:** Broiler Chickens, Peas, Phosphorus Bioavailability)

INTRODUCTION

About three quarters of phosphorus (P) in most plant feed ingredients is in the form of phytate P which is poorly utilized by poultry (Ravindran et al., 1995b). Consequently, poultry diets are routinely supplemented with inorganic sources of P to meet requirements (NRC, 1994). This practice not only adds considerably to the cost of poultry diets but also increases the risk of environmental pollution due to excess P excretion, especially under conditions of intensive production (Honeyman, 1993). Therefore, it is critical that poultry diets are formulated to accurately match supply with requirements.

Field pea can be utilized in poultry diets as a source of protein and energy. Brenes et al. (1993) reported that

inclusion levels up to 48% raw field pea can be used as an alternative protein and energy source to replace soybean meal and corn in broiler chicken diets. Similarly, Thacker et al. (2013) reported that pea can be incorporated at levels as high as 30% in diets fed to broiler chicks with no negative effects on performance. At such high inclusion levels and considering the relatively high levels of P in pea, utilizing pea in poultry diets can make a significant contribution to total phosphorus content. However, information on bioavailability of P in pea for broiler chickens is scarce. Therefore, the objective of the current experiment was to determine the bioavailability of P in brown and yellow cultivars of peas by broiler chicks.

MATERIALS AND METHODS

The experimental protocol was reviewed and approved by the Animal Care Protocol Management and Review Committee of the University of Manitoba and birds were

---

* Corresponding Author: C. M. Nyachoti.
E-mail: martin_nyachoti@umanitoba.ca

cared for according to the guidelines of the Canadian Council on Animal Care (CCAC, 2009).

One hundred and sixty-eight 1-d old male broiler chicks were used in the present study. On d 1, birds were weighed and divided based on body weight (BW) into 42 uniform groups each with 4 birds. The birds (Ross 308, Aviagen) were obtained from a local hatchery (Carlton Hatchery, Grunthal, MB, Canada) and were housed in Petersime battery brooders (Petersime Incubator Co., Gettysburg, OH, USA) in a room with continuous fluorescent lighting. Room temperature was maintained at 32°C, 28°C, and 24°C during wk 1, 2, and 3, respectively. The control diet was based on a corn and soybean meal and was formulated to meet or exceed the nutrient requirements (Table 1).

Six replicate cages were randomly assigned to one of 7 dietary treatments for a 21-d period. The 7 dietary treatments consisted of three reference diets formulated by adding 0, 0.5, and 1.0 g/kg total P from monosodium phosphate (MSP; that contained 24% total P) to a corn-soybean meal basal diet and by adding 0.5 and 1.0 g/kg total P from either brown- (that contained 0.47% total P) or yellow-seeded (that contained 0.38% total P) pea to the basal diet at the expense of cornstarch. Birds had free access to feed and water throughout the study period.

Growth performance and bone measurements

Birds were weighed on d 1, 7, 14, and 21 to calculate BW gain and feed intake was measured on d 7, 14, and 21 to calculate average daily feed intake. The two measurements were used to calculate feed conversion ratio. On the last day of the experiment (day 22), two birds were randomly selected from each cage and killed by cervical dislocation. Left and right tibiae were excised, placed in sealed plastic bags, and stored frozen at –20°C until subsequent analysis of tibia ash and bone mineral density, respectively.

The tibiae were de-fleshed after autoclaving at 121°C for 1 min. Tibia for determination of tibia ash were dried in an oven at 45°C for 2 d. They were then fat-extracted using hexane for 2 d, dried in a fume hood for 2 d to allow the hexane to evaporate and ashed at 550°C in a muffle furnace for 12 h. Bone mineral density was estimated from the right tibia using a dual energy x-ray absorptiometry x-ray densitometer (GE Healthcare, Lunar Prodigy Advance PA+130472, Small Animal Software, Diegem, Belgium).

Chemical analysis

Feed and pea samples were finely ground through a 1-mm screen in a Thomas Wiley Mill (Thomas model 4 Wiley Mill; Thomas Scientific, Swedesboro, NJ, USA) and thoroughly mixed before analysis. All analyses were performed in duplicates. Analyses for dry matter (DM) in feed and pea samples were carried out according to AOAC

Table 1. Composition of the basal diet

Items	%
Ingredients	
Corn	29.00
Soybean meal (48% CP)	30.70
Corn starch	26.00
Monosodium phosphate	0.00
Peas	0.00
Casein	7.00
Limestone	2.50
Vegetable oil	4.00
Iodized salt	0.40
Mineral/vitamin premix[1]	0.30
DL-methionine	0.10
Lysine HCl	0.00
Threonine	0.00
Tryptophan	0.00
Total	100.00
Calculated nutrient content	
CP (%)	23.20
ME (kcal/kg)	3,484
Ca (%)	1.07
Total P (%)	0.35

CP, crude protein; ME, metabolizable energy.
[1] Provided per kilogram of diet: vitamin A, 8,255.0 IU; vitamin D_3, 3,000.0 IU; vitamin E, 30.0 IU; vitamin B_{12}, 0.013 mg; vitamin K_3, 2.0 mg; niacin, 41.2 mg; choline, 1,300.5 mg; folic acid, 1.0 mg; biotin, 0.25 mg; pyridoxine, 4.0 mg; thiamine, 4.0 mg; calcium pantothenic acid, 11.0 mg; riboflavin, 6.0 mg; manganese, 70.0 mg; zinc, 80.0 mg; iron, 80.0 mg; iodine, 0.5 mg; copper, 10 mg; and selenium, 0.3 mg.
Assumptions: Monosodium phosphate contained 22.4% P and field pea contain 0.39% total P.

(1990; method 934.01). Ingredients and feed samples for Ca and P analysis were ashed at 600°C for 12 h, digested according to AOAC (1990; method 985.01) procedures and read on a Varian Inductively Coupled Plasma Mass Spectrometer (Varian Inc., Palo Alto, CA, USA). Crude protein (N×6.25) was determined using a Leco NS 2000 Nitrogen analyzer (LECO Corporation, St. Joseph, MI, USA). Gross energy was measured using a Parr adiabatic oxygen bomb calorimeter (Parr Instrument co., Moline, IL, USA).

Statistical analysis

Data were subjected to analysis of variance as a completely randomized design using the general linear models (GLM) procedure of SAS (SAS Institute, Inc. Cary, NC, USA). The cage was the experimental unit for growth performance variables and the 2 birds per cage for bone variables. Linear and quadratic contrasts were used to examine the relationship between growth performance response criteria or bone measurements and supplemental P from MSP, brown-seeded pea or yellow-seeded pea. Relative bioavailability of P in brown- and yellow-seeded pea was estimated using the slope-ratio technique (Finney 1978). The validity of the following 3 assumptions of the slope-ratio technique were first tested as described by Littell et al. (1997) using Proc GLM of SAS: the linearity and lack of curvature of response curve for each nutrient source, the equality of intercepts of the two regression lines, and the equality of the common-intercept and the 'zero level' of the basal treatment mean. Multiple regression analyses were then conducted using Proc GLM of SAS as outlined by Littell et al. (1997) to estimate the relative bioavailability of P in brown- and yellow-seeded pea.

RESULTS AND DISCUSSION

The analyzed composition of the two types of field pea tested in the present study is shown in Table 2. The crude protein, Ca and P contents were 23.35% and 24.26%, 0.07% and 0.12%, and 0.47% and 0.38%, respectively, for the brown- and yellow-seeded peas. These values are within the range of values reported for field pea for these nutrients (Hickling, 2003; NRC, 2012). Ravindran et al. (2010) reported the P content in five pea cultivars to range from 0.38% to 0.53%, with an average value of 0.45%, which is comparable with the values reported in the current study. Earlier, Igbasan et al. (1997) reported the P content in 12 pea cultivars to range from 2.9 to 5.6 g/kg DM with an average of 4.3±0.88 g/kg DM.

The growth performance of broiler chicks fed diets with increasing P content from MSP, brown- and yellow-seeded is shown in Table 3. The final BW was linearly increased (p<0.001) with the addition of P from all three sources,

Table 2. Proximate composition of brown- and yellow-seeded pea

	Brown pea	Yellow pea
Moisture (%)	12.47	11.13
Dry matter (%)	87.53	88.87
Crude protein (%)	23.35	24.26
Calcium (%)	0.07	0.12
Total phosphorus (%)	0.47	0.38
Gross energy (kcal/kg)	3,824	3,891

although the performance of birds fed the diet with 0.5% additional P from yellow-seeded pea was not different from that of chicks fed the control diet. Feed intake and feed conversion ratio responses followed the same trend as weight gain (Table 3). These observations are consistent with those of Sands et al. (2003) showing a linear increase in performance of broiler chicks fed diets with increasing P content from MSP and low phytate soybean meal. Increasing supplemental P from pea from 0.5% to 1.0% increased growth responses but such improvements were not significant, except for average daily feed intake for the brown-seeded pea, which was significantly increased (p<0.05) from 25.95 to 31.72 g/bird. For all the performance response criteria, responses to P from MSP were much higher (p<0.05) compared with P from peas, which is likely a reflection of the fact that P from MSP is more readily available than P from pea. Indeed, it has been reported that P from plant-based source is of low bioavailability compared to inorganic sources such as MSP (NRC, 1994; 2012).

Tibia ash and bone mineral density of broiler chickens fed diets with increasing P from different sources are shown

Table 3. Effect of dietary treatment on performance of broilers from day 1 to day 21 of age

Item	Dietary treatments							SEM	Contrasts	
	0%	MSP		Brown pea		Yellow pea			Linear	Quadratic
		0.5%	1.0%	0.5%	1.0%	0.5%	1.0%			
Initial BW (g/b)	47.4	46.9	46.6	46.8	46.8	46.8	46.8	0.83		
Final BW (g/b)	206.5[c]	884.4[a]	898.4[a]	384.9[b]	421.9[b]	309.9[bc]	435.9[b]	0.67	0.006	0.121
ADFI (g/b)	22.84[c]	42.07[a]	42.92[a]	25.95[c]	31.72[b]	25.74[c]	26.32[c]	2.50	0.013	0.252
ADG (g/b)	7.51[c]	39.88[a]	40.56[a]	16.10[b]	17.86[b]	12.53[bc]	18.55[b]	3.22	0.006	0.118
FCR (g/g)	3.55[a]	1.06[c]	1.07[c]	1.78[bc]	2.27[b]	2.17[b]	1.50[bc]	0.52	<0.001	0.009

MSP, monosodium phosphate; SEM, standard error of the mean; BW, body weight; ADFI, average daily feed inatke; ADG, average daily gain; FCR, feed conversion ratio.
[a-c] Mean values within a row with unlike superscript letters were significantly different between groups (p<0.05).

Table 4. Effect of dietary treatment on bone mineralization in broilers at 21 days of age

Item	Dietary treatments							SEM	Contrasts	
	0%	MSP		Brown pea		Yellow pea			Linear	Quadratic
		0.5%	1.0%	0.5%	1.0%	0.5%	1.0%			
Tibia ash (%)	27.3[c]	48.8[a]	48.8[a]	31.7[b]	32.6[b]	29.0[bc]	30.0[bc]	1.33	0.022	0.137
Tibia ash (g/bone)	0.384[c]	1.859[a]	1.970[a]	0.544[bc]	0.705[b]	0.469[c]	0.598[bc]	0.070	0.022	0.299
Bone density (g/cm²)	0.071[d]	0.168[a]	0.168[a]	0.079[cd]	0.110[b]	0.087[bcd]	0.097[bc]	0.008	0.013	0.177

MSP, monosodium phosphate; SEM, standard error of the mean.
[a-d] Mean values within a row with unlike superscript letters were significantly different between groups (p<0.05).

in Table 4. Birds fed the basal diet had a low (p<0.001) tibia ash content compared with those fed the diets with supplemental P from MSP and brown-seeded pea; values for yellow-seeded pea were similar. Tibia ash contents were not different between the birds fed the pea-containing diets. There was an increase (p<0.001) in tibia ash content as the additional P content from MSP and brown-seeded pea increased from 0% to 0.5%; no additional improvement was observed when the supplemental P content was further increased to 1.0%. The response of tibia ash content to increasing supplemental P from yellow-seeded pea was not significant (p>0.10). In the study of Sands et al. (2003), tibia ash content was shown to respond linearly to increased supplemental P from MSP and low and normal phytate soybean meal; the results of the current study are consistent with these observations. Dietary addition of P from MSP at either 0.5% or 1.0% and from pea at the 1.0% level increased (p<0.001) bone density compared with the basal diet (Table 4). Values for birds fed diets with 0.5% supplemental P from pea were not different from those of values for birds fed the basal diet.

Figures 1 to 5 shows the common-intercept multiple linear regression of average (mean±standard error) final BW daily gain (ADG), tibia ash (%), tibia ash (g) and bone mineral density on supplemental P intake in broiler chickens fed diets supplemented P from MSP, brown- and yellow-seeded pea. When using final BW as response criterion, the regression equation was Final BW = 259.35±31.58+1793.75±148.25MSP+565.22±202.67BP+649.30±238.83YP, R2 = 0.77; where BP and YP are brown- and yellow-seeded pea, respectively. The slope for MSP differed (p<0.0001) from that for brown-seed pea or yellow-seeded pea; however, the slope for brown-seed pea did not differ (p = 0.628) from that for yellow-seeded pea. The estimated relative bioavailability values obtained with the slope-ratio assay were 31.5% and 36.2% for BP and YP, respectively.

When using ADG as response criterion, the regression equation was ADG = 8.83±1.44+89.02±7.11MSP+31.72±9.73BP+33.22±11.40YP, R2 = 0.76. The slope for MSP differed (p<0.0001) from that for brown-seed pea or yellow-seeded pea; however, the slope for brown-seed pea did not differ (p = 0.825) from that for yellow-seeded pea. The estimated relative bioavailability values obtained with the slope-ratio assay were 35.6% and 37.3% for BP and YP, respectively.

The common-intercept, multiple-linear regression equation obtained with percent tibia ash was Tibia ash (%) = 29.06±1.03+54.40±5.04MSP+12.49±6.92BP+3.05±8.09YP, R2 = 0.72. The slope for MSP differed (p<0.0001) from that for brown-seed pea or yellow-seeded pea; however, the slope for brown-seed pea did not differ (p = 0.250) from that for yellow-seeded pea. The relative bioavailabilities estimated by slope-ratio were 23.0% and 5.60% for BP and YP, respectively.

The common-intercept, multiple-linear regression equation obtained with absolute amount of tibia ash was Tibia ash (g) = 0.468±0.052+4.127±0.253MSP+0.757±0.347BP+0.381±0.406YP, R2 = 0.86. The slope for MSP differed (p<0.0001) from that for brown-seed pea or yellow-seeded pea; however, the slope for brown-seed pea did not differ (p = 0.246) from that for yellow-seeded pea. The relative bioavailabilities estimated by slope-ratio were 18.3% and 9.23% for BP and YP, respectively.

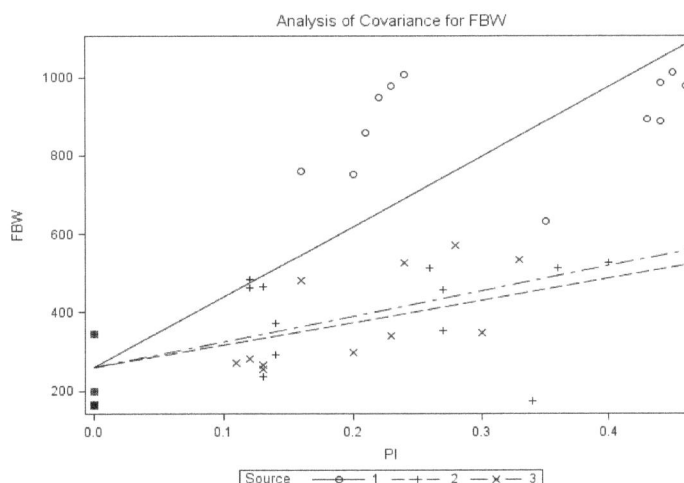

Figure 1. Common-intercept, multiple-linear regression of final body weight (FBW; g) on supplemental P intake of chicks fed diets supplemented with P from monosodium phosphate (MSP; Source 1), brown-seeded pea (BP; Source 2), and Yellow-seeded pea (YP; Source 3). The circular marker represents MSP, the plus marker represents BP, and the multiplication marker represents YB. The common-intercept, multiple-linear regression equation (mean±standard error) was Y = 259.35±31.58+1,793.75±148.25MSP+565.22±202.67BP+649.30±238.83YP, R2 = 0.77, p<0.0001; the relative bioavailabilities estimated by slope-ratio were 31.5% and 36.2% for BP and YP, respectively.

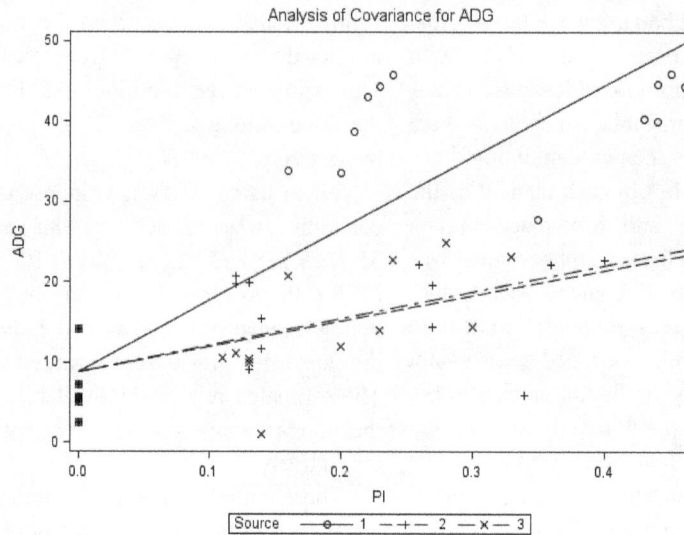

Figure 2. Common-intercept, multiple-linear regression of average daily gain (ADG; g) on supplemental P intake of chicks fed diets supplemented with P from monosodium phosphate (MSP; Source 1), brown-seeded pea (BP; Source 2), and Yellow-seeded pea (YP; Source 3). The circular marker represents MSP, the plus marker represents BP, and the multiplication marker represents YB. The common-intercept, multiple-linear regression equation (mean±standard error) was Y = 8.83±1.44+89.02±7.11MSP+31.72±9.73BP+ 33.22±11.40YP, R2 = 0.76, p<0.0001; the relative bioavailabilities estimated by slope-ratio were 35.6% and 37.3% for BP and YP, respectively.

The common-intercept, multiple-linear regression equation obtained with bone mineral density (BMD) was BMD = 0.077±0.006+0.242±0.028MSP+0.097±0.039BP +0.074±0.045YP, R2 = 0.60. The slope for MSP differed (p<0.0001) from that for brown-seed pea or yellow-seeded pea; however, the slope for brown-seed pea did not differ (p = 0.596) from that for yellow-seeded pea. The relative bioavailabilities estimated by slope-ratio were 40.3% and

30.3% for BP and YP, respectively.

Based on the results in Figures 1 to 5, the relative bioavailability of P values for brown- and yellow-seeded peas did not differ regardless of the response criterion used to estimate the bioavailability. In the current study, the brown-seeded pea contained more P than the yellow-seeded pea. Igbasan et al. (1997) also reported greater content of tannin (which reduces nutrient digestibility) in brown-

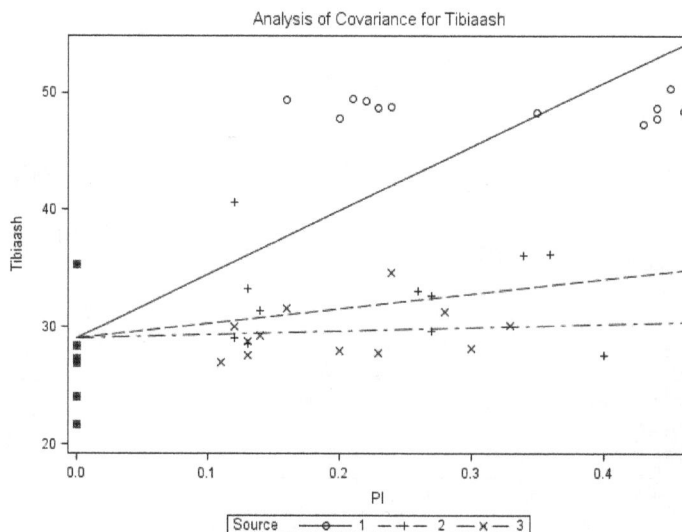

Figure 3. Common-intercept, multiple-linear regression of relative amount of tibia ash (tibiaash; %) on supplemental P intake of chicks fed diets supplemented with P from monosodium phosphate (MSP; Source 1), brown-seeded pea (BP; Source 2), and Yellow-seeded pea (YP; Source 3). The circular marker represents MSP, the plus marker represents BP, and the multiplication marker represents YB. The common-intercept, multiple-linear regression equation (mean±standard error) was Y = 29.06±1.03+54.40±5.04MSP+12.49±6.92BP +3.05±8.09YP, R2 = 0.72, p<0.0001; the relative bioavailabilities estimated by slope-ratio were 23.0% and 5.60% for BP and YP, respectively.

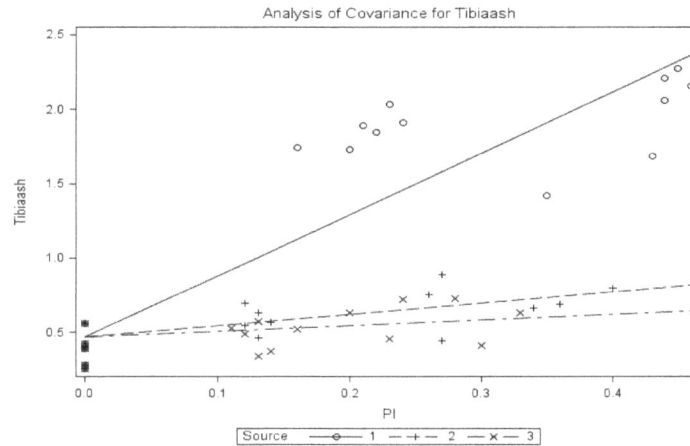

Figure 4. Common-intercept, multiple-linear regression of absolute amount of tibia ash (tibiaash; g) on supplemental P intake of chicks fed diets supplemented with P from monosodium phosphate (MSP; Source 1), brown-seeded pea (BP; Source 2), and Yellow-seeded pea (YP; Source 3). The circular marker represents MSP, the plus marker represents BP, and the multiplication marker represents YB. The common-intercept, multiple-linear regression equation (mean±standard error) was Y = 0.468±0.052+4.127±0.253MSP+0.757±0.347BP+ 0.381±0.406YP, R2 = 0.86, p<0.0001; the relative bioavailabilities estimated by slope-ratio were 18.0% and 9.23% for BP and YP, respectively.

seeded pea than in yellow-seeded pea, and hence lower amino acid digestibility for the former than the latter in poultry. Thus, it appears that the differences in the composition between the 2 pea cultivars (brown vs yellow seeds) does not affect P bioavailability in pea.

We could not find any report on the bioavailability of P in pea for poultry in the literature and therefore the results of the current study cannot be compared. Tibia ash is routinely used as a sensitive criterion for assessing responses to dietary P in poultry (Ravindran et al., 1995a; Sands et al., 2003; Kim et al., 2008; Shastak et al., 2012).

However, the value of P bioavailability obtained with tibia ash as response criteria was lower than the value obtained with final body weight, ADG or bone mineral density as response criterion. Also, the bioavailability of P in regular and low-phytate soybean meals for broilers estimated using tibia ash as response criterion was lower than that estimated using bone mineral density as response criterion (Sands et al., 2003). Additionally, the bioavailability of P in meat and bone meal for pigs estimated using metacarpal ash as response criterion was lower than that estimated using metacarpal strength as response criterion (Taylor et al.,

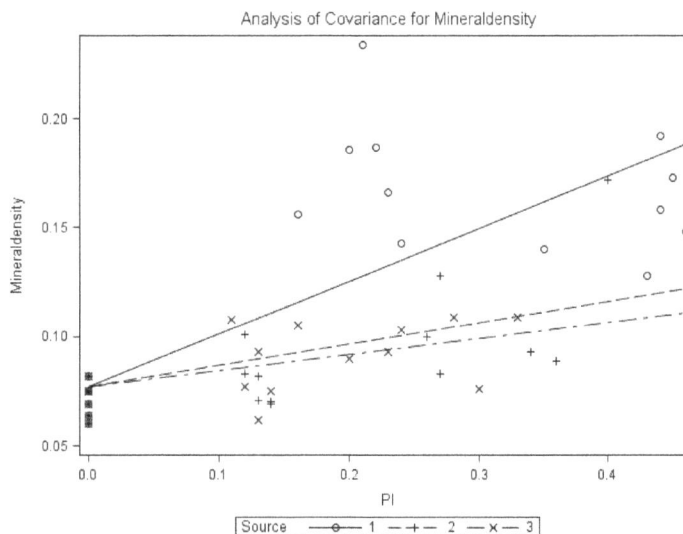

Figure 5. Common-intercept, multiple-linear regression of bone mineral density (mineraldensity; g/cm^2) on supplemental P intake of chicks fed diets supplemented with P from monosodium phosphate (MSP; Source 1), brown-seeded pea (BP; Source 2), and Yellow-seeded pea (YP; Source 3). The circular marker represents MSP, the plus marker represents BP, and the multiplication marker represents YB. The common-intercept, multiple-linear regression equation (mean±standard error) was Y = 0.077±0.006+0.242±0.028MSP+0.097± 0.039BP+0.074±0.045YP, R2 = 0.60, p<0.0001; the relative bioavailabilities estimated by slope-ratio were 40.3% and 30.3% for BP and YP, respectively.

Table 5. Bioavailability of P in two pea cultivars for broilers based on body weight gain, feed intake, feed conversion ratio, tibia ash and bone mineral density

Item	Bioavailability coefficient (%)	Total P content (%)	Bioavailable P content (%)
Brown-seeded pea			
Final BW	31.5	0.47	0.15
BW gain	35.6	0.47	0.17
Tibia ash	23.0	0.47	0.11
Tibia ash weight	18.3	0.47	0.09
Bone density	40.3	0.47	0.19
Yellow-seeded pea			
Final BW	36.2	0.38	0.14
BW gain	37.3	0.38	0.14
Tibia ash	5.60	0.38	0.02
Tibia ash weight	9.23	0.38	0.04
Bone density	30.3	0.38	0.12

BW, body weight.

2005). Thus, the relative bioavailability values can vary depending on the response criterion used. The proportion of digestible dietary P that is deposited in soft tissues of pigs ranges from 20% to 40% (Crenshaw, 2001), and it is affected by several factors including lean deposits (Baker et al., 2013). This variation in proportion of digestible dietary P that is deposited in soft tissues of pigs could partly explain the variation in relative P bioavailability values due to response criterion used.

Table 5 shows the estimates of bioavailability, total P content and bioavailable P content in the tested pea and as obtained using the three different response criteria. When averaged, the relative bioavailabilities in brown-seeded and yellow-seeded pea were 29.7% and 23.7%, respectively. Respective values for bioavailable P content were 0.14% and 0.09%.

In conclusion, the relative bioavailability of P in pea did not differ depending on the cultivar (brown- vs yellow-seed). However, the relative bioavailability of P in pea estimated using tibia ash as response criteria was lower than the value estimated using final body weight, ADG or bone mineral density as response criterion. Thus, relative bioavailability of P in pea may vary depending on the response criterion used to measure the bioavailability.

REFERENCES

AOAC. 1990. Official Methods of Analysis. 15th ed. Assoc. Off. Anal. Chem., Washington, DC, USA.

Baker, S. R., B. G. Kim, and H. H. Stein. 2013. Comparison of values for standardized total tract digestibility and relative bioavailability of phosphorus in dicalcium phosphate and distillers dried grains with solubles fed to growing pigs. J. Anim. Sci. 91:203-210.

Brenes A., B. A. Rotter, R. R. Marquardt, and W. Guenter. 1993. The nutritional value of raw, autoclaved and dehulled peas (*Pisum sativum* L.) in chicken diets as affected by enzyme supplementation. Can. J. Anim. Sci. 73:605-614.

CCAC. 2009. Guidelines on the care and use of farm animals in research, teaching and testing. Canadian Council on Animal Care, Ottawa, ON, Canada.

Crenshaw, T. D. 2001. Calcium, phosphorus, vitamin D, and vitamin K in swine nutrition. In: Swine Nutrition (Eds. A. J. Lewis and L. L. Southern). 2nd ed. CRC Press, Boca Raton, FL, USA. pp. 187-212.

Finney, D. J. 1978. Statistical Methods in Biological Assay. 3rd ed. Charles Griffin and Co. Ltd., High Wycombe, UK.

Hickling, D. 2003. Canadian Feed Pea Industry Guide. Pulse Canada, Winnipeg, MB, Canada.

Honeyman, M. S. 1993. Environment-friendly swine feed formulation to reduce nitrogen and phosphorus excretion. Am. J. Altern. Agric. 8:128-132.

Igbasan, F. A., W. Guenter, and B. A. Slominski. 1997. Field peas: chemical composition and energy and amino acid availabilities for poultry. Can. J. Anim. Sci. 77:293-300.

Littell, R. C., P. R. Henry, A. J. Lewis, and C. B. Ammerman. 1997. Estimation of relative bioavailability of nutrients using SAS procedures. J. Anim. Sci. 75:2672-2683.

Kim, E. J., C. M. Amezcua, P. L. Utterback, and C. M. Parsons. 2008. Phosphorus bioavailability, true metabolizable energy, and amino acid digestibilities of high protein corn distillers dried grains and dehydrated corn germ. Poult. Sci. 87:700-705.

NRC. 2012. Nutrient Requirements of Swine. 11th rev. ed. Natl. Acad. Press, Washington, DC, USA.

NRC. 1994. Nutrient Requirements of Poultry. 9th rev. ed. Natl. Acad. Press, Washington, DC, USA.

Onyango, E. M., P. Y. Hester, R. Stroshine, and O. Adeola. 2003. Bone densitometry as an indicator of percentage tibia ash in broiler chicks fed varying dietary calcium and phosphorus levels. Poult. Sci. 82:1787-1791.

Potter, L. M., M. Potchanakorn, V. Ravindran, and E. T. Kornegay. 1995. Bioavailability of phosphorus in various phosphate sources using body weight and toe ash as response criteria. Poult. Sci. 74:813-820.

Ravindran, G., C. L. Nalle, A. Molan, and V. Ravindran. 2010. Nutritional and biochemical assessment of field peas (*Pisum sativum* L.) as a protein source in poultry diets. J. Poult. Sci. 47:48-52.

Ravindran, V., E. T. Kornrgay, L. M. Potter, B. O. Ogunabameru, M. K. Welten, J. H. Wilson, and M. Potchanakorn. 1995a. An evaluation of various response criteria in assessing biological availability of phosphorus for broilers. Poult. Sci. 74:1820-1830.

Ravindran, V., W. L. Bryden, and E. T. Kornegay. 1995b. Phytates: Occurrence, bioavailability and implications in poultry nutrition. Poult. Avian Biol. Rev. 6:125-143.

Sands, J. S., D. Ragland, J. R. Wilcox, and O. Adeola. 2003.

Relative bioavailability of phosphorus in low-phytate soybean meal for broiler chicks. Can. J. Anim. Sci. 83:95-100.

Shastak, Y., M. Witzig, K. Hartung, W. Bessei, and M. Rodehutscord. 2014. Comparison and evaluation of bone measurements for the assessment of mineral phosphorus sources in broilers. Poult. Sci. 91:2210-2220.

Thacker, P., A. Deep, D. Petri, and T. Warkentin. 2013. Nutritional evaluation of low-phytate peas (*Pisum sativum* L.) for young broiler chicks. Arch. Anim. Nutr. 67:1-14.

Traylor, S. L., G. L. Cromwell, and M. D. Lindemann. 2005. Bioavailability of phosphorus in meat and bone meal for swine. J. Anim. Sci. 83:1054-1061.

8

Effect of Feeding Date Pits on Milk Production, Composition and Blood Parameters of Lactating Ardi Goats

S. B. AL-Suwaiegh*

Animal and Fish Production Department, King Faisal University, Al-Ahsa, 31982, Kingdom of Saudi Arabia

ABSTRACT: Twenty Ardi lactating goats were used to investigate the effect of substituting 10%, 15%, and 20% of concentrate feed with date pits on milk production, composition, and blood parameters. Four isocaloric and isonitrogenous dietary treatments were used. Four levels (0% [control], 10%, 15%, and 20%) of date pits were used to replace concentrate feed. The forages to concentrate ratio used was 60 to 40. Dry matter intake (DMI) of goats fed diets containing 10% and 15% date pits was significantly (p<0.05) higher than those fed diets containing 0% and 20%. However, goats fed a diet containing 20% date pits were significantly (p<0.05) lower in DMI compared to those fed control diet. The protein percent was significantly higher for goats fed control diet compared to the other dietary treatments. Total solids percent was significantly the lowest for goats fed diet supplemented with 10% date pits. Goats fed a diet containing 20% date pits was significantly (p<0.05) higher in the total protein compared to those fed a diet containing 10%. In addition, goats fed a diet containing 15% date pits exhibited no significant differences in the total protein percent compared to those fed a diet containing 20% date pits. Triglyceride was significantly higher for goats fed diets containing 10% and 20% date pits compared to those fed 15% date pits. Results obtained in the present study suggest that date pits can be added up to 20% of the concentrate feeds into lactating Ardi goat diets without negative effects on their productive performance. (**Key Words:** Goats, Date Pits, Dietary Treatments, Milk Production, Albumin, Triglyceride)

INTRODUCTION

Saudi Arabia is considered as one of the largest producer of dates in the world (Haider et al., 2014). The number of date palm trees in Saudi Arabia is more than 25 million and this number is expected to increase gradually (Ministry of Agriculture, 2013). Date palm trees produce large quantity of agricultural waste. Although the quantity of date palm trees by-products produced in Saudi Arabia is relatively large, most of these are disposed by dumping, burning or burying. Date seeds are one of these wastes that are the product of many date processing plants producing pitted dates, date syrup, and date confectionery. Date pits can be utilized efficiently by animals as non-traditional feed. However, the feeding value of the date pits are lower than the traditional concentrate feed because date pits have a hard seed coat that makes the seed components difficult to be digested. Thus, to increase their nutritive value, it is necessary to process the seeds before feeding them to ruminant animals. Grinding of date pits makes nutrients more available for animals by breaking and removing the seed coat. However, one of grinding drawbacks is its energy-consuming and may cause wear and tear to the machine used in grinding (Barreveld, 1993). Many feeders search for alternatives to high priced traditional feeds that provide will the nutrient requirements of the animal in a cost-effective manner. Thus, regional agricultural by-products, especially those with a similar nutritional value, can be a viable alternative to high cost conventional feed. (Costa et al., 2009).

Goats belong to the small ruminant group of animals and they are important livestock for meat and milk production in many developing countries. The number of goats in Saudi Arabia in 2012 was estimated to be 1,075,394 (Ministry of Agriculture, 2013). Furthermore, milk from goats is more digestible and has longer shelf life

* Corresponding Author: S. B. AL-Suwaiegh.
E-mail: shaker300@hotmail.com

compared to cow milk. Nagura (2004) found that goat milk is very beneficial for diabetic patients. Also, people that have some diseases such as stomach ulcers, some types of allergy, and various gastro intestinal disorders that develop from intolerance to cow milk may benefit from consumption of goat milk (Haenlein, 2004). These facts also favors goats for adoption as dairy animals, especially in the developing world where a majority of the goat population is found with people having low economic status. To this end, the objective of this study was to investigate the effect of partial substitution of date pits with conventional concentrate feeds on lactation performance and blood parameters profiles of Ardi goats.

MATERIALS AND METHODS

Experimental site

The trial was conducted at Agricultural Research and Training Station at King Faisal University in Al-Ahsa, Saudi Arabia from July to September, 2013. The local environment in Al-Ahsa is characterized by its high temperature and humidity in summer, particularly in July and August. The average temperature and humidity in these two months are 48°C and 90%, respectively.

Procurement and processing of date pits

A total of 500 kg of date pits was purchased from local date market in Al-Ahsa, Saudi Arabia. Date pits were transported to AL-Ghadeer Feed Mill in Al-Ahsa to be milled through a date pit grinder and then used in the experimental diets.

Animals and experimental diets

Twenty Ardi goats of almost the same age and body weight were assigned randomly to one of four diets with five animals each. All goats were subjected to regular health inspection before the beginning of the trial to ensure that they were in good condition. Chemical composition of the experimental diets was determined before the beginning of

Table 1. Ingredients of the experimental diets

Ingredient (% of DM)	Inclusion of date pits (%)			
	0	10	15	20
Alfalfa hay	60.0	60.0	60.0	60.0
Date pit	0.0	10.0	15.0	20.0
Barley	6.2	11.07	9.55	5.74
Soybean	1.9	4.0	5.0	7.6
Wheat bran	28.12	11.18	6.7	2.9
Dicalcium phosphate	1.0	1.0	1.0	1.0
Limestone	1.99	1.97	1.97	1.96
Salt	0.69	0.68	0.68	0.70
Vit. and Min. premix[1]	0.1	0.1	0.1	0.1

DM, dry matter.
[1] Vit. and Min. premix contains vitamin A 10,000,000 IU, vitamin D_3 1,000,000 IU, vitamin E 10,000 mg, magnesium 100,000 mg, manganese 50,000 mg, zinc 45,000 mg, iron 80,000 mg, copper 6,000 mg, cobalt 800 mg, iodine 2,500 mg, selenium 100 mg (per kg premix).

the trial. Four isocaloric and isonitrogenous treatments were formulated according to the recommendation for goats (National Research Council, 1981). The total digestible nutrients (TDN) and crude protein of the four treatments were around 60% and 14%, respectively. Ingredients used in formulating the experimental and control diets and chemical composition of date pits and dietary treatments are presented in Table 1 and 2. Group 1 was fed a control diet which contained 60% alfalfa hay and 40% concentrate feed (soybean meal, wheat bran, barley, and vitamins and minerals premix). Group 2, 3, and 4 were fed experimental diets, in which 10%, 15%, and 20% of date pits were substituted for the concentrate portion of the concentrate feed. All ingredients in concentrate feed were mixed thoroughly through a 100-kg feed mixer. Forages and concentrates were mixed carefully by hand before feeding in order to ensure uniformity and diets were offered as a total mixed ration. Animals in each group were fed individually in metabolic crates located under a semi-shed and the trial lasted for 90 days. Goats were milked daily at 8 a.m. and milk samples of each goat were collected weekly and analyzed by infrared spectroscopy (Milko-Scan FT 110,

Table 2. Chemical composition of date pits and experimental diets

Chemical composition (% of dry matter)	Date pits	Inclusion of date pits (%)			
		0	10	15	20
Dry matter	89.0	89.3	88.8	88.6	88.4
Crude protein	6.4	14.9	14.2	14.1	14.6
Total digestible nutrients	63.0	61.2	61.5	61.1	60.6
Fat	7.0	2.29	2.38	2.52	2.66
Ash	2.0	7.1	5.7	6.2	6.1
Crude fiber	28.9	21.6	23.0	23.9	24.9
Neutral detergent fiber	66.1	42.3	42.2	43.1	43.9
Acid detergent fiber	45.5	26.1	28.9	30.5	32.3
Calcium	0.35	1.87	1.87	1.89	1.91
Total phosphorus	0.23	0.7	0.6	0.5	0.5

Foss Electric, Hillerod, Denmark) to determine the milk components (fat, protein, lactose, total solids, and solid-not-fat). Rations were offered once daily, after milking. Mineral blocks and water were available to goats all the time. Feeds and orts of each treatment were weighed daily to determine the feed intake. The amount of ration offered was adapted so refusals were maintained between 1% and 2%. Goats were weighed every two weeks before the feeding time from the beginning of the trial until the end. Feed intake, feed conversion ratio, and body weight gain were calculated to evaluate goat performance.

Sample collection and analysis

Alfalfa hay and experimental diets were composited every two weeks for analysis. These composited samples were oven-dried (60°C), ground through a Wiley mill (1-mm screen), and analyzed for dry matter [DM] (105°C for 24 h). Crude protein (CP) was determined by macro-Kjeldahl procedure (method 955.04; AOAC International, 2002). Crude fibre (CF) analysis was determined by filter bag technique (ANKOM Technology, 2052 O'Neil Rd, Macedon, NY, USA) (method 978.10; AOAC International, 2002). Ether extract was determined with petroleum ether as the solvent (method 920.39; AOAC International, 2002). Ash was analyzed according to AOAC procedures (AOAC International, 2002). Acid detergent fiber (ADF) and neutral detergent fiber (NDF) (Van Soest and Robertson, 1985). TDN of the experimental diets were calculated based on the TDN values of ingredients.

Blood collection and biochemical analyses

Blood samples, about 10 mL, were withdrawn every two weeks via jugular venipuncture into vacuum glass tubes containing no anticoagulant. Samples were kept for one hour after collection and then plasma samples were obtained by centrifugation at 3,500 rpm for 20 min. The plasma samples were stored at –25°C until chemical analysis. Plasma total proteins (g/dL), creatinine (mg/dL), blood urea nitrogen (mg/dL), plasma glucose (mg/dL), triglycerides (mg/dL), albumin (g/dL), alanine amino transferase (ALT; IU/L), and aspartate amino transferase (AST; IU/L) activities were analyzed using available kits supplied by Bio Mérieux SA, F-69280 Marcy l'Etoile, France.

Statistical analysis

A completely randomized design was proposed with the experiment composed of four treatments with five animals each. The traits studied were milk production, milk composition, and blood parameters. Data were analyzed using SPSS version 14.0 (SPSS Inc., 2007) according the following model:

$$Y_{ij} = \mu + t_i + e_{ij}$$

Where Y_{ij} is the trait studied, u is the overall mean; t_i is the ith treatment; and e_{ij} is the error term which is assumed to be randomly and normally distributed with 0 mean and variance $\delta^2 e$. Duncan's multiple range test was used to compare differences among treatment means using 0.05 level of significance (Steel and Torrie, 1980).

RESULTS AND DISCUSSION

Chemical composition of the experimental and control groups shows that the dry matter, crude protein, fat, and TDN were relatively close in values (Table 2). Crude fiber, NDF, and ADF were higher in the diets contained date pits because of high concentration of fiber in the date pits.

Effect of experimental diets on milk yield and composition

Effects of the experimental diets on the initial and final body weight, milk yield, and milk composition are presented in Table 3. Goats fed the experimental and control diets showed no significant difference in terms of initial body weight, final body weight, and body weight gain which indicated that the experimental diets supplied goats with their nutrient requirements for maintenance and production. These results were in disagreement with Suliman and Mustafa (2014) who found that weight gain of lambs increased as the percentage of ground date seeds increased up to 30%. In another study, the addition of ground date seeds to lamb diets improved animal performance (Soliman et al., 2006). In addition, results obtained in this study were not comparable with Abdel-Fattah et al. (2012) who found that total gain of lambs reduced when concentrate feed contained 30% ground date seeds. The differences observed in this study compared to others in term of total body gain may be attributed to the differences of date varieties used, the percent of date pits included in the diets, the breed of animals, and physiological response of the animal to date pits. Milk yield was not significantly affected by the dietary treatments. This indicated that inclusion of date pits up to 20% in lactating dairy goat diets had no detrimental effect on milk production and the energy availability requirements of the animals was met. Dry matter intake (DMI) of goats fed 10% and 15% of date pits was significantly (p<0.05) higher than the goat fed 20% and the control group. In addition, goats fed 20% date pits showed significantly (p<0.05) lower DMI compared to the control group. The reduction in DMI for the goats fed 20% date pits may be attributed to higher content of fiber in date pits that may affect digestibility of the diet and therefore decrease the rumen passage rate. The result of this study was consistent with the finding of

Table 3. Dry matter intake, live body weight and daily milk production and composition of Ardi lactating goats fed diets supplemented with date pits

Item	Inclusion of date pits			
	0	10	15	20
Initial body weight (kg)	59.6 ±3.61	57.60 ±1.87	57.9 ±0.81	57.6 ±1.08
Final body weight (kg)	62.8 ±3.15	60.90 ±2.15	63.8 ±1.46	60.2 ±1.08
Body gain (kg)	3.20 ±0.97	3.30 ±0.70	5.90 ±1.44	2.60 ± 0.19
Milk production (kg/d)	1.40 ±0.02	1.41 ±0.02	1.43 ±0.01	1.39 ± 0.01
DMI (gm/d)	1,468.6 ±2.29[b]	1,479.8 ±2.35[a]	1,483.2 ±1.20[a]	1,449.8 ±0.86[c]
MY/DMI (feed conversion ratio)	1.22 ±0.01	1.16 ±0.01	1.17 ±0.01	1.21 ±0.01
Fat (%)	3.69 ±0.33	3.28 ±0.28	3.29 ±0.37	2.90 ±0.28
Fat yield (gm/d)	51.71 ±4.62	45.73 ±3.89	46.45 ±5.32	40.22 ±3.84
Protein (%)	2.57 ±0.03[a]	2.46 ±0.05[b]	2.43 ±0.02[b]	2.47 ±0.03[b]
Protein yield (gm/d)	36.00 ±0.83	34.79 ±0.95	34.83 ±0.11	34.60 ±0.44
Lactose (%)	4.25 ±0.09	4.05 ±0.04	4.07 ±0.13	4.08 ±0.12
Lactose yield (gm/d)	59.52 ±0.89	56.98 ±1.14	58.04 ±1.71	56.67 ±1.56
Total solids (%)	12.55 ±0.09[a]	12.27 ±0.08[b]	12.61 ±0.03[a]	12.62 ±0.01[a]
Solid not fat (%)	7.32 ±0.10	7.21 ±0.11	7.20 ±0.20	7.26 ±0.18

DMI, dry matter intake; MY, milk yield.
[a,b] Means within a row with different superscripts differ (p<0.05).

Suliman and Mustafa (2014) who showed that as the percentage of ground date seeds increased in lamb diets up to 15.57%, the DMI was significantly increased compared to other groups fed lower percentage of date pits but it was in contrast with the finding of Al-Shanti et al. (2013) who found that DMI of lambs was not affected when crushed date seeds replaced 50%, 75%, and 100% of corn and barley of concentrate feeds.

Feed conversion ratio was not significantly different among all the dietary treatments. Al-Shanti et al. (2013) showed that feed conversion ratio of Assaf lambs fed crushed date seeds was reduced when the ratio of date seeds was 45% and 60% of the concentrate feed mixture. Suliman and Mustafa (2014) indicated that the inclusion of date pits in lamb diets should not exceed 20% of the corn grain to avoid an adverse effect on nutritive value of feed intake and consequently on animal performance. The fat percent and yield was not significantly different among all the dietary treatments which indicated that goats fed diets containing date pits produced an adequate amount of acetate required for milk fat synthesis. In this study, the percentage of fiber and NDF of date pits was 29% and 66%, respectively (Table 2). However, Al-Farsi and Lee (2008) indicated that the total dietary fiber of date pits was much higher than this study. They found that date pits contained 58% crude fiber and 53% of it was insoluble dietary fiber mainly as hemicellulose, cellulose, and lignin. The fat percentage obtained in this study is lower than the values obtained by Al-Dobaib et al. (2009), but it was consistent with those of Strzalkowska et al. (2009) who found that the fat percentage in the first stage of lactation of goats was similar to the result of this study. In terms of percentage protein, goats in

the control group showed significantly (p<0.05) higher protein percent compared to the treatment groups. In fact, the inclusion of date pits at 20% of concentrate feed reduced protein yield by 1.4 g/d compared with the control group. The reduction of percentage protein from goats fed date pits may be attributed to lower protein availability for ruminal microbes, which would limit microbial protein synthesis. The protein percentages observed in this study for all the dietary treatments were comparable and fall into the range of those reported by Strzalkowska et al. (2009). Al-Dobaib et al. (2009) indicated that feeding discarded dates up to 30% to lactating Ardi goats increased protein percent compared to the control group. They attributed the higher protein percent in the milk to the higher non-fiber carbohydrates of dates. Goats fed the control and experimental diets showed no significant difference in terms of lactose percent, lactose yield and solid not fat percent. Total solids percent was significantly (p<0.05) lower for goats fed 10% diet compared to the other treatments. The results obtained in this study for Ardi goats were higher than that reported by Suranindyah and Astuti (2012), but it was comparable with the values found by Strzalkowska et al. (2009). The differences in composition of goat milk observed in this study compared to the others depended on many factors, but the most important among them was diet (Min et al., 2005).

Effect of experimental diets on blood parameters

Hematological data was used as an indication of the health status of the goats fed the experimental diets. The effects of experimental diets on serum biochemical values of Ardi lactating goats are presented in Table 4. Goats fed

Table 4. Serum biochemical values of Ardi lactating goats

Item	Inclusion of date pits			
	0	10	15	20
Total protein (g/dL)	8.21 ± 0.30^{ab}	7.33 ± 0.18^{b}	7.97 ± 0.37^{ab}	8.57 ± 0.28^{a}
Albumin (g/dL)	3.02 ± 0.05	2.72 ± 0.08	2.87 ± 0.18	3.01 ± 0.10
Urea (mg/dL)	18.80 ± 1.55	12.89 ± 1.51	14.93 ± 2.23	16.70 ± 1.77
Triglycerides (mg/dL)	8.24 ± 0.56^{ab}	9.45 ± 1.22^{a}	7.40 ± 1.10^{b}	9.46 ± 1.79^{a}
Creatinine (mg/dL)	0.87 ± 0.03	0.69 ± 0.05	0.81 ± 0.04	0.84 ± 0.04
Glucose (mg/dL)	45.09 ± 7.63	43.05 ± 8.81	53.80 ± 8.97	54.06 ± 5.89
ALT (IU/L)	16.86 ± 0.88	16.30 ± 0.68	16.92 ± 0.99	14.87 ± 1.02
AST (IU/L)	61.60 ± 5.28	67.60 ± 3.55	62.56 ± 4.81	57.81 ± 1.19

ALT, alanine amino transferase; AST, aspartate amino transferase.
[a,b] Means within a row with different superscripts differ ($p < 0.05$).

the 20% date pits was significantly ($p < 0.05$) higher in total protein compared to goats fed 10%, but no significant differences were found between the control and the other experimental groups. However, goats fed the 15% date pits was not significantly different from the goats fed the 20%. In term of albumin, no significant differences were found among all dietary treatments. The results of total protein and albumin were in the normal ranges reported for goats (Kahn and Line, 2010) and it was in agreement with the finding of Opara et al. (2010). The values of total protein and albumin observed in this study might indicate that date pits contain low level of tannins, known to reduce nutrient permeability in the digestive system as well as decrease protein utilization in the body, which are then subsequently excreted in feces without alteration to their metabolism. Turner et al. (2005) studied the effect of feeding growing goat diets containing lespedeza or alfalfa hay. They showed that concentration of condensed tannins in lespedeza hay had a negative effect on the uptake of protein from the small intestine causing the muscles to release more amino acids to support gluconeogenesis since forage quality of lespedeza was lower than alfalfa hay. The blood urea level of both the experimental and control groups were not significant different and it was within the normal ranges. The result in this study was inconsistent with the finding of Njidda et al. (2013) who found high serum urea of goats. The values of urea observed in the present study showed that goats fed the date pits diets had adequate amounts of protein for maintenance and production since high serum urea is a good indicator of protein deficiency (Oduye and Adadevoh, 1976). However, it might indicate that the level of glucose in blood of goats fed the experimental diets was normal since catabolic activity was increased for gluconeogenesis which increased serum urea level (Radostits et al., 1994). Triglyceride was significantly ($p < 0.05$) higher for the goats fed 10% and 20% compared to those fed 15% date pits, but no significant differences were found between the control and the experimental diets. The triglyceride values observed in this study were

inconsistent with the values found by Elitok (2012) in Saannen goats. The normal range of triglycerides observed in this study might indicate that goats fed the experimental diets did not experience negative energy balance which was reflected in the level of blood urea. The diets in this study did not significantly affect the creatinine levels in the serum of goats. Creatinine levels were within the normal range and they were comparable with the finding of Elitok (2012). Goats fed the experimental and the control groups showed no significant difference in glucose level. The value of blood glucose concentration obtained in this study was within the normal range from 48 to 76 mg/dL reported for goats (Kahn and Line, 2010). Since blood glucose concentration is an important indicator of dietary energy intake, the values reported in this study confirmed that the experimental diets supplied the animals with their requirement for energy. Date pits are recognized as a rich source of carbohydrates (Hamada et al., 2002) and might increase glucose circulation in blood and consequently the body responds by releasing insulin and inhibiting glucagon to stimulate the uptake of glucose by body cells. This study showed no significant difference in the concentrations of both AST and the ALT and the values of both enzymes were within the normal ranges reported for goats (Kahn and Line, 2010). The AST and ALT concentrations in this study were in agreement with Njidda et al. (2013) while were in disagreement with the values reported by Abdel-Fattah et al. (2012) who fed ground date seeds to male lambs. Higher liver enzymes were associated with lower nutrient intake (Oni et al., 2006) which indicated that goats fed the experimental diets had adequate amounts of nutrients to sustain their maintenance and milk production.

CONCLUSION

A big concern to ruminant breeders is the increasing prices of concentrate feeds around the world which is driving the search for an alternative traditional concentrates. Date pits can be considered as a good alternative feed ingredient that can be utilized efficiently by ruminants. Taking into account that the feed mills must have grinding machines that operate efficiently in milling date pits without the occurrence of any problems during the milling process, the finding of this study may indicate that date pits can be added to local lactating goat diets up to 20% of the concentrate feeds with no adverse effects on animal health or production.

ACKNOWLEDGMENTS

I am pleased to extend my sincere thanks and appreciation to the Deanship of Scientific Research at King Faisal University for funding this research. Thanks and appreciation are also extended to the Agricultural Experimental Station of King Faisal University for facilitating the research process and provide all the necessary materials needed for the trial. I am also pleased to thank AL-Gadeer Feed Mill in Al-Ahsa for their help in grinding the date pits.

REFERENCES

Abdel-Fattah, M. S., A. A. Abdel-Hamid, A. M. Ellamie, M. M. El-Sherief, and M. S. Zedan. 2012. Growth rate, some plasma biochemical and amino acid concentrations of Barki lambs fed ground date palm at Siwa oasis, Egypt. Am. Eurasian J. Agric. Environ. Sci. 12:1166-1175.

Al-Dobaib, S. N., M. A. Mehaia, and M. H. Khalil. 2009. Effect of feeding discarded dates on milk yield and composition of Aradi goats. Small Rumin. Res. 81:167-170.

Al-Farsi, M. A. and C. Y. Lee. 2008. Optimization of phenolics and dietary fibre extraction from date seeds. Food Chem. 108:977-985.

Al-Shanti, H. A., A. M. Kholif, K. J. Al-Shakhrit, M. F. Al-Banna, and I. E. Abu Showayb. 2013. Use of crushed date seeds in feeding growing Assaf lambs. Egypt. J. Sheep Goat Sci. 8:65-73.

AOAC International. 2002. Official Methods of Analysis of AOAC International. 17th edn. Association of Official Analytical Chemists, Gaithersburg, MD, USA.

Barreveld, W. H. 1993. Date Palm Products: FAO Agricultural Services Bulletin No. 101. Food and Agriculture Organization of the United Nations, Rome, Italy.

Costa, R. G., E. M. Beltrao Filho, R. C. R. Queiroga, A. N. Medeiros, M. O. Maia, and S. E. S. Cruz. 2009. Partial replacement of soybean meal by urea on production and milk physicochemical composition in Saanen goats. Rev. Bras. Saude Prod. Anim. 10:596-603.

Elitok, B. 2012. Reference Values for Hematological and Biochemical Parameters in Saanen Goats Breeding in Afyonkarahisar Province. Kocatepe Vet. J. 5:7-11.

Haenlein, G. F. W. 2004. Goat milk in human nutrition. Small Rumin. Res. 51:155-163.

Haider, M. S., I. A. Khan, M. J. Jaskani, S. A. Naqvi, and M. M. Khan. 2014. Biochemical attributes of dates at three maturation stages. Emir. J. Food Agric. 26:953-962.

Hamada, J. S., I. B. Hashim, and F. A. Sharif. 2002. Preliminary analysis and potential uses of date pits in foods. Food Chem. 76:135-137.

Kahn, C. M. and S. Line. 2010. 10th ed. The Merck Veterinary Manual. Merck and Co. Inc., Whitehouse Station, NJ, USA. pp. 905-908.

Min, B. R., S. P. Hart, T. Sahlu, and L. D. Satter. 2005. The effect of diets on milk production and composition, and on lactation curves in pastured dairy goats. J. Dairy Sci. 88:2604-2615.

Ministry of Agriculture. 2013. Annual Report. The Annual Agricultural Statistical Book. Department of Studies, Planning and Statistics. Vice Ministry for Research Affairs and Agricultural Development, Ministry of Agriculture, Riyadh, Kingdom of Saudi Arabia, No.22 (In Arabic).

Nagura, Y. 2004. Utilization of goat milk and meat in Japan. Farming Jpn. 36:9-13.

Njidda, A. A., T. Hassan, and E. A. Olatunji. 2013. Haematological and biochemical parameters of goats of semi arid environment fed on natural grazing rangeland of Northern Nigeria. IOSR J. Agric. Vet. Sci. 3:1-8.

National Research Council, 1981. Nutrient Requirements of Domestic Animals. National Academic Press, Washington, DC, USA. 91 p.

Oduye, O. O. and B. K. Adadevoh. 1976. Biochemical values in apparently normal Nigerian goats. J. Nig. Vet. Med. Assoc. 5:51-55.

Oni, A. O., C. F. I. Onwuka, O. O. Oduguwa, O. S. Onifade, O. M. Arigbede, and J. E. N. Olatunji. 2006. Utilization of citrus pulp based diets and Enterolobium cyclocarpum foliage (Jacq. Griseb) by West African Dwarf goats. J. Anim. Vet. Adv. 5:814-818.

Opara, M. N., N. Udevi, and I. C. Okoli. 2010. Haematological parameters and blood chemistry of apparently healthy West African Dwarf (Wad) goats in Owerri, South Eastern Nigeria. New York Sci. J. 3:68-72.

Radostits, O. M., D. C. Blood, and C. C. Gay. 1994. Veterinary Medicine: A Textbook of the Diseases of Cattle, Sheep, Pigs, Goats and Horses. 8th edn. Bailliere Tindall, London, England.

Soliman, A. A. M., A. I. A. Suliman, and A. H. A. Morsy. 2006. Productive performance of growing lambs fed on unconventional diets based on ground date palm seeds. Egypt. J. Anim. Poult. Manag. 1:101-119.

SPSS Inc. 2007. SPSS Version 14. SPSS Inc., Chicago, IL, USA.

Steel, R. G. D. and J. H. Torrie. 1980. Principles and Procedures of Statistics: A Biometrical Approach. 2nd edn. McGraw-Hill Book Company, New York, NY, USA.

Strzalkowska, N., A. Jozwik, E. Bagnicka, J. Krzyzewski, K. Horbanczuk, B. Pyzel, and J. O. Horbanczuk, 2009. Chemical composition, physical traits and fatty acid profile of goat milk as related to the stage of lactation. Anim. Sci. Pap. Rep. 27:311-320.

Suliman, A. I. A. and S. M. S. Mustafa. 2014. Effects of ground date seeds as a partial replacer of ground maize on nitrogen metabolism and growth performance of lambs. Egypt J. Sheep Goat Sci. 9:23-31.

Suranindyah, Y. and A. Astuti. 2012. The effects of feeding dried fermented cassava peel on milk production and composition of Etawah Crossedbred goat. World Acad. Sci. Eng. Technol. 6:10-21.

Turner, K. E., S. Wildeus, and J. R. Collins. 2005. Intake, performance, and blood parameters in young goats offered high forage diets of lespedeza or alfalfa hay. Small Rumin. Res. 59:15-23.

Turner, K. E., S. Wildeus, and J. R. Collins. 2005. Intake, performance, and blood parameters in young goats offered high forage diets of lespedeza or alfalfa hay. Small Rumin. Res. 59:15-23.

Van Soest, P. J. and J. B. Robertson. 1985. Analysis of Forages and Fibrous Feeds: A Laboratory Manual for Animal Science 613. Cornell University, Ithaca, NY, USA.

9

Growth Performance, Relative Meat and Organ Weights, Cecal Microflora, and Blood Characteristics in Broiler Chickens Fed Diets Containing Different Nutrient Density with or without Essential Oils

Sang-Jin Kim[1,2,a], Kyung-Woo Lee[1,a], Chang-Won Kang[1], and Byoung-Ki An[1,*]

[1] Laboratory of Poultry Science, Department of Animal Science and Technology,
College of Animal Bioscience and Technology, Konkuk University, Seoul 05029, Korea

ABSTRACT: The present study was conducted to investigate whether dietary essential oils could affect growth performance, relative organ weights, cecal microflora, immune responses and blood profiles of broiler chickens fed on diets containing different nutrient densities. A total of eight hundred-forty 1-d-old male broiler chicks were randomly allotted into twenty-eight pens (7 pens per treatment, 30 chicks per pen). There were four experimental diets containing two different nutrient densities and supplemented with or without essential oils. Experimental period lasted for 35 days. No clear interaction between nutrient density and essential oils on any of growth performance-related parameters was observed. Live body weights were affected ($p<0.05$) by nutrient density at 21 days and by dietary essential oils at 35 days. Essential oils significantly ($p<0.05$) increased daily body weight gain and feed conversion ratio during the periods of 22 to 35 and 1 to 35 days, but failed to affect feed intake during the entire experimental period. Daily weight gain at 1 to 21 days and feed intake at 1 to 21 and 1 to 35 days were significantly impaired ($p<0.05$) by nutrient density. There were significant treatment interactions ($p<0.05$) on relative weights of bursa of Fabricius and abdominal fat contents. Finally, either essential oil or nutrient density did not influence the relative percentages of breast and leg meats, the population of cecal microflora, blood parameters and antibody titers against Newcastle disease and infectious bronchitis in broiler chickens. It was concluded that dietary essential oils, independent to nutrient density, failed to stimulate feed intake, but increased growth performance in broiler chickens. (**Key Words:** Growth Performance, Nutrient Density, Essential Oils, Broiler Chickens)

INTRODUCTION

Essential oils (EO) derived from spices and herbs, as single components or as mixed preparations, can play significant roles in supporting both performance and health status of poultry (Huyghebaert et al., 2011; Lee et al., 2011; Khattak et al., 2014). Beneficial effects of active plant substances in poultry nutrition include the stimulation of appetite and feed intake, the improvement of endogenous digestive enzyme secretion, the activation of immune response and antibacterial, antiviral, antioxidant and antihelminthic actions (Lee et al., 2004; Jamroz et al., 2005; Zeng et al., 2015b). The aforementioned biological effects of EO could then lead to efficient nutrient utilization in poultry. Indeed, the supplementation of EO into low energy or nutrient density diet significantly increased daily weight gain and enhanced the digestibilities of nutrients compared with the low energy or nutrient density based control diet and exhibited comparable performance compared with a standard energy or nutrient density diet in pigs (Yan et al., 2010; Zeng et al., 2015a). In line with the latter studies, it was reported that dietary EO exhibited identical growth performance of broiler chickens fed the low-nutrient density (LND) diets compared with those fed the high-nutrient density (HND) diet (Scheuermann et al., 2009). Unfortunately, the absence of the LND control diet in the latter experiment made it difficult to prove the observed

* Corresponding Author: Byoung-Ki An.
E-mail: abk7227@hanmail.net
[2] Yuhan Corp, Seoul 06927, Korea.
[a] These authors contributed equally to this work.

effect. Buchanan et al. (2008) reported that improvement of feed conversion ratio was only observed when EO was added into a HND vs LND diet. In contrast, no significant effect of EO on growth performance was observed in broiler chickens fed a threonine-deficient diet (Muhl and Liebert, 2007). In contrast to pig trials (Yan et al., 2010; Zeng et al., 2015a), the effect of EO on growth performance in broiler chickens fed a diet containing LND vs HND diet is inconclusive. Thus, the present experiment was intended to validate whether there is interaction between nutrient density and EO on growth performance in broiler chickens. In addition to production traits, various parameters such as blood metabolites, blood immune responses, organ weights, and cecal microflora were evaluated to see the presence or absence of the interaction between EO and nutrient density.

MATERIALS AND METHODS

Experimental design

The experimental facility was thoroughly cleaned and disinfected before the initiation of the experiment. A total of eight hundred-forty 1-d-old male broiler chicks (Ross 308) were obtained from local hatchery. Upon arrival, they were individually weighed and randomly placed into one of twenty-eight pens (1.8 m×1.8 m) with fresh rice husks as a bedding material and stocking density was set at 0.108 m^2 per bird. The present experiment consisted of four dietary treatments which were given for 35 days (starter 1 to 21 days, finisher 22 to 35 days). Each treatment consisted of seven pens and each pen had 30 chickens (n = 210 chickens/treatment). There were four dietary experimental diets with two nutrient density diets (HND vs LND) supplemented with or without EO at 150 mg/kg of diet. The commercially available EO preparation (Biostrong 510, Delacon, Steyregg, Austria) consisted of a mixture of EO with thyme and star anise, Quillaja extracts, and bulking and anti-caking agents. Corn-soybean meal based HND and LND based diets were formulated (Table 1). The LND vs HND diet contained less energy and nutrients by one percent point of crude protein (CP), 50 kcal of nitrogen-corrected true metabolizable energy, 0.05 percent point of available phosphate, and 0.05 percent point of lysine. These nutrient reduction can be considered moderate as compared with the previous studies (Brickett et al., 2007; Zhao et al., 2009; Li et al., 2010; Mirshekar et al., 2013), in which the LND vs HND diet was formulated to reduce CP levels by 2 percent points, and energy levels by more than 150 kcal/kg of diet. Diet and water were provided *ad libitum*. Continuous lighting program was used and the temperature of facility was maintained at 32°C during the first week posthatch and gradually decreased to reach 25°C at 3 weeks and kept thereafter. At days 14 and 28, all broiler chicks used in this study were vaccinated against Newcastle

Table 1. Ingredient and chemical composition of the starter and finisher diets

Items	Starter (1 to 21 d)		Finisher (22 to 35 d)	
	HND	LND	HND	LND
Ingredients (%)				
Yellow corn	55.12	59.83	59.92	63.78
Soybean meal	33.23	30.97	29.22	27.18
Corn gluten meal	3.36	2.89	2.95	2.99
Tallow	4.48	2.57	4.35	2.59
DL-methionine (98%)	0.17	0.16	0.05	0.03
Salt	0.33	0.33	0.33	0.33
Limestone	1.19	1.38	1.56	1.74
Dicalcium phosphate	1.64	1.38	1.12	0.86
Vitamin+mineral mixture[1]	0.40	0.40	0.40	0.40
Choline chloride (50%)	0.08	0.09	0.10	0.10
Total	100.0	100.0	100.0	100.0
Calculated values				
Crude protein (%)	21.50	20.50	20.00	19.00
Crude fat (%)	7.00	5.22	6.96	5.31
Crude fiber (%)	3.36	3.33	3.22	3.19
Crude ash (%)	6.00	5.84	5.66	5.51
Ca (%)	1.00	1.00	1.00	1.00
Available P (%)	0.40	0.35	0.30	0.25
Lysine (%)	1.13	1.07	1.02	0.97
Cys+met (%)	0.90	0.86	0.73	0.70
TMEn (kcal/kg)	3,100	3,045	3,150	3,100

HND, high-nutrient density; LND, low-nutrient density; TMEn, nitrogen-corrected true metabolizable energy.

[1] Vit.+Min. mixture provided the following nutrients per kg of diet: vitamin A, 18,000 IU; vitamin D$_3$, 3,750 IU; vitamin E, 30 IU; vitamin K$_3$, 2.7 mg; vitamin B$_1$, 3.0 mg; vitamin B$_2$, 9.0 mg; vitamin B$_6$, 4.5 mg; vitamin B$_{12}$, 30.0 mg; niacin, 37.5 mg; pantothenic acid, 15 mg; folic acid, 1.5 mg; biotin, 0.07 mg; Fe, 75.0 mg; Zn, 97.5 mg; Mn, 97.5 mg; Cu, 7.5 mg; I, 1.5 mg; Se, 0.2 mg.

disease (ND) and infectious bronchitis (IB) by eye drop using the attenuated live mixed vaccine (Nobilis Ma5+Clone 30, MSD AH Korea Ltd., Seoul, Korea). All experimental protocols were approved by the Animal Care Committee of KonKuk University.

Sampling

Feed intake and body weight per pen were measured on a weekly basis and used to calculate feed conversion ratio. At 35 days, eight chickens per pen were randomly selected for blood sampling. Blood was collected from cardiac puncture immediately after cervical dislocation. Sera were obtained by gentle centrifugation (600 g for 15 min) and stored at –20°C prior to use. Immediately after blood sampling, organs such as liver, spleen, bursa of Fabricius and abdominal fat were excised, weighed and expressed as relative weight to live body weight. In addition, right breast and thigh meats were sampled and weighed. Finally, cecal contents were removed aseptically, placed into sterile tubes and kept on ice until used for gut microbiota analysis on the same day of the sampling.

Measurement of total cholesterol, glutamic-oxaloacetic transaminase and glutamic-pyruvic transaminase in serum samples

Blood parameters were measured as described elsewhere (Lee et al., 2010). In brief, total cholesterol concentration and activities of glutamic-oxaloacetic transaminase (GOT) and glutamic-pyruvic transaminase (GPT) in sera sampled at 35 days were measured according to the colorimetric method using cholesterol diagnostic kit (Cholesterol E kit, Asan Phamaceutical Co., Seoul, Korea) and GOT-GPT assay kits (Asan Phamaceutical Co., Korea).

Measurement of infectious bronchitis- and Newcastle disease-reactive antibody response

Viral antibodies against IB and ND in sera sampled at 35 days were determined (Lee et al., 2012) using commercial enzyme-linked immunosorbent assay kits (IDEXX Laboratory, Westbrook, ME, USA) according to the manufacturer's instructions. The results were expressed as antibody titer that was calculated using the formula provided by IDEXX.

Enumeration of intestinal microflora

Individual cecal contents were subjected to serial 10-fold (w/v) dilution with ice-cold phosphate-buffered saline as suggested by Noh et al. (2014). Total microbes were counted after grown on total plate agar (Difco, Detroit, MI, USA), lactic acid bacteria on Man Rosa-Sharpe agar (Difco,

USA), and total coli forms on MacConkay agar (Difco, USA) at 37°C for 24 h. Results obtained were presented as base-10 logarithm colony forming unit (cfu) per gram of cecal digesta.

Statistical analysis

Pen was considered an experimental unit. Two-way analysis of variance was performed to define the effect of diet (LND vs HND), EO (0, 150 mg/kg of diet) and the diet by EO interaction on the variables using the general linear model procedure of SAS program (SAS, 2000). A statistical significance was preset at $p<0.05$ unless otherwise stated.

RESULTS

There was no interaction between nutrient density and EO on the measured parameters except for bursa of Fabricius and abdominal fat in broiler chickens. Body weights were affected ($p<0.05$) by nutrient density at 21 days and by EO supplementation at 35 days. HND vs LND diet significant increased ($p<0.05$) daily body weight gain at 1 to 21 days (Table 2). Dietary EO significantly increased ($p<0.05$) daily body weight gain during the periods of 22 to 35 and 1 to 35 days. Daily feed intake was not affected ($p>0.05$) by dietary EO, but significantly reduced ($p<0.05$) by the LND vs HND diet, especially during the periods of 1 to 21 and 1 to 35 days. Feed conversion ratio was significantly improved ($p<0.05$) by EO addition during the periods of 22 to 35 days and 1 to 35 days (Table 2). Either nutrient density or EO did not influence the relative weights of liver, spleen, breast and leg meat yields (Table 3), cecal

Table 2. Effect of dietary essential oils on growth performance in broiler chickens fed diets containing different nutrient densities[1]

	HND	LND	HND+EO	LND+EO	Pooled SEM	p-values		
						Diet	EO	Diet×EO
Initial BW (g/bird)	40.3	40.3	40.3	40.3	0.033	0.880	0.880	0.880
1 to 21 d BW (g/bird)	708.0	670.1	714.9	688.5	11.7	0.011	0.289	0.624
Final BW (g/bird)	1,698.7	1,656.4	1,754.5	1,708.9	25.5	0.084	0.036	0.947
Feed intake (g/d/bird)								
1 to 21 d	52.2	49.3	52.8	50.6	0.759	0.002	0.237	0.649
22 to 35 d	154.3	151.8	152.6	145.5	2.410	0.057	0.112	0.344
1 to 35 d	91.9	89.1	91.7	87.7	0.865	0.001	0.358	0.467
BW gain (g/d/bird)								
1 to 21 d	32.5	30.0	32.1	30.2	0.644	0.002	0.866	0.668
22 to 35 d	71.3	70.5	74.3	73.2	1.321	0.472	0.042	0.925
1 to 35 d	48.0	47.5	50.4	48.8	0.732	0.162	0.023	0.440
Feed/gain (g/g)								
1 to 21 d	1.61	1.64	1.64	1.68	0.018	0.064	0.064	0.784
22 to 35 d	2.17	2.16	2.06	1.99	0.058	0.496	0.024	0.609
1 to 35 d	1.92	1.88	1.82	1.80	0.029	0.307	0.005	0.731

HND, high-nutrient density; LND, low-nutrient density; HND+EO, HND diet added with 150mg/kg of essential oils (EO); LND+EO, LND diet added with 150 mg/kg of EO; SEM, pooled standard error of the mean; BW, body weight.
[1] Values are expressed as means of seven replicates per dietary group.

Table 3. Effect of dietary essential oils on relative organ and meat weights in broiler chickens fed diets containing different nutrient densities[1]

	HND	LND	HND+EO	LND+EO	Pooled SEM	p-values Diet	p-values EO	p-values Diet×EO
Liver	2.36	2.29	2.28	2.35	0.081	0.999	0.903	0.398
Spleen	0.12	0.14	0.13	0.13	0.010	0.327	0.999	0.327
Bursa of Fabricius	0.23	0.19	0.21	0.23	0.013	0.457	0.457	0.033
Abdominal fat	1.53	1.87	2.16	1.53	0.188	0.448	0.447	0.016
Breast meat	7.70	7.62	8.02	7.85	0.228	0.522	0.166	0.817
Leg meat	8.87	9.00	9.25	9.19	0.166	0.838	0.105	0.579

HND, high-nutrient density; LND, low-nutrient density; HND+EO, HND diet added with 150 mg/kg of essential oils (EO); LND+EO, LND diet added with 150 mg/kg of EO; SEM, pooled standard error of the mean.
[1] Values (g/100g of body weight) are expressed as means of seven replicates per dietary group.

microflora (Table 4) and blood parameters such as total cholesterol, GOT, GPT, and ND- and IB-specific antibodies (Table 5).

DISCUSSION

The present study was designed to investigate whether dietary EO added into different nutrient density diets could improve growth performance of broiler chickens. Initially, a positive effect was expected in the light of previous reports showing that dietary EO increased growth performance and nutrient utilization in broiler chickens (Hernandez et al., 2004; Mountzouris et al., 2011; Bravo et al., 2014; Cho et al., 2014). In this study, we confirmed the previous reports (Buchanan et al., 2008; Leeson, 2012) on the well-established negative effect of nutrient density on growth

performance (e.g., decrease in live body weight, feed intake, and daily weight gain) in broiler chickens. It was observed that EO supplementation, regardless of nutrient density, significantly increased (p<0.05) daily body weight gain during the periods of 22 to 35 and 1 to 35 days, and also improved feed conversion ratio during the periods of 22 to 35 and 1 to 35 days. The consistent increase in daily weight gain and decrease in feed conversion ratio by dietary EO would be likely the consequence of increased nutrient utilization as reported elsewhere (Hernandez et al., 2004; Lee et al., 2004; Mountzouris et al., 2011; Bravo et al., 2014; Cho et al., 2014). However, we did not measure nutrients digestibility or digestive enzyme activities in this study.

It is of note that EO-mediated increase in growth performance was equally effective in chickens fed diet containing different nutrient densities. In previous studies

Table 4. Effect of dietary essential oils on cecal microflora in broiler chickens fed diets containing different nutrient densities[1]

	HND	LND	HND+EO	LND+EO	Pooled SEM	p-values Diet	p-values EO	p-values Diet×EO
Total microbes (log cfu/g)	6.25	6.59	6.59	6.65	0.228	0.818	0.491	0.927
Lactic acid bacteria (log cfu/g)	6.56	6.76	7.04	6.61	0.228	0.644	0.413	0.608
Coli forms (log cfu/g)	6.01	5.39	5.63	5.92	0.241	0.742	0.328	0.566

HND, high-nutrient density; LND, low-nutrient density; HND+EO, HND diet added with 150 mg/kg of essential oils (EO); LND+EO, LND diet added with 150 mg/kg of EO; SEM, pooled standard error of the mean.
[1] Values are expressed as means of seven replicates per dietary group.

Table 5. Effects of dietary essential oils on blood characteristics and antibody titers against Newcastle disease (ND) virus and infectious bronchitis (IB) in broiler chickens fed diets containing different nutrient densities[1]

	HND	LND	HND+EO	LND+EO	Pooled SEM	p-values Diet	p-values EO	p-values Diet×EO
Total cholesterol (mg/dL)	92.05	90.00	90.47	90.97	4.59	0.867	0.948	0.784
GOT (U/L)	277.20	281.62	278.31	278.33	3.86	0.571	0.780	0.574
GPT (U/L)	8.86	8.90	8.87	8.98	0.40	0.853	0.911	0.931
Viral antibody titer (log_{10})								
ND titer	2.29	2.00	2.57	2.43	0.300	0.480	0.248	0.805
IB titer	3.00	4.00	3.00	3.29	0.392	0.113	0.375	0.375

HND, high-nutrient density; LND, low-nutrient density; HND+EO, HND diet added with 150 mg/kg of essential oils (EO); LND+EO, LND diet added with 150 mg/kg of EO; SEM, pooled standard error of the mean; GOT, glutamic-oxaloacetic transaminase; GPT, glutamic-pyruvic transaminase.
[1] Values are expressed as means of seven replicates per dietary group.

with EO supplemented into different nutrient density, conflicted results have been reported. For example, Buchanan et al. (2008) observed an increase in feed efficiency in broiler chickens fed the EO-added, HND diet, but this effect was not seen when the LND diet was used. Bozkurt et al. (2012) reported that body weights were increased in broiler chickens fed the EO-fortified, wheat-based diet, but were significantly decreased when EO was added into the corn-based diet. Finally, dietary EO did not improve overall growth performance in broiler chickens fed a threonine-deficient diet (Muhl and Liebert, 2007). In this study, the LND vs HND diet contained less energy and CP, but formulated to keep the ratios of lysine to limiting amino acids constant. Whether that the effect of dietary EO on growth performance is more effective in low energy/nutrient density diets with balanced amino acids needs to be addressed. Nonetheless, our study provides evidence that dietary EO could improve growth performance of broiler chicken and the EO-mediated effect was independent to nutrient density. The latter finding is considered of practical relevance in feed formulation in poultry production. According to Leeson (2012), the LND vs HND diet will be feasible in poultry production as the feed prices are currently on the increase.

None of parameters measured in this study were affected by either nutrient density or dietary EO. However, it was found that there were significant interactions ($p < 0.05$) between density and EO on the relative weight of bursa of Fabricius and abdominal fat. Dietary EO when added into a LND diet increased relative weight of bursa of Fabricius, which reached to the value shown in birds fed a HND diet alone. Whether this indicates the consequence of altered immune status by EO needs to be verified as no difference in humoral responses against ND and IB were observed in this study. To our surprise, addition of EO into the HND, but not LND diet, tended to increase the relative abdominal fat, thus leading to significant treatment interaction. At this stage, clear explanation on the confounding result is not readily available.

In conclusion, the present study showed that dietary EO, independent to nutrient density, effectively increased body weight, daily weight gains and feed conversion ratio in broiler chickens. In addition, there were significant treatment interactions on relative bursal of Fabricius and abdominal fats. Finally, none of parameters measured, i.e., relative organ weights, percentage of meat yields, cecal microflora, blood characteristics were affected by either nutrient density or dietary EO.

ACKNOWLEDGMENTS

This paper was supported by the KU Research Professor Program of KonKuk University, and Delacon Biotechnik Ges.m.b.H., and also partially supported by the grant "Investigation of black rice extract on functional mechanism modulating bone health and obesity using in vivo method (PJ01009001)" of Rural Development Administration, Republic of Korea. We thank In-Sook An in the Poultry Science Laboratory, Department of Animal Science and Technology, College of Animal Bioscience and Technology, KonKuk University for the technical assistance.

REFERENCES

Bozkurt, M., K. Kucukyilmaz, A. U. Catli, Z. Ozyildiz, M. Cinar, M. Cabuk, and F. Coven. 2012. Influences of an essential oil mixture supplementation to corn versus wheat-based practical diets on growth, organ size, intestinal morphology and immune response of male and female broilers. Ital. J. Anim. Sci. 11:e54.

Bravo, D., V. Pirgozliev, and S. P. Rose. 2014. A mixture of carvacrol, cinnamaldehyde, and capsicum oleoresin improves energy utilization and growth performance of broiler chickens fed maize-based diet. J. Anim. Sci. 92:1531-1536.

Brickett, K. E., J. P. Dahiya, H. L. Classen, and S. Gomis. 2007. Influence of dietary nutrient density, feed form, and lighting on growth and meat yield of broiler chickens. Poult. Sci. 86:2172-2181.

Buchanan, N. P., J. M. Hott, S. E. Cutlip, A. L. Rack, A. Asamer, and J. S. Moritz. 2008. The effects of a natural antibiotic alternative and a natural growth promoter feed additive on broiler performance and carcass quality. J. Appl. Poult. Res. 17:202-210.

Cho, J. H., H. J. Kim, and I. H. Kim. 2014. Effects of phytogenic feed additive on growth performance, digestibility, blood metabolites, intestinal microbiota, meat color and relative organ weight after oral challenge with *Clostridium perfringens* in broilers. Livest. Sci. 160:82-88.

Hernandez, F., J. Madrid, V. Garcia, J. Orengo, and M. D. Megias. 2004. Influence of two plant extracts on broilers performance, digestibility, and digestive organ size. Poult. Sci. 83:169-174.

Huyghebaert, G., R. Ducatelle, and F. Van Immerseel. 2011. An update on alternatives to antimicrobial growth promoters for broilers. Vet. J. 187:182-188.

Jamroz, D., A. Wiliczkiewicz, T. Wertelecki, J. Orda, and J. Skorupinska. 2005. Use of active substances of plant origin in chicken diets based on maize and locally grown cereals. Br. Poult. Sci. 46:485-493.

Khattak, F., A. Ronchi, P. Castelli, and N. Sparks. 2014. Effects of natural blend of essential oil on growth performance, blood biochemistry, cecal morphology, and carcass quality of broiler chickens. Poult. Sci. 93:132-137.

Lee, K. W., H. Everts, and A. C. Beynen. 2004. Essential oils in broiler nutrition. Int. J. Poult. Sci. 3:738-752.

Lee, S. H., H. S. Lillehoj, S. I. Jang, K. W. Lee, M. S. Parks, D. Bravo, and E. P. Lillehoj. 2011. Cinnamaldehyde enhances *in vitro* parameters of immunity and reduces *in vivo* infection against avian coccidiosis. Br. J. Nutr. 106:862-869.

Lee, K. W., H. S. Lillehoj, S. I. Jang, and S. H. Lee. 2012. Effects of various field coccidiosis control programs on host innate and adaptive immunity in commercial broiler chickens. Korean J. Poult. Sci. 39:17-25.

Lee, D. W., J. H. Shin, J. M. Park, J. C. Song, H. J. Shu, U. J. Chang, B. K. An, C. W. Kang, and J. M. Kim. 2010. Growth performance and meat quality of broiler chicks fed germinated and fermented soybeans. Korean J. Food Sci. Anim. Resour. 30:938-945.

Leeson, S. 2012. Future considerations in poultry nutrition. Poult. Sci. 941:1281-1285.

Li, W. B., Y. L. Guo, J. L. Chen, R. Wang, Y. He, and D. G. Su. 2010. Influence of lighting schedule and nutrient density in broiler chickens: effect on growth performance, carcass traits and meat quality. Asian Australas. J. Anim. Sci. 23:1510-1518.

Mirshekar, R., B. Dastar, B. Shabanpour, and S. Hassani. 2013. Effect of dietary nutrient density and vitamin premix withdrawal on performance and meat quality of broiler chickens. J. Sci. Food Agric. 93:2979-2985.

Mountzouris, K. C., V. Paraskevas, P. Tsirtsikos, I. Palamidi, T. Steiner, G. Schatzmayr, and K. Fegeros. 2011. Assessment of a phytogenic feed additive effect on broiler growth performance, nutrient digestibility and caecal microflora composition. Anim. Feed Sci. Technol. 168:223-231.

Muhl, A. and F. Liebert. 2007. Growth, nutrient utilization and threonine requirement of growing chicken fed threonine limiting diets with commercial blends of phytogenic feed additives. J. Poult. Sci. 44:297-304.

Noh, H. S, S. L. Ingale, S. H. Lee, K. H. Kim, I. K. Kwon, Y. H. Kim, and B. J. Chae. 2014. Effects of citrus pulp, fish by-product and *Bacillus subtilis* fermentation biomass on growth performance, nutrient digestibility, and fecal microflora of weanling pigs. J. Anim. Sci. Technol. 56:10.

SAS. 2002. SAS User's Guide. Statistics, Version 8, SAS Institute Inc., Cary, NC, USA.

Scheuermann, G. N., A. C. Junior, L. Cyprioano, and A. M. Gabbi. 2009. Phytogenic additive as an alternative to growth promoters in broiler chickens. Cienc. Rural 39:522-527.

Yan, L., J. P. Wang, H. J. Kim, Q. W. Meng, X. Ao, S. M. Hong, and I. H. Kim. 2010. Influence of essential oil supplementation and diets with different nutrient densities on growth performance, nutrient digestibility, blood characteristics, meat quality and fecal noxious gas content in grower-finisher pigs. Livest. Sci. 128:115-122.

Zeng, Z. K., X. Xu, Q. Zhang, P. Li, P. Zhao, Q. Li, J. Liu, and X. Piao. 2015a. Effects of essential oil supplementation of a low-energy diet on performance, intestinal morphology and microflora, immune properties and antioxidant activities in weaned pigs. Anim. Sci. J. 86:279-285.

Zeng, Z., S. Zhang, H. Wang, and X. Piao. 2015b. Essential oil and aromatic plants as feed additives in non-ruminant nutrition: A review. J. Anim. Sci. Biotechnol. 6:7.

Zhao, J. P., J. L. Chen, G. P. Zhao, M. Q. Zheng, R. R. Jiang, and J. Wen. 2009. Live performance, carcass composition, and blood metabolite responses to dietary nutrient density in two distinct broiler breeds of male chickens. Poult. Sci. 88:2575-2584.

10

A Comparison of Natural (D-α-tocopherol) and Synthetic (DL-α-tocopherol Acetate) Vitamin E Supplementation on the Growth Performance, Meat Quality and Oxidative Status of Broilers

K. Cheng, Y. Niu, X. C. Zheng, H. Zhang, Y. P. Chen, M. Zhang[1], X. X. Huang[1],
L. L. Zhang, Y. M. Zhou, and T. Wang*

College of Animal Science and Technology, Nanjing Agricultural University, Nanjing 210095, China

ABSTRACT: The present study was conducted to compare the supplementation of natural (D-α-tocopherol) and synthetic (DL-α-tocopherol acetate) vitamin E on the growth performance, meat quality, muscular antioxidant capacity and genes expression related to oxidative status of broilers. A total of 144 1 day-old Arbor Acres broiler chicks were randomly allocated into 3 groups with 6 replicates of 8 birds each. Birds were given a basal diet (control group), and basal diet supplemented with either 20 IU D-α-tocopherol or DL-α-tocopherol acetate for 42 days, respectively. The results indicated that treatments did not alter growth performance of broilers (p>0.05). Compared with the control group, concentration of α-tocopherol in the breast muscle was increased by the supplementation of vitamin E (p<0.05). In the thigh, α-tocopherol content was also enhanced by vitamin E inclusion, and this effect was more pronounced in the natural vitamin E group (p<0.05). Vitamin E supplementation increased the redness of breast (p<0.05). In the contrast, the inclusion of synthetic vitamin E decreased lightness of thigh (p<0.05). Dietary vitamin E inclusion reduced drip loss at 24 h of thigh muscle (p<0.05), and this effect was maintained for drip loss at 48 h in the natural vitamin E group (p<0.05). Broilers given diet supplemented with vitamin E showed decreased malondialdehyde (MDA) content in the breast (p<0.05). Additionally, natural rather than synthetic vitamin E reduced MDA accumulation in the thigh (p<0.05). Neither natural nor synthetic vitamin E supplementation altered muscular mRNA abundance of genes related to oxidative stress (p>0.05). It was concluded that vitamin E supplementation, especially the natural vitamin E, can enhance the retention of muscular α-tocopherol, improve meat quality and muscular antioxidant capacity of broilers. (**Key Words:** D-α-tocopherol, DL-α-tocopherol Acetate, Meat Quality, Oxidative Status, Broiler)

INTRODUCTION

Lipid oxidation is one of the major causes of quality deterioration in meat and meat products. This oxidation can lead to serious consequences for meat quality, including the production of off-flavors and odors, reduction of polyunsaturated fatty acids, fat-soluble vitamins and pigments, lower consumer acceptability, and the generation of compounds such as peroxides and aldehydes which may be toxic (Morrissey et al., 1994). Vitamin E as a lipid component of biological membranes is known to be a major chain-breaking antioxidant (Sahin et al., 2002). Various studies have demonstrated that vitamin E can improve meat quality and inhibit lipid oxidation of muscles in broilers (O'neill et al., 1998; Voljč et al., 2011; Rey et al., 2015).

Eight naturally occurring substances have been found to have vitamin E activity: α-, β-, γ-, and δ-tocopherols and α-, β-, γ-, and δ-tocotrienols (Voljč et al., 2011). Of these the α-tocopherol is the most biologically active and most widely distributed form (Halliwell and Gutteridge, 2000). This molecule possesses three centers where stereoisomers can occur, and the naturally occurring molecule is the D-α-tocopherol (RRR-α-tocopherol) configuration that has the highest vitamin activity (McDonald et al., 2011). Synthetic DL-α-tocopherol acetate (all racemic α-tocopherol acetate)

* Corresponding Author: T. Wang.
E-mail: tianwangnjau@163.com
[1] Jiangsu Wilmar Spring Fruit Nutrition Products Co., Ltd. Taixing 225434, China.

consisting of all eight possible stereoisomers is commonly used as a vitamin E supplement in poultry feeds. Previous studies have shown that D-α-tocopherol exhibited a superior bioavailability than the synthetic DL-α-tocopherol acetate in retaining in tissues (Lauridsen et al., 2002), alleviating lipopolysaccharide-induced inflammatory response (Kaiser et al., 2012), and improving meat quality and muscular antioxidant capacity (Boler et al., 2009; Voljč et al., 2011; Rey et al., 2015). However, in broilers, limit was known about the use of D-α-tocopherol as an alternative for the synthetic vitamin E on the muscular oxidative status, and especially meat quality and muscular mRNA abundance of genes involved in antioxidant system. The present study was therefore conducted to compare the supplementation of D-α-tocopherol and DL-α-tocopherol acetate on the growth performance, meat quality, muscular antioxidant capacity and genes expression related to oxidative status of broilers.

MATERIAL AND METHODS

Experimental design, diets and management

All experimental conditions and animal procedures were approved by Nanjing Agricultural University Institutional Animal Care and Use Committee. A total of 144 1 day-old Arbor Acres broiler chicks obtained from a commercial hatchery (Hewei Co., Ltd, Anhui, China) were randomly allocated into 3 groups with 6 replicates of 8 birds each (one replicate per cage, 4 males and 4 females per cage). Birds were given a basal diet (control group), and basal diet supplemented with either 20 IU D-α-tocopherol or DL-α-tocopherol acetate for 42 days, respectively. D-α-tocopherol was provided by Jiangsu Wilmar Spring Fruit Nutrition Products Co., Ltd. (Taixing, Jiangsu, China) with the purity of 6%. DL-α-tocopherol was purchased from Zhejiang NVB Co., Ltd (Xinchang, Zhejiang, China), and the purity was 50%. The basal diet was formulated according to the NRC (1994) to meet the nutrient requirements of the broiler. The analyzed α-tocopherol in the basal grower (1 to 21 day) and finisher (22 to 42 day) diet was 7.12 and 8.96 mg/kg, respectively. Formulation and nutrient level of basal diet are presented in Table 1. Birds had free access to mash feed and water in 3-layer cages in a temperature-controlled room with continuous lighting. The temperature of the room was maintained at 32°C to 34°C for the first 3 day and then reduced by 2°C to 3°C per week to a final temperature of 20°C. At 21 and 42 day of age, birds were weighed after feed deprivation for 12 h with water being provided *ad libitum*. Feed intake was recorded by replicate (cage) to calculate average daily feed intake (ADFI), average daily gain (ADG), and feed/gain ratio (F:G). Birds that died during the experiment were weighed, and the data were included in the calculation of F:G.

Table 1. Composition and nutrient level of basal diet (g/kg, as fed basis unless otherwise stated)

Items	1 to 21 d	22 to 42 d
Ingredients		
Corn	576.1	622.7
Soybean meal	310	230
Corn gluten meal	32.9	60
Soybean oil	31.1	40
Limestone	12	14
Dicalcium phosphate	20	16
L-Lysine	3.4	3.5
DL-Methionine	1.5	0.8
Sodium chloride	3	3
Premix[1]	10	10
Calculated nutrient levels[2]		
Apparent metabolizable energy (MJ/kg)	12.56	13.19
Crude protein	211	196
Calcium	10.00	9.50
Available phosphorus	4.60	3.90
Lysine	12.00	10.50
Methionine	5.00	4.20
Methionine+cystine	8.50	7.60
Analyzed composition[3]		
Crude protein	208	192
Ash	57.2	56.5

[1] Premix provided per kilogram of diet: transretinyl acetate, 3.44 mg; cholecalciferol, 0.075 mg; menadione, 1.3 mg; thiamin, 2.2 mg; riboflavin, 8 mg; nicotinamide, 40 mg; choline chloride, 400 mg; calcium pantothenate, 10 mg; pyridoxine·HCl, 4 mg; biotin, 0.04 mg; folic acid, 1 mg; vitamin B_{12} (cobalamin), 0.013 mg; Fe (from ferrous sulfate), 80 mg; Cu (from copper sulphate), 8.0 mg; Mn (from manganese sulphate), 110 mg; Zn (from zinc oxide), 60 mg; I (from calcium iodate), 1.1 mg; Se (from sodium selenite), 0.3 mg.
[2] The nutrient levels were as fed basis.
[3] Values based on analysis of triplicate samples of diet.

Sample collection

At the end of experiment (42 day), 18 male broilers (1 bird per cage) were randomly selected and weighed after withdrawal of feed for 12 h. Birds were killed by cervical dislocation after bleeding from jugular veins. About 3 g breast and thigh samples were then immediately collected and stored at −80°C for further mRNA abundance determination and antioxidant parameters. After that, breast and thigh muscles were excised for meat quality measurement. The left-side muscle was used for the assay of pH value and meat color, and the right-side muscle was used to determine drip loss and cooking loss.

Meat quality assay

The meat color was measured at 24 h postmortem by a colorimeter (Minolta CR-10, Konica Minolta Sensing, Osaka, Japan) according to CIELAB system (L* = lightness; a* = redness; b* = yellowness). The pH values of muscles at 45 min ($pH_{45\ min}$) and 24 h ($pH_{24\ h}$) postmortem were measured at a 1 cm depth using a pH meter (HI9125, HANNA Instruments, Clujnapoca, Romania). Drip loss of muscle was measured according to method of Zhang et al. (2015). Briefly, around size of 3 cm (length)×2 cm (width)×1 cm (thickness) muscle samples trimmed of adjacent fat and connective tissues were weighed, suspended in plastic bag, sealed, and stored at 4°C for 24 h and 48 h. After 24 h and 48 h at 4°C, the samples were reweighed to calculate drip loss. For the determination of cooking loss, approximately 15 g muscle sample was weighed, held in plastic bags, and immersed in a water bath at 80°C until the internal temperature reached 75°C. After that, samples were reweighed to calculate the drip loss of muscle after cooling to room temperature.

Preparation of muscular homogenate

Approximately 0.3 g of muscle samples were homogenized (1:4, wt/vol) with ice-cold 154 mmol/L sodium chloride solution using an Ultra-Turrax homogenizer (Tekmar Co., Cicinati, OH, USA), and then centrifuged at 4,550 g for 15 min and at 4°C. The supernatant was used for determination of α-tocopherol and antioxidant parameters.

Measurement of α-tocopherol

Muscular α-tocopherol was determined according to the method described by Zhang et al. (2009) with modifications. In detail, a 2% solution pyrogallol in ethanol was slowly added to (5 mL) muscular homogenate (0.5 mL) while mixing, and the mixture was then heated for 2 min in a 70°C shaking water bath. Then, 0.25 mL of saturated KOH was added to the mixture. The mixture was heated again in a 70°C shaking water bath for 30 min and next placed in an ice bath after saponification. Hexane (2 mL) and water (0.5 mL) were added to the saponified samples, which were then shaken vigorously for 2 min. After that, 1 mL of the hexane layer was transferred to a 4-mL glass test tube for analysis. Standards solutions of α-tocopherol were also prepared. A 0.2% bathophenanthroline solution (200 μL) was then added to all the samples and thoroughly mixed. 200 μL of 1 mmol/L $FeCl_3$ solution was added and samples were vortexed. After 1 min, 200 μL of an H_3PO_4 solution was added and vortexed again. Absorbance of final solutions was read on a spectrophotometer at 534 nm (Mapada Instruments Co., Ltd., Shanghai, China). The standard curve was used to calculate the concentration of α-tocopherol in each sample. The concentrations of α-tocopherol were expressed as microgram per gram of fresh muscle.

Determination of antioxidant parameters

Muscular homogenate was used to determine the activities of total antioxidant capacity (T-AOC), total superoxide dismutase (T-SOD) and glutathione peroxidase (GSH-PX), and the content of malondialdehyde (MDA) and total protein by commercial kits according to the instructions of the manufacture (Nanjing Jiancheng Institute of Bioengineering, Nanjing, Jiangsu, China). T-AOC was measured based on the method of Benzie and Strain (1996). One unit of T-AOC was defined as the amount of enzyme per milligram protein which would increase the absorbance by 0.01 at 37°C in 1 min. The activity of SOD was determined using the hydroxylamine method (Ōyanagui, 1984), and one unit of SOD activity was defined as the amount of enzyme per milligram protein of muscle that produced 50% inhibition of the rate of nitrite generation at 37°C. GSH-PX activity was determined by dithio-nitro benzene method (Hafeman et al., 1974), and one unit of GSH-PX activity was defined as the amount of enzyme per milligram protein that would catalyze the conversion of 1 μmol/L of reduced glutathione to oxidized glutathione at 37°C in 5 min. MDA concentration was measured using thiobarbituric acid method (Placer et al., 1966). All results were normalized against total protein concentration in each sample for inter-sample comparison.

Total RNA isolation and mRNA quantification

The abundance of muscular mRNA was measured according to method of Zhang et al. (2014). RNA was isolated by a TRIzol reagent (TaKaRa Biotechnology, Dalian, Liaoning, China) from muscle sample (breast and thigh) according to protocol of manufacturer. The concentration of RNA and its purity were determined from OD260/280 readings (ratio>1.8) using a NanoDrop ND-1000 UV spectrophotometer (NanoDrop Technologies, Wilmington, DE, USA). After that, 1 μg of total RNA was reverse-transcribed into cDNA using the PrimeScript RT reagent kit (TaKaRa Biotechnology, China) according to the instructions of manufacturer. Real-time polymerase chain reaction (PCR) was carried out on an ABI StepOnePlus Real-Time PCR system (Applied Biosystems, Grand Island, NY, USA) according to the manufacturer's guidelines. The primer sequences for the target and reference genes (nuclear factor erythroid 2-related factor 2 [*Nrf2*], heme oxygenase 1 [*HO-1*], superoxide dismutase 1 [*SOD1*], glutathione peroxidase 1 [*GPX1*], and β-actin) are shown in Table 2. In detail, the reaction mixture was prepared using 2 μL of cDNA, 0.4 μL of forward primer, 0.4 μL of reverse primer, 10 μL of SYBR Premix Ex Taq (TaKaRa Biotechnology, China), 0.4 μL of ROX reference dye (TaKaRa Biotechnology, China) and 6.8 μL of double-distilled water. Each sample was tested in duplicate. PCR consisted of a pre-run at 95°C for 30 s and 40 cycles of denaturation at 95°C for 5 s, followed by a 60°C annealing step for 30 s.

Table 2. Sequences of real-time PCR primers

Gene	GeneBank ID	Primer sequence (5'→3')	Product size (bp)
Nrf2	NM_205117.1	GATGTCACCCTGCCCTTAG CTGCCACCATGTTATTCC	215
HO-1	HM237181.1	GGTCCCGAATGAATGCCCTTG ACCGTTCTCCTGGCTCTTGG	138
SOD1	NM_205064.1	CCGGCTTGTCTGATGGAGAT TGCATCTTTTGGTCCACCGT	124
GPX1	NM_001277853.1	GACCAACCCGCAGTACATCA GAGGTGCGGGCTTTCCTTTA	205
β-actin	NM_205518.1	TGCTGTGTTCCCATCTATCG TTGGTGACAATACCGTGTTCA	150

PCR, polymerase chain reaction; *Nrf2*, nuclear factor erythroid 2-related factor 2; *HO-1*, heme oxygenase 1; *SOD1*, superoxide dismutase 1; *GPX1*, glutathione peroxidase 1.

The conditions of the melting curve analysis were as follows: one cycle of denaturation at 95°C for 10 s, followed by an increase in temperature from 65°C to 95°C at a rate of 0.5°C/s. The relative levels of mRNA expression were calculated using the $2^{-\Delta\Delta C_T}$ method (Livak and Schmittgen, 2001) after normalization against the reference gene β-actin. The values of broilers in the control group were used as a calibrator.

Statistical analysis

Data were analyzed by one-way analysis of variance using SPSS statistical software (Ver.16.0 for windows, SPSS Inc., Chicago, IL, USA). Differences among treatments were examined using Duncan's multiple range tests. The differences were considered to be significant at $p<0.05$. Data are presented as means and their pooled standard errors.

RESULTS

Growth performance

As indicated in Table 3, broilers fed diets supplemented with vitamin E exhibited similar growth performance (ADG, ADFI, and F:G) to those given basal diet during the 42-day study ($p>0.05$).

Muscular α-tocopherol retention

Compared with the control group (Table 4), the concentration of α-tocopherol in the breast muscle was significantly increased by the supplementation of vitamin E ($p<0.05$), but the content of α-tocopherol in the breast was not affected by vitamin E source ($p>0.05$). In the thigh, the concentration of α-tocopherol was also significantly enhanced by vitamin E inclusion, and this effect was more pronounced in the natural vitamin E group ($p<0.05$).

Meat quality

Vitamin E supplementation (Table 5) in either natural or synthetic form significantly increased the redness of breast ($p<0.05$) when compared with the control group, whereas no difference was observed in the redness of breast between the natural and synthetic vitamin E group ($p>0.05$). Synthetic vitamin E supplementation also significantly decreased lightness of thigh ($p<0.05$), but lightness of thigh was not altered by the form of vitamin E ($p>0.05$). Dietary vitamin E supplementation significantly reduced drip loss at 24 h of thigh muscle ($p<0.05$), and this effect was maintained for drip loss at 48 h in the natural vitamin E group ($p<0.05$). However, supplementation of vitamin E did not alter pH value, cooking loss and yellowness of muscles (breast and thigh), drip loss and lightness of breast, and the

Table 3. Effects of natural and synthetic vitamin E on the growth performance of broilers

Items	Period (d)	Control[1]	SVE	NVE	SEM[3]	p-value
ADG (g/bird/per d)	1-21	25.6	25.5	26.1	0.3	0.756
	22-42	70.1	75.9	75.1	1.3	0.122
	1-42	48.4	51.3	51.3	0.6	0.077
ADFI (g/bird/per d)	1-21	40.2	39.1	39.9	0.5	0.660
	22-42	134	143	133	2	0.092
	1-42	88.0	91	87	1	0.132
F:G (g:g)	1-21	1.57	1.53	1.54	0.02	0.654
	22-42	1.92	1.88	1.91	0.02	0.791
	1-42	1.82	1.78	1.79	0.01	0.604

SEM, pooled standard error of the means; ADG, average daily gain; ADFI, average daily feed intake; F:G, feed/gain ratio.
[1] Control, basal diet; SVE, basal diet supplemented with 20 IU/kg synthetic vitamin E; NVE, basal diet supplemented with 20 IU/kg natural vitamin E.

Table 4. Effects of natural and synthetic vitamin E on the concentration of α-tocopherol in muscles f broilers (μg/g tissue)

Items	Control[1]	SVE	NVE	SEM	p-value
Breast					
α-Tocopherol	0.57[b]	1.81[a]	2.32[a]	0.28	0.001
Thigh					
α-Tocopherol	1.11[c]	3.11[b]	4.22[a]	0.36	<0.001

SEM, pooled standard error of the means.
[1] Control, basal diet; SVE, basal diet supplemented with 20 IU/kg synthetic vitamin E; NVE, basal diet supplemented with 20 IU/kg natural vitamin E.
[a-c] Means within a row with different superscripts are different at p<0.05.

redness of thigh (p>0.05).

Muscular antioxidant capacity

Broilers given diet supplemented with (Table 6) either natural or synthetic vitamin E showed decreased MDA content in the breast (p<0.05) than those given the basal diet. Additionally, natural rather than synthetic vitamin E reduced MDA accumulation in the thigh (p<0.05). However, treatments did not alter activities of T-SOD, T-AOC, and GSH-PX (p>0.05).

Muscular mRNA abundance

Neither natural nor synthetic vitamin E (Table 7) supplementation altered muscular mRNA abundance including Nrf2, HO-1, GPX1, and SOD1 as compared with the control group (p>0.05).

DISCUSSION

Many factors would affect vitamin E requirement, including feed composition, dietary selenium, antioxidant occurred in the feed, fat content and its profile. In this study, neither natural nor synthetic vitamin E supplementation altered the growth performance of broilers, which was in agreement with the results of Sheehy et al. (1991), and

Table 5. Effects of natural and synthetic vitamin E on meat quality of broilers

Items	Control[1]	SVE	NVE	SEM	p-value
Breast					
pH$_{45min}$	6.52	6.43	6.42	0.04	0.541
pH$_{24h}$	5.79	5.85	5.84	0.01	0.218
Lightness	46.9	47.6	46.7	0.3	0.568
Redness	3.03[b]	3.90[a]	4.29[a]	0.17	0.001
Yellowness	15.0	14.6	15.0	0.3	0.840
Drip loss at 24 h (g/kg)	104.1	92.8	86.5	4.5	0.269
Drip loss at 48 h (g/kg)	133.4	120.2	114.0	4.1	0.122
Cooking loss (g/kg)	214.0	201.9	206.0	4.8	0.552
Thigh					
pH$_{45min}$	6.32	6.37	6.42	0.04	0.576
pH$_{24h}$	6.17	6.31	6.17	0.03	0.072
Lightness	51.1[b]	48.1[b]	49.2[ab]	0.5	0.047
Redness	6.67	7.11	6.31	0.21	0.308
Yellowness	14.0	15.4	13.8	0.4	0.182
Drip loss at 24 h (g/kg)	102.5[a]	67.3[b]	72.9[b]	6.3	0.023
Drip loss at 48 h (g/kg)	120.1[a]	96.9[ab]	85.2[b]	5.7	0.027
Cooking loss (g/kg)	273.3	232.6	263.4	8.3	0.109

SEM, pooled standard error of the means.
[1] Control, basal diet; SVE, basal diet supplemented with 20 IU/kg synthetic vitamin E; NVE, basal diet supplemented with 20 IU/kg natural vitamin E.
[a-b] Means within a row with different superscripts are different at p<0.05.

Bartov and Frigg (1992). Recently, Rey et al. (2015) have also demonstrated that performance parameters were not modified by source (natural vs synthetic) and dosage of vitamin E in turkeys. It has been reported that supplementation of vitamin E (α-tocopherol) from 0 to 100 mg/kg did not alter growth performance of chicks given a maize-soya bean meal-soya oil type diet, and it may be due to the fact that vitamin E content in the basal diet (α-tocopherol content in grower and finisher diet was 7 and 6 mg/kg, respectively) would meet the requirement of broilers (Guo et al., 2001). The similar growth performance among

Table 6. Effects of natural and synthetic vitamin E on antioxidant status of meat in broilers

Items	Control[1]	SVE	NVE	SEM	p-value
Breast					
MDA (nmol/mg protein)	0.70[a]	0.30[b]	0.23[b]	0.07	0.022
GSH-PX (U/mg protein)	12.8	14.1	16.3	0.8	0.163
T-AOC (U/mg protein)	0.26	0.24	0.24	0.01	0.786
T-SOD (U/mg protein)	51.2	52.7	50.5	1.6	0.873
Thigh					
MDA (nmol/mg protein)	0.32[a]	0.27[a]	0.14[b]	0.03	0.014
GSH-PX (U/mg protein)	18.4	25.1	22.5	1.39	0.136
T-AOC (U/mg protein)	2.80	3.14	2.64	0.18	0.565
T-SOD (U/mg protein)	39.8	43.0	40.1	2.00	0.795

SEM, pooled standard error of the means; MDA, malondialdehyde; GSH-PX, glutathione peroxidase; T-AOC, total antioxidant capacity; T-SOD, total superoxide dismutase.
[1] Control, basal diet; SVE, basal diet supplemented with 20 IU/kg synthetic vitamin E; NVE, basal diet supplemented with 20 IU/kg natural vitamin E.
[a-b] Means within a row with different superscripts are different at p<0.05.

Table 7. Effects of natural and synthetic vitamin E on the muscular mRNA abundance of genes related to oxidative status of broilers

Items[1]	Control[2]	SVE	NVE	SEM	p-value
Breast					
Nrf2	1.00	0.95	1.02	0.03	0.733
HO-1	1.00	1.11	1.25	0.05	0.172
GPX1	1.00	0.98	1.02	0.03	0.898
SOD1	1.00	1.12	1.19	0.04	0.189
Thigh					
Nrf2	1.00	1.03	1.04	0.04	0.911
HO-1	1.00	1.19	1.02	0.04	0.137
GPX1	1.00	1.01	0.97	0.02	0.669
SOD1	1.00	1.18	0.99	0.04	0.173

SEM, pooled standard error of the means; Nrf2, nuclear factor erythroid 2-related factor 2; HO-1, heme oxygenase 1; SOD1, superoxide dismutase 1; GPX1, glutathione peroxidase 1.

[1] Control, basal diet; SVE, basal diet supplemented with 20 IU/kg synthetic vitamin E; NVE, basal diet supplemented with 20 IU/kg natural vitamin E.

[2] Expressed in arbitrary units. The mRNA level of each target gene in the control treatment was assigned a value of 1 and normalised to that of β-actin.

treatments observed in this study may also suggest that vitamin E in the basal diet would meet the requirement and therefore no change was detected with extra vitamin E supplementation.

Dietary vitamin E supplementation, irrespective of sources, increased its retention in tissues (liver and breast muscle) and plasma (Voljč et al., 2011). In this study, vitamin E supplementation in either natural (D-α-tocopherol) or synthetic (DL-α-tocopherol acetate) form increased α-tocopherol accumulation in the muscles (breast and thigh), and it was consistent with the finding of Nam et al. (1997) who reported that α-tocopherol contents of breast and thigh muscles were significantly increased by vitamin E supplementation. Bioavailability of vitamin E is related to its form. The vitamin E values of feed ingredients are always interpreted in terms of international units, one IU of vitamin E being defined as the specific activity of 1 mg of synthetic DL-α-tocopherol acetate, and it is generally accepted that 1 mg of D-α-tocopherol is equivalent to 1.49 IU vitamin E (McDonald et al., 2011). In this study, the capacity to increase muscular α-tocopherol accumulation was more pronounced in the D-α-tocopherol group. Similar results were also observed by Rey et al. (2015) in turkeys. Also, more efficient accumulation of the natural form of vitamin E than the synthetic form has been reported in swine (Yang et al., 2009; Amazan et al., 2014). The less efficient adsorption of the synthetic DL-α-tocopherol acetate may account for the difference in the muscular retention (Brigelius-Flohé and Traber, 1999).

Visual appearance of fresh meat influences a consumer's decision to buy a particular meat product, and consumers usually discriminate against surface discoloration of meat. Enhanced meat color through dietary vitamin E supplementation has been demonstrated in swine (Kim et al., 2015), beef (Sherbeck et al., 1995) and poultry (Rey et al., 2015). In the present study, vitamin E supplementation increased redness of meat whereas decreased lightness of thigh. Discoloration of meat is known to be closely related with lipid peroxidation (Renerre et al., 1990; Morrissey et al., 1994; Jensen et al., 1998). The activity of the metmyoglobin reducing systems is believed to be retained for longer periods in meats with reduced lipid oxidation that resulted from vitamin E supplementation, and a positive relationship between dietary vitamin E and improved color stability has also been clearly demonstrated in animals (Sherbeck et al., 1995; Jensen et al., 1998). Drip loss is considered one of the major quality deterioration factors within the meat industry. Consistent with the results of O'neill et al. (1998) and Li et al. (2009), vitamin E supplementation also decreased the drip loss of thigh muscle in this study. Protection of membranal lipids against lipid oxidation by vitamin E has been suggested to be the mechanism responsible for the positive influence of dietary vitamin E on the water holding capacity since the integrity of the cell membrane is thought to be associated with drip loss (Asghar et al., 1991). MDA, the main end product of lipid peroxidation by radical oxygen species, is an important indication of lipid peroxidation. Vitamin E supplementation decreased muscular MDA accumulation. The reduced MDA accumulation was in agreement with the simultaneously enhanced α-tocopherol retention and improved meat quality, which in turn suggested that vitamin E may improve the meat quality by enhancing muscular α-tocopherol content which could inhibit lipid peroxidation as evidenced by reduced generation of MDA. Boler et al. (2009) and Li et al. (2009) also found that vitamin E could prevent lipid oxidation of meat by reducing the generation of thiobarbituric acid reactive substance. In this study, vitamin E supplementation did not alter the T-AOC (an indicator of total antioxidant capacity) and activities of antioxidant enzymes (SOD and GPX) that are key constituents of enzymatic antioxidant system. Similar results were also found by Voljč et al. (2011) in which neither synthetic nor natural vitamin E supplementation affected antioxidant enzymes including SOD, GPX, and glutathione reductase, and the total antioxidant status. Thus, the reduced lipid oxidation in the muscles in this study was most likely to result from enhanced α-tocopherol retention rather than improved enzymatic antioxidant capacity. Nrf2 as a basic leucine zipper transcription factor protects the cell against oxidative stress through antioxidant response elements-mediated induction of several phase 2 detoxifying and anti-oxidant enzymes, including the HO-1, GPX1, and SOD1 (Cho et al., 2002). However, neither synthetic nor natural vitamin E supplementation altered the muscular mRNA abundances of Nrf2, HO-1, GPX1, and SOD1 in this study. Li et al. (2015) also observed that dietary vitamin E, ferulic acid or their combination supplementation did not alter mRNA expression of genes that involved in Nrf2 path way including nuclear factor erythroid 2-related factor 2-antioxidant response element and GPX1 in pigs.

Vitamin E is a highly lipophilic molecule that cannot be directly dispersed into aqueous solutions. Instead, it must be incorporated into an appropriate colloidal delivery system prior to dispersion (Yang and McClements, 2013). In addition to its biological activity after consumption, encapsulation and carrier selection of vitamin E have also been reported to improve its physicochemical stability during feed manufacture and storage (Reboul et al., 2006; Yang and McClements, 2013). However, the details regarding the carrier, encapsulation and other manufacture processes cannot be obtained due to commercial sensitivity. Further studies that incorporated both manufacture process and animal trials would be worthy to be conducted when evaluating the bioavailability of nature and synthetic vitamin E.

In conclusion, vitamin E supplementation had no effect on the growth performance and mRNA abundance of genes related to oxidative status of broilers. Vitamin E inclusion especially in the natural form can enhance the retention of muscular α-tocopherol, improve meat quality and muscular antioxidant capacity.

REFERENCES

Amazan, D., G. Cordero, C. J. López-Bote, C. Lauridsen, and A. I. Rey. 2014. Effects of oral micellized natural vitamin E (d-α tocopherol) vs. synthetic vitamin E (dl-α-tocopherol) in feed on α-tocopherol levels, stereoisomer distribution, oxidative stress and the immune response in piglets. Animal 8:410-419.

Asghar, A., J. I. Gray, A. M. Booren, E. A. Gomaa, M. M. Abouzied, E. R. Miller, and D. J. Buckley. 1991. Effects of supranutritional dietary vitamin E levels on subcellular deposition of α-tocopherol in the muscle and on pork quality. J. Sci. Food Agric. 57:31-41.

Bartov, I. and M. Frigg. 1992. Effect of high concentrations of dietary vitamin E during various age periods on performance, plasma vitamin E and meat stability of broiler chicks at 7 weeks of age. Br. Poult. Sci. 33:393-402.

Benzie, I. F. and J. J. Strain. 1996. The ferric reducing ability of plasma (FRAP) as a measure of "antioxidant power": The FRAP assay. Anal. Biochem. 239:70-76.

Boler, D. D., S. R. Gabriel, H. Yang, R. Balsbaugh, D. C. Mahan, M. S. Brewer, F. K. McKeith, and J. Killefer. 2009. Effect of different dietary levels of natural-source vitamin E in grow-finish pigs on pork quality and shelf life. Meat Sci. 83:723-730.

Brigelius-Flohé, R. and M. G. Traber. 1999. Vitamin E: Function and metabolism. FASEB J. 13:1145-1155.

Cho, H. Y., A. E. Jedlicka, S. P. Reddy, T. W. Kensler, M. Yamamoto, L. Y. Zhang, and S. R. Kleeberger. 2002. Role of NRF2 in protection against hyperoxic lung injury in mice. Am. J. Respir. Cell Mol. Biol. 26:175-182.

Guo, Y., Q. Tang, J. Yuan, and Z. Jiang. 2001. Effects of supplementation with vitamin E on the performance and the tissue peroxidation of broiler chicks and the stability of thigh meat against oxidative deterioration. Anim. Feed Sci. Technol. 89:165-173.

Hafeman, D. G., R. A. Sunde, and W. G. Hoekstra. 1974. Effect of dietary selenium on erythrocyte and liver glutathione peroxidase in the rat. J. Nutr. 104:580-587.

Halliwell, B. and J. M. C. Gutteridge. 2000. Free Radicals in Biology and Medicine. 3rd ed. Oxford Univ. Press, New York, NY, USA.

Jensen, C., C. Lauridsen, and G. Bertelsen. 1998. Dietary vitamin E: Quality and storage stability of pork and poultry. Trends Food Sci. Technol. 9:62-72.

Kaiser, M. G., S. S. Block, C. Ciraci, W. Fang, M. Sifri, and S. J. Lamont. 2012. Effects of dietary vitamin E type and level on lipopolysaccharide-induced cytokine mRNA expression in broiler chicks. Poult. Sci. 91:1893-1898.

Kim, J. C., C. G. Jose, M. Trezona, K. L. Moore, J. R. Pluske, and B. P. Mullan. 2015. Supra-nutritional vitamin E supplementation for 28 days before slaughter maximises muscle vitamin E concentration in finisher pigs. Meat Sci. 110: 270-277.

Lauridsen, C., H. Engel, S. K. Jensen, A. M. Craig, and M. G. Traber. 2002. Lactating sows and suckling piglets preferentially incorporate RRR-over all-rac-α-tocopherol into milk, plasma and tissues. J. Nutr. 132:1258-1264.

Li, W. J., G. P. Zhao, J. L. Chen, M. Q. Zheng, and J. Wen. 2009. Influence of dietary vitamin E supplementation on meat quality traits and gene expression related to lipid metabolism in the Beijing-you chicken. Br. Poult. Sci. 50:188-198.

Li, Y. J., L. Y. Li, J. L. Li, L. Zhang, F. Gao, and G. H. Zhou. 2015. Effects of dietary supplementation with ferulic acid or vitamin E individually or in combination on meat quality and antioxidant capacity of finishing pigs. Asian Australas. J. Anim. Sci. 28:374-381.

Livak, K. J. and T. D. Schmittgen. 2001. Analysis of relative gene expression data using real-time quantitative PCR and the $2^{-\Delta\Delta C}{}_T$ method. Methods 25:402-408.

McDonald, P., R. A. Edwards, J. F. D. Greenhalgh, C. A. Morgan, L. A. SinclaiL, and R. G. Wilkinson. 2011. Animal Nutrition, 7th ed. Pearson, Harlow, England.

Morrissey, P. A., D. J. Buckley, P. J. Sheehy, and F. J. Monahan. 1994. Vitamin E and meat quality. Proc. Nutr. Soc. 53:289-295.

Nam, K. T., H. A. Lee, B. S. Min, and C. W. Kang. 1997. Influence of dietary supplementation with linseed and vitamin E on fatty acids, α-tocopherol and lipid peroxidation in muscles of broiler chicks. Anim. Feed Sci. Technol. 66:149-158.

NRC. 1994. Nutrient Requirements of Poultry. 9th rev. ed. National Academy Press, Washington, DC, USA.

O'neill, L. M., K. Galvin, P. A. Morrissey, and D. J. Buckley. 1998. Comparison of effects of dietary olive oil, tallow and vitamin E on the quality of broiler meat and meat products. Br. Poult. Sci. 39:365-371.

Ōyanagui, Y. 1984. Reevaluation of assay methods and establishment of kit for superoxide dismutase activity. Anal. Biochem. 142:290-296.

Placer, Z. A., L. L. Cushman, and B. C. Johnson. 1966. Estimation of product of lipid peroxidation (malonyl dialdehyde) in biochemical systems. Anal. Biochem. 16:359-364.

Renerre, M. 1990. Factors involved in the discoloration of beef meat. Int. J. Food Sci. Technol. 25:613-630.

Rey, A. I., J. Segura, A. Olivares, A. Cerisuelo, C. Piñeiro, and C. J. López-Bote. 2015. Effect of micellized natural (D-α-tocopherol) vs. synthetic (DL-α-tocopheryl acetate) vitamin E supplementation given to turkeys on oxidative status and breast meat quality characteristics. Poult. Sci. 94:1259-1269.

Reboul, E., M. Richelle, E. Perrot, C. Desmoulins-Malezet, V. Pirisi, and P. Borel. 2006. Bioaccessibility of carotenoids and vitamin E from their main dietary sources. J. Agric. Food Chem. 54:8749-8755.

Sahin, K., N. Sahin, and M. F. Gursu. 2002. Effects of vitamins E and A supplementation on lipid peroxidation and concentration of some mineral in broilers reared under heat stress (32°C). Nutr. Res. 22:723-731.

Sheehy, P. J. A., P. A. Morrissey, and A. Flynn. 1991. Influence of dietary α-tocopherol on tocopherol concentrations in chick tissues. Br. Poult. Sci. 32:391-397.

Sherbeck, J. A., D. M. Wulf, J. B. Morgan, J. D. Tatum, G. C. Smith, and S. N. Williams. 1995. Dietary supplementation of vitamin E to feedlot cattle affects beef retail display properties. J. Food Sci. 60:250-252.

Voljč, M., T. Frankič, A. Levart, M. Nemec, and J. Salobir. 2011. Evaluation of different vitamin E recommendations and bioactivity of α-tocopherol isomers in broiler nutrition by measuring oxidative stress in vivo and the oxidative stability of meat. Poult. Sci. 90:1478-1488.

Yang, H., D. C. Mahan, D. A. Hill, T. E. Shipp, T. R. Radke, and M. J. Cecava. 2009. Effect of vitamin E source, natural versus synthetic, and quantity on serum and tissue α-tocopherol concentrations in finishing swine. J. Anim. Sci. 87:4057-4063.

Yang, Y. and D. J. McClements. 2013. Encapsulation of vitamin E in edible emulsions fabricated using a natural surfactant. Food Hydrocoll. 30:712-720.

Zhang, H., Y. Chen, Y. Li, L. Yang, J. Wang, and T. Wang. 2014. Medium-chain TAG attenuate hepatic oxidative damage in intra-uterine growth-retarded weanling piglets by improving the metabolic efficiency of the glutathione redox cycle. Br. J. Nutr. 112:876-885.

Zhang, J., Z. Hu, C. Lu, K. Bai, L. Zhang, and T. Wang. 2015. Effect of various levels of dietary curcumin on meat quality and antioxidant profile of breast muscle in broilers. J. Agric. Food Chem. 63:3880-3886.

Zhang, X. H., X. Zhong, Y. M. Zhou, H. M. Du, and T. Wang. 2009. Effect of RRR-α-tocopherol succinate on the growth and immunity in broilers. Poult. Sci. 88:959-966.

11

Effects of Supplemental Levels of *Saccharomyces cerevisiae* Fermentation Product on Lactation Performance in Dairy Cows under Heat Stress

W. Zhu, B. X. Zhang, K. Y. Yao, I. Yoon[1], Y. H. Chung[1], J. K. Wang, and J. X. Liu*

Institute of Dairy Science, College of Animal Sciences, Zhejiang University, Hangzhou 310058, China

ABSTRACT: The objectives of this study were to evaluate the effects of different supplemental levels of *Saccharomyces cerevisiae* fermentation product (SCFP; Original XP; Diamond V) on lactation performance in Holstein dairy cows under heat stress. Eighty-one multiparous Holstein dairy cows were divided into 27 blocks of 3 cows each based on milk yield (23.6±0.20 kg/d), parity (2.88±0.91) and day in milk (204±46 d). The cows were randomly assigned within blocks to one of three treatments: 0 (control), 120, or 240 g/d of SCFP mixed with 240, 120, or 0 g of corn meal, respectively. The experiment was carried out during the summer season of 2014, starting from 14 July 2014 and lasting for 9 weeks with the first week as adaption period. During the experimental period, average daily temperature-humidity index (measured at 08:00, 14:00, and 20:00) was above 68, indicating that cows were exposed to heat stress throughout the study. Rectal temperatures tended to decrease linearly (p = 0.07) for cows supplemented with SCFP compared to the control cows at 14:30, but were not different at 06:30 (p>0.10). Dry matter intake was not affected by SCFP supplementation (p>0.10). Milk yield increased linearly (p<0.05) with increasing levels of SCFP. Feed efficiency (milk yield/ dry matter intake) was highest (p<0.05) for cows fed 240 g/d SCFP. Cows supplemented with SCFP gained (p<0.01) body weight, while cows in the control lost body weight. Net energy balance also increased linearly (p<0.01) with increasing levels of SCFP. Concentrations of milk urea nitrogen (p<0.01) decreased linearly with increasing levels of SCFP, while no difference (p>0.10) was observed among the treatments in conversion of dietary crude protein to milk protein yield. In summary, supplementation of SCFP alleviated the negative effect of heat stress in lactating Holstein dairy cows and allowed cows to maintain higher milk production, feed efficiency and net energy balance. Effects of SCFP were dose-dependent and greater effects were observed from higher doses. (**Key Words:** Heat Stress, *Saccharomyces cerevisiae*, Lactation Performance, Dairy Cow)

INTRODUCTION

Heat stress is detrimental to dairy cows. The comfortable ambient temperatures for dairy cows are between 5°C and 25°C, and a temperature-humidity index (THI) above 68 typically affects dairy production parameters negatively (Burgos-Zimbelman and Collier, 2011). During warm summer months, milk production decreases by 10% to 35%, which represents a significant cost to the global dairy industry (St-Pierre et al., 2003). The deficit in energy and nutrient availability in heat stressed

cows is thought to limit milk production during a thermal load (Shwartz et al., 2009). Methods of increasing digestion efficiency and providing additional energy include supplemental dietary modifiers.

Cows under heat stress are at a higher risk for suboptimal rumen function (Baumgard et al., 2006). Increased respiration rate (causing increased secretion of bicarbonate by the kidneys), reduced feed intake (causing reduced rumination and saliva production) and altered feeding behavior (sorting, slug feeding, etc.) are among the contributing factors (Berman et al., 1985; Collier et al., 2006). Optimizing rumen function of heat stressed cows could mitigate the negative effect of heat stress on lactation performance of dairy cows.

Feed additives such as *Saccharomyces cerevisiae*

* Corresponding Author: J. X. Liu.
E-mail: liujx@zju.edu.cn
[1] Diamond V, Cedar Rapids, IA 52405, USA.

fermentation product (SCFP; Original XP; Diamond V, Cedar Rapids, IA, USA) are widely used as ruminant fermentation modifiers to optimize rumen health and improve lactation performance in dairy cows. Arambel and Kent (1990) suggested that yeast products might be more effective under heat stress than in normal conditions. Schingoethe et al. (2004) reported a significant improvement in feed efficiency when mid-lactation dairy cows were supplemented with SCFP during summer months. Optimum feeding rate of SCFP may differ under heat stress condition. However, optimum level of supplementary SCFP under heat stress has not been determined.

Therefore, we hypothesized that SCFP would improve lactation performance of dairy cows exposed to heat stress and a higher feeding rate of SCFP could be more effective under such conditions. To address this hypothesis, the effects of different levels of SCFP on dry matter intake (DMI), lactation performance, rectal temperature, and energy status in dairy cows during heat stress were evaluated.

MATERIAL AND METHODS

Animals, diets, and experimental design

The use of animals was approved by the Animal Care Committee of Zhejiang University (Hangzhou, China). The experiment was conducted at Hangjiang Dairy farm (Hangzhou, China). Eighty-one multiparous Holstein dairy cows were allocated into 27 blocks based on day in milk (DIM) (204±46; mean±standard deviation), parity (2.88±0.91) and milk yield (23.6±0.20 kg/d). Cows were randomly assigned within block to one of three dietary treatments: 0 (control), 120, or 240 g of SCFP per head per day (Diamond V XP, USA) mixed with 240, 120, or 0 g of corn meal, respectively (Table 1). The SCFP, a fully fermented yeast culture containing fermentation metabolites, residual yeast cells, and growth media was provided by Diamond V (USA). The supplement was top-dressed daily at the time of feeding individually to each cow. Each cow was observed for 20 min after the feeding to ensure complete consumption of the SCFP. Total mixed ration (TMR) were formulated to meet or exceed the nutrient requirements (MOA, 2004) for lactating Holstein

cows weighing 600 kg and producing 25 kg/d of milk. The ingredient and nutrient content of the feed components are presented in Table 2.

Cows were housed in an individual tie-stall barn and fed the TMR daily at 0630, 1330 and 2000 h with free access to drinking water. Feed was offered *ad libitum* to yield 10% residues. The barn contained 105-cm diameter fans over feeding alleys every 8 m, and the fans were operated once ambient temperature reached 25°C. The experiment commenced from July 14 to September 14, 2014, which is the typical hot season in Hangzhou, China. The feeding trial was composed of a 1-week covariate period and 8-week experimental period. Cows were milked 3 times daily at 0700, 1400, and 2030 h. The basal TMR was mixed on site three times daily, with the grain mix prepared every two weeks.

Sampling, measurement, and analyses

To measure the environmental conditions inside the barn, temperature and relatively humidity (RH) were recorded by calibrated data logging equipment (Ming Gao, Mingle Instruments Co. Ltd., Shenzhen, China) three times

Table 1. Experimental design

Ingredient	SCFP supplementation (g/d)		
	0	120	240
No. of cows	27	27	27
SCFP[1] (g/d)	0	120	240
Corn meal[2] (g/d)	240	120	0

SCFP, *Saccharomyces cerevisiae* fermentation product (Diamond V Original XP, Cedar Rapids, IA, USA).

[1] Corn meal (% of dry matter; n = 5): Organic matter 92.1, Crude protein 8.4, Neutral detergent fiber 9.5, and Acid detergent fiber 3.4.

Table 2. Ingredients and chemical composition of basal diets used in the experiment (n = 8)

Items	Contents
Ingredient (% of DM)	
Corn silage	20.7
Alfalfa hay	12.2
Wild ryegrass	8.1
Ground corn	14.2
Steam-flaked corn	4.9
Ground barley	4.9
Soybean meal	8.1
Whole cotton seed	6.1
Cottonseed meal	3.2
Wheat bran	2.0
Brewers dried grain	4.3
Beet pulp	8.1
Minerals and vitamins[1]	3.2
Chemical composition (% of DM)	
CP	16.7
NDF	36.4
ADF	22.6
Ca[2]	0.63
P[2]	0.46
NE_L[2] (Mcal/kg DM)	1.58

DM, dry matter; CP, crude protein; NDF, neutral detergent fiber; ADF, acid detergent fiber; NE_L, net energy for lactation.

[1] Formulated to provide (per kg of DM): 10 g of CP, 150 g of EE, 60 g of crude fiber, 70 g of Ca, 13 g of P, 100 g of salt, 30 g of Mg, 15 g of K, 10 g of Met, 260 mg of Cu, 260 mg of Fe, 1,375 mg of Zn, 500 mg of Mn, 112,000 IU of vitamin A, 29,500 IU of vitamin D_3, and 700 IU of vitamin E.

[2] Calculated based on Ministry of Agriculture of P.R. China (MOA, 2004).

daily (08:00, 14:00, 20:00). Recorders were set at the east and west of the study pen, and placed at a height of 1.9 m from the floor. Temperature and relative humidity were recorded within±0.2°C and ±2%, respectively. The THI was calculated as: THI = dew point temperature (TD)–(0.55–0.55 RH/100) (TD–58), where TD was the dry bulb temperature in °F (°F = 32°C+1.8°C) and RH was expressed as a percentage (NOAA, 1976). The average daily temperature and RH were determined using the recording data, and mean THI were calculated.

Feed offered and residues were weighed on the fourth day of each week to determine the individual DMI throughout the experiment. The representative samples of the TMR, dietary ingredients (corn silage and concentrate mixes) and residues were collected also on the fourth day of every week and stored at –20°C until analyses. All samples were then dried at 60°C for 48 hours, ground through a 1 mm-mesh screen (Tecator 1093, Hoganas, Sweden), and stored in closed plastic bottles at room temperature (approximately 25°C) until further analysis. The dry matter, crude protein, and crude ash contents in the test samples (Tables 1 and 2) were determined according to AOAC methods as described by procedures 934.01, 976.05, and 927.02 (AOAC, 2012), respectively. The neutral detergent fiber and acid detergent fiber were determined according to method described by Van Soest et al. (1991).

Cows were milked 3 times daily and individual milk production was recorded daily using a milk-sampling device (Waikato Milking Systems NZ Ltd., Waikato, Hamilton, New Zealand) throughout the experiment. Milk (50 mL) samples were collected at the second day of each week from each cow at each milking and 0.06% potassium dichromate was added as the preservative (milk preservative, D&F Control Systems, San Ramon, CA, USA). Samples from each milking per day were pooled in a proportion of 4:3:3. The samples were send to Shanghai Dairy Herd Improvement testing center (Shanghai, China) for analysis of milk protein, fat, lactose, somatic cell count (SCC), and milk urea nitrogen (MUN) using Combi Foss FT+instrument (Foss Electric, Hillerød, Denmark).

Animal body weights were estimated at the first and last day of the experiment based on the measurement of heart girth and body length using the follow equation: body weight (BW, kg) = heart girth[2] (m)×body length (m)×90 (Wang, 2006). Cows were scored for body condition according to Edmonson et al. (1989) using 5-point scale where 1 = thin and 5 = fat on the seventh day of the first, fourth, and eighth week by 2 experienced investigators blind to the treatments. The daily mean body condition score (BCS) (average of BCS scores of the two investigators) was used for statistical analysis of BCS.

Rectal temperatures were recorded within±0.1°C on the fifth of each week using clinical veterinary thermometers at 06:30 and 14:30.

Statistical analysis

Statistical analyses of data were carried out using SAS software (SAS, 2000). All data except for BW gain were analyzed through the PROC MIXED program of SAS with the covariance type analytical reagent (AR) (1) for repeated measures. A randomized block design with repeated measurements was used, with week, treatment, interaction of treatment×week and block as the fixed effects. Cow was included as a random effect. Means were separated using the PDIFF option in the LSMEANS statement. Data on BW gain were analyzed using the PROC general linear model of SAS. The statistical model was the same as indicated above except that week and treatment×week were omitted. Results are reported as the least squares means. Linear and quadratic effects of treatment were tested for all data using orthogonal polynomial contrasts. Significance was considered at p≤0.05 and tendency was declared at 0.05<p≤0.10.

RESULTS AND DISCUSSION

Environmental temperature-humidity index and rectal temperatures

The average of mean daily THI during the entire experiment was 76.6±3.69. The heat wave was moderate during the second half of the study with decreasing THI (Figure 1). Diurnal variation of THI was not large, with the range from 68 to 86 through all days. Mean daily THI throughout the study was above 68, which characterizes exposure to heat stress for lactating dairy cows (Burgos-Zimbelman and Collier, 2011). Even with cooling enhancements such as fans in the experiment barn, there were days when cows showed signs of suffering from heat stress, such as a decreased DMI and lack of movement. Rectal temperatures tended to decrease linearly (p = 0.07) with increasing amount of SCFP compared to the control at 14:30, but were not different at 06:30 (p>0.10) (Table 3). This suggests that higher dose of SCFP is more effective in alleviating metabolic heat load of a cow during the day when heat stress reached a peak, although mechanism of action is not clear at this time.

Dry matter intake and lactation performance

Supplementation of SCFP did not affect (p>0.10) the DMI (Table 4). According to Poppy et al. (2012) effect of SCFP supplementation on DMI was dependent on the stage of lactation. Cows with <70 DIM increased DMI with SCFP, while cows with >70 DIM decreased DMI (Poppy et al., 2012). In this study, mid to late lactation cows with an average DIM of 204 were used and a numerical decrease in DMI was observed with SCFP supplementation. Similar

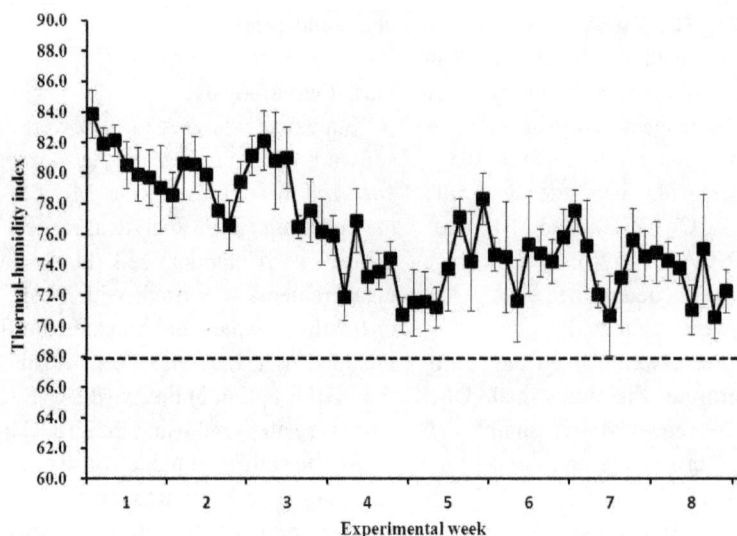

Figure 1. Daily thermal-humidity index (THI) during the trial period (mean±standard deviation). The dashed line represents THI = 68, when cows are expected to begin suffering heat stress (Burgos-Zimbelman and Collier, 2011).

results were observed when SCFP was supplemented to mid-lactation dairy cows during the summer (Schingoethe et al., 2004).

Milk yield increased linearly (p<0.05) with increasing levels of SCFP (Table 4), which is in agreement with the results of a meta-analyses by Poppy et al. (2012). Pattern of milk yield was partially attributed to the thermal-humidity index fluctuations (Figure 2). Overall effect as well as effect of SCFP level on milk yield was more pronounced from the fourth to eighth week of the study, suggesting that cows need time to adapt to the supplement before demonstrating significant production response. Although the effect of SCFP on rumen fermentation characteristics was not detected in the present study, and it is likely that the positive responses to SCFP supplementation in milk production resulted from the stabilizing effect on rumen fermentation (Callaway and Martin, 1997). The stabilized rumen condition allows increased growth of fiber-digesting bacteria (Harrison et al., 1988), resulting in improved fiber-digestion and rumen fermentation (Yoon and Garrett, 1998; Mao et al., 2013). Feed efficiency (milk yield/DMI) was highest (p<0.05) for cows fed 240 g/d SCFP. Enhancement of feed efficiency in response to SCFP supplementation was also observed in other studies (Schingoethe et al., 2004; Zhang et al., 2013), and such effects might be attributable to the increased milk production in the SCFP-

supplementation cows. Yields of 3.5% fat-corrected milk and energy corrected milk were not affected (p>0.10) by SCFP supplementation.

The percentage and yield of milk composition are also presented in Table 4. According to the meta-analysis (Poppy et al., 2012), SCFP-supplementation increased yield of milk fat and protein, while no changes in percentage of milk fat or protein was reported. In the present study, neither milk

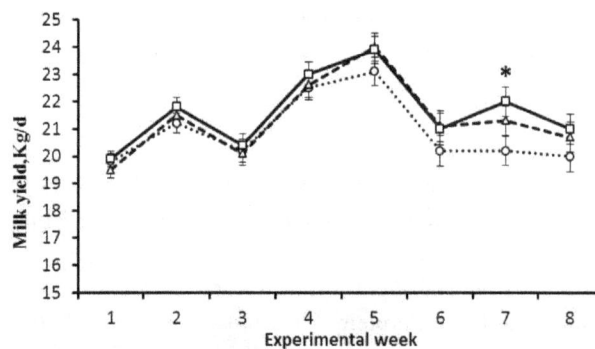

Figure 2. Weekly mean of milk yield of cows during heat stress with supplementation of a *Saccharomyces cerevisiae* fermentation product (SCFP) at level of 0 (○), 120 (Δ), or 240 (□) g/d. Pattern of milk yield was partially attributed to the thermal-humidity index fluctuations. Bars indicated standard error mean. * In the marked week (week 7), milk yield for cows fed 240 g/d SCFP tended to be greater than control cows (p = 0.07).

Table 3. Effect of SCFP supplementation on rectal temperatures in lactation dairy cow during heat stress

Parameters	SCFP supplementation (g/d)			SEM	p-value		
	0	120	240		T	L	Q
Morning (06:30)	39.3	39.2	39.2	0.10	0.66	0.46	0.60
Afternoon (14:30)	39.2	39.1	38.9	0.16	0.15	0.07	0.54

SCFP, *Saccharomyces cerevisiae* fermentation product (Diamond V Original XP, Cedar Rapids, IA, USA); SEM, standard error of the mean; T, treatment effect; L, linear effect; Q, quadratic effect.

Table 4. Effect of SCFP supplementation on dry matter intake and lactation performance in dairy cows during heat stress

Parameters	SCFP supplementation (g/d)			SEM	p-value		
	0	120	240		T	L	Q
DMI (kg/d)	17.2	16.9	16.9	0.23	0.60	0.33	0.77
Yield (kg/d)							
Milk	20.8[b]	21.3[ab]	21.5[a]	0.19	0.04	0.02	0.50
3.5% FCM[1]	24.3	24.9	24.6	0.25	0.23	0.32	0.16
ECM[2]	24.9	25.4	25.2	0.24	0.33	0.41	0.21
Milk protein	0.718	0.722	0.718	0.0077	0.94	0.98	0.72
Milk fat	0.939	0.973	0.955	0.0112	0.10	0.32	0.07
Milk composition (%)							
Fat	4.55	4.65	4.54	0.085	0.34	0.93	0.14
Protein	3.44	3.44	3.41	0.029	0.63	0.41	0.61
Lactose	4.77	4.74	4.80	0.019	0.18	0.43	0.09
Total solids	13.8	13.8	13.7	0.09	0.82	0.79	0.57
SCC ($\times 10^4$) /mL	19.8	22.4	21.3	2.51	0.75	0.66	0.54
MUN (mg/dL)	15.5[a]	15.3[a]	14.6[b]	0.21	0.02	<0.01	0.25
BW gain (g/d)	−13.0[c]	17.8[a]	11.1[b]	0.61	<0.01	<0.01	<0.01
BCS	2.82[b]	3.05[a]	2.84[b]	0.071	0.04	0.83	0.01
Feed efficiency[3]	1.28[b]	1.29[b]	1.32[a]	0.012	0.04	0.07	0.72
Nitrogen conversion[4]	0.269	0.272	0.275	0.0035	0.55	0.28	0.97
Net energy balance[5]	2.81[c]	3.12[b]	4.13[a]	0.047	<0.01	<0.01	<0.01

SCFP, *Saccharomyces cerevisiae* fermentation product (Diamond V Original XP, Cedar Rapids, IA, USA); SEM, standard error of the mean; T, treatment effect; L, linear effect; Q, quadratic effect; DMI, dry matter intake; FCM, fat-corrected milk; ECM, energy corrected milk; SCC, somatic cell count; MUN, milk urea nitrogen; BW, body weight; BCS, body condition score; NE_L, net energy for lactation.
[1] 3.5% FCM = (milk kg×0.432)+(fat kg×16.216) (Dairy Records Management Systems, 2006).
[2] ECM = 0.3246×milk yield (kg)+13.86×milk fat (kg)+7.04×milk protein (kg) (Orth, 1992).
[3] Feed efficiency = milk yield/DMI.
[4] Nitrogen conversion = milk protein yield/dietary crude protein intake.
[5] Net energy balance = (DMI×NE_L diet)−[(0.08×$BW^{0.75}$)+{(0.0929×fat+0.0563×protein+0.0395×lactose)×milk yield}] (NRC, 2001).
[a-c] Means within a row with different superscripts differ (p<0.05, n = 27).

fat percentage (p>0.10) nor milk protein percentage (p>0.10) was affected by SCFP-supplementation. The positive effect on milk production resulted in 3.6% greater (p = 0.10) milk fat yield in cows fed 120 g/d SCFP than that of the control cows and supported the results reported in the meta-analysis. No differences (p>0.10) among the groups were observed in contents of milk lactose, total solids, and SCC, similar with the results reported by Schingoethe et al. (2004), where the SCFP products were fed to mid-lactation dairy cows during hot season. Concentrations of MUN decreased linearly (p<0.01) with increasing levels of SCFP, but no difference was observed among the treatments in conversion of dietary N to milk N. Lower concentration of MUN with 240 g/d SCFP supplementation in dairy cows might indicate higher amino acid utilization for productive uses.

Net energy balance

Cows supplemented with SCFP gained (p<0.01) BW, but control cows lost BW during the study (Table 4). Body condition score of cows fed 120 g/d SCFP were higher (p<0.05) than that of the control cows and cows fed 240 g/d SCFP. Net energy balance, calculated based on DMI, milk yield and composition, and estimated BW (NRC, 2001), increased linearly (p<0.01) with increasing levels of SCFP. Improved BW gain, BCS and milk yield without affecting DMI supports the improved net energy balance with SCFP supplementation. Such results in the present study suggest that SCFP supplementation dosage dependently improves dietary energy utilization or absorption in heat-stressed dairy cows.

CONCLUSION

Supplementation of SCFP alleviated the negative effect of heat stress in lactating Holstein dairy cows and allowed cows to maintain higher milk production, feed efficiency and net energy balance. Effects of SCFP were dose-dependent and greater effects were observed from higher doses.

ACKNOWLEDGMENTS

This work was supported partly by funds from Diamond V (Cedar Rapids, USA) and from the China Agriculture

(Dairy) Research System (CARS-37). The authors gratefully thank the personnel of the Hangjiang Dairy Farm (Hangzhou, China) for their assistance in milking and care of the animals, the members of the Institute of Dairy Science Zhejiang University (Hangzhou, China) for their assistance in the sampling and analysis of the feed stuff.

REFERENCES

AOAC. 2012. Official Methods of Analysis. 17th edn. Association of Official Analytical Chemists, Arlington, VA, USA.

Arambel, M. J. and B. A. Kent. 1990. Effect of yeast culture on nutrient digestibility and milk yield response in early- to midlactation dairy cows. J. Dairy Sci. 73:1560-1563.

Baumgard, L. H., J. B. Wheelock, and G. Shwartz, M. O'Brien, M. J. VanBaale, R. J. Collier, M. L. Rhoads, and R. P. Rhoads. 2006. Effects of heat stress on nutritional requirements of lactating dairy cattle. In: Proceedings of the 5th Annual Arizona Dairy Production Conference. The University of Arizona Arizona, UT, USA. 8-16.

Berman, A., Y. Folman, M. Kaim, M. Mamen, Z. Herz, D. Wolfenson, A. Arieli, and Y. Graber. 1985. Upper critical temperatures and forced ventilation effects for high-yielding dairy cows in a subtropical climate. J. Dairy Sci. 68:1488-1495.

Burgos-Zimbelman, R. and R. J. Collier. 2011. Feeding strategies for high-producing dairy cows during periods of elevated heat and humidity. Tri-State Dairy Nutrition Conference, Fort Wayne, IN, USA. 111-126.

Callaway, E. S. and S. A. Martin. 1997. Effects of a *Saccharomyces cerevisiae* culture on ruminal bacteria that utilize lactate and digest cellulose. J. Dairy Sci. 80:2035-2044.

Collier, R. J., G. E. Dahl, and M. J. VanBaale. 2006. Major advances associated with environmental effects on dairy cattle. J. Dairy Sci. 89:1244-1253.

Edmonson, A. J., I. J. Lean, L. D. Weaver, T. Farver, and G. Webster. 1989. A body condition scoring chart for Holstein dairy cows. J. Dairy Sci. 72:68-78.

Harrison, G. A., R. W. Hemken, K. A. Dawson, R. J. Harmon, and K. B. Barber. 1988. Influence of addition of yeast culture supplement to diets of lactating cows on ruminal fermentation and microbial populations. J. Dairy Sci. 71:2967-2975.

Mao, H. L., H. L. Mao, J. K. Wang, J. X. Liu, and I. Yoon. 2013. Effects of *Saccharomyces cerevisiae* fermentation product on *in vitro* fermentation and microbial communities of low-quality forages and mixed diets. J. Anim. Sci. 91:3291-3298.

MOA (Ministry of Agriculture, China). 2004. Feeding Standard of Dairy Cattle (NY/T 34-2004). Beijing, China.

NOAA (National Oceanic and Atmospheric Administration). 1976. Livestock hot weather stress. US Dept. Commerce, Natl. Weather Serv. Central Reg., Reg. Operations Manual Lett. C-31-76.

NRC (Nutrient Requirents of Dairy Cattle). 2001. Nutrient Requirements of Dairy Cattle. 7th rev. ed. Natl. Acad. Sci. Washington, DC, USA.

Poppy, G. D., A. R. Rabiee, I. J. Lean, W. K. Sanchez, K. L. Dorton, and P. S. Morley. 2012. A meta-analysis of the effects of feeding yeast culture produced by anaerobic fermentation of *Saccharomyces cerevisiae* on milk production of lactating dairy cows. J. Dairy Sci. 95:6027-6041.

SAS Institute. 2000. SAS User's Guide. Statistics, Version 8.01. SAS Inst., Inc., Cary, NC, USA.

Schingoethe, D. J., K. N. Linke, K. F. Kalscheur, A. R. Hippen, D. R. Rennich, and I. Yoon. 2004. Feed efficiency of mid-lactation dairy cows fed yeast culture during summer. J. Dairy Sci. 87:4178-4181.

Shwartz, G., M. L. Rhoads, M. J. VanBaale, R. P. Rhoads, and L. H. Baumgard. 2009. Effects of a supplemental yeast culture on heat-stressed lactating Holstein cows. J. Dairy Sci. 92:935-942.

St-Pierre, N. R., B. Cobanov, and G. Schnitkey. 2003. Economic losses from heat stress by US livestock industries. J. Dairy Sci. 86 (E Suppl.):E52-E77.

Van Soest, P. J., J. B. Robertson, and B. A. Lewis. 1991. Methods for dietary fiber, neutral detergent fiber, and nonstarch polysaccharides in relation to animal nutrition. J. Dairy Sci. 74:3583-3597.

Wang, J. Q. 2006. Modern Dairy Production Science. China Agricultural Press, Beijing, China. (In Chinese).

Yoon, I. and J. E. Garrett. 1998. Yeast culture and processing effects on 24-hour *in situ* ruminal degradation of corn silage. Proc. 8th World Conf. Anim. Prod., Seoul, Korea. 1:322-323. Seoul National University, Seoul, South Korea.

Zhang, R. Y., I. Yoon, W. Y. Zhu, and S. Y. Mao. 2013. Effect of *Saccharomyces cerevisiae* fermentation product on lactation performance and lipopolysaccharide concentration of dairy cows. Asian Australas. J. Anim. Sci. 26:1137-1143.

12

The Effects of Dietary Phosphorus on the Growth Performance and Phosphorus Excretion of Dairy Heifers

B. Zhang, C. Wang[1], Z. H. Wei, H. Z. Sun, G. Z. Xu[2], J. X. Liu, and H. Y. Liu*

Institute of Dairy Science, College of Animal Science, Zhejiang University, Hangzhou 310058, China

ABSTRACT: The objective of this study was to investigate the effects of reducing dietary phosphorus (P) on the frame size, udder traits, blood parameters and nutrient digestibility coefficient in 8- to 10-month-old Holstein heifers. Forty-five heifers were divided into 15 blocks according to the mo of age and were randomly assigned one of three dietary treatments: 0.26% (low P [LP]), 0.36% (medium P [MP]), or 0.42% (high P [HP]) (dry matter basis). Samples were collected at the wk 1, 4, 8. The results show that low dietary P had no effect on body measurement. The blood P concentration decreased with decreasing dietary P (p<0.05), while the blood calcium content of LP was higher than that of the MP and HP groups (p<0.05), though still in the normal range. The serum contents of alkalinephosphatase, potassium, and magnesium were similar among the treatments. No differences were found in all nutrients' apparent digestibility coefficients with varied dietary P. However, with P diet decreased from HP to LP, the total fecal P and urine P concentration declined significantly, as did fecal water soluble P (p<0.05). In conclusion, reducing the dietary P from 0.42% to 0.26% did not negatively affect the heifers' growth performance but did significantly lessen manure P excretion into the environment. (**Key Words:** Heifers, Growth Performance, Phosphorus, Phosphorus Excretion)

INTRODUCTION

Phosphorus (P) is known to play an important role in various events of animal production and reproduction, including adenosine triphosphate (ATP), buffering systems, cell signaling, structure and strength of bones, and synthesis of cell walls, RNA and DNA (Hill et al., 2008; Geisert et al., 2010). Currently, the most critical environment problem in China is water pollution, and the excessive concentration of P has been recognized as a major cause of eutrophication in surface water (Correll, 1998; Imura, 2013). Previous studies demonstrated that the average amount of total P was 0.077 mg/L in Taihu Lake, of which animal and poultry manure P account for 46% (Li et al., 2000; Imura, 2013). Therefore, improving the efficiency of P utilization and lessening manure P excretion are the important ways to reduce the

* Corresponding Author: H. Y. Liu.
E-mail: hyliu@zju.edu.cn
[1] College of Animal Science and Technology, Zhejiang A & F University, Hangzhou, 311300, China.
[2] Institute of Shanghai Dairy Science, Shanghai 200032, China.

potential P pollution of freshwater.

A recent survey demonstrated that livestock producers in eastern China fed heifers 0.46% to 0.72% dietary P on a dry matter (DM) basis (Guo, 2013), which was higher than the amount recommended by the NRC (2001) (0.25% to 0.29% diet P [DM basis], body weight [BW] from 250 to 350 kg, average daily gain [ADG] = 1,000 g). The high content of dietary P was enabling heifers to reach puberty and pregnancy faster (Hill et al., 2007). However, many studies indicated that overfeeding P is not necessary, because increase P in manure and augments the expense of the producer (Wu et al., 2000; Valk et al., 2002; Bjelland et al., 2011). Tallam et al. (2005) reported no effect on ovarian activity, reproduction, or lactation benefit when feeding multiparous Holstein cows 0.35% to 0.47% of dietary P. In addition, excess dietary P was simply excreted, suggesting that the supplied additional mineral P in grain-based diets is not necessary. There was a significant amount of information on P excretion reduction from lactating cows, but few data on heifers are available. Therefore, the objective of this study was to assess whether lessening dietary P affects the growth performance of dairy heifers in China.

MATERIAL AND METHODS

Animals and experimental diets

The use of heifers was approved by the Animal Care Committee of Zhejiang University, Hangzhou, China. Forty-five Holstein heifers were divided into 15 blocks according to the mo of age (9.3±0.8) and were randomly assigned one of the following treatments: 0.26% (low P [LP]), 0.36% (medium P [MP]), and 0.42% (high P [HP]) (Table 1). The LP diet contained no supplemental P, whereas the other 2 diets were obtained by adding different amounts of dicalcium phosphate. The amount of P in the LP diet was near the level recommended by the NRC (BW from 250 to 350 kg, ADG = 1,000 g), the MP dietary P level was close to the Chinese dairy cow feeding standard recommendations (NY/T 34-2004, BW from 250 to 350 kg,

ADG = 1,000 g), and the HP diet contained P commonly fed by livestock producers in the east region of China (Guo, 2013). In order to avoid excessive fattening, all heifers were limited feeding, 2.1% of BW, and average dry matter intake (DMI) of each group was measured weekly. Heifers were housed in a tie-stall barn with free access to water and fed 3 times daily at 06:30, 14:00, and 20:30 h. At each feeding time, mixed concentrates were offered first, and then corn silage and grass hay were provided. The experiment was conducted from October until December 2013.

Measurements and analytical methods

A proportional amount of feed offered was collected weekly. Samples were dried in a forced oven at 55°C for 48 h and then milled through a Wiley mill with1-mm screen, and analyzed for crude protein (CP), Ca, P, ash, ether extract (AOAC, 1990), fecal water soluble P (Dou et al., 2002), neutral detergent fiber (NDF; Van Soest et al., 1991), acid detergent fiber (ADF; Van Soest et al., 1991). Spot fecal and urinary samples collected at 07:00, 14:30, and 20:30 h on the d 3, 27, and 53 were mixed across hours with a day as described by Wang et al. (2014), and sampled for later analysis. The record of hip height, body high, body length, heart girth, and teat length were taken at the d 2, 26, and 52. Blood samples (5 mL) were collected from the coccygeal vein on the d 1, 25, and 51, and centrifuged at $3,000\times g$ for 10 min to collect serum, which were frozen at –20°C. Alkalinephosphatase (ALP), Ca, P, magnesium (Mg) and potassium (K) serum concentrations were analyzed using a HITACHI (7020) Automatic Analyzer, and kits were provided by NINGBO MEDICALSYSTEM BIOTECHNOLOGY CO., LTD (Zhejiang, China).

Table 1. Ingredients and nutrient composition of the diet

Items	Dietary treatment		
	HP	MP	LP
Ingredient, % DM basis			
Chinese wild rye	39.8	40.0	39.9
Corn silage	25.1	25.1	25.2
Corn	13.9	14.0	14.0
Barley	6.1	6.3	6.3
Rapeseed meal	4.6	4.7	4.7
Soybean meal	1.7	1.7	1.7
DDGS(corn)	4.8	4.9	4.9
Peptide protein	0.9	0.9	0.9
Mineral-Vitamin premix[1]	-	-	2.4
P mineral-Vitamin premix[2]	2.2	2.4	-
Di-calcium phosphate	0.99	-	-
Analyzed chemical composition			
CP	11.2	11.2	11.1
NDF	55.3	54.2	56.0
ADF	29.0	28.4	28.7
P	0.42	0.36	0.26
Ca	1.0	0.9	0.9
EE	2.5	2.3	2.4
Ash	6.9	6.9	6.8
NE$_G$ (Mcal/kg)[3]	0.95	0.96	0.96

LP, low phosphorus (P); MP, medium P; HP, high P; DM, dry matter; DDGS, distillers dried grains with solubles; CP, crude protein; NDF, neutral detergent fiber; ADF, acid detergent fiber; EE, ether extract; NE$_G$, net energy for gain.

[1] Mineral-vitamin premix per kg containing: Ca (g) 166; P (g) 0; Fe (mg) 1,800; Cu (mg) 630; Mn (mg) 630; Zn (mg) 2,940; Se (mg) 21; I (mg) 38; Co (mg) 8; Vitamin A (IU) 240,000; Vitamin D (IU) 60,000; Vitamin E (IU) 1,200.

[2] P Mineral-vitamin premix per kg containing: Ca (g) 166; P (g) 70; Fe (mg) 1,800; Cu (mg) 630; Mn (mg) 630; Zn (mg) 2,940; Se (mg) 21; I (mg) 38; Co (mg) 8; Vitamin A (IU) 240,000; Vitamin D (IU) 60,000; Vitamin E (IU) 1,200.

[3] As calculated by NRC (2001).

Calculations and statistical analysis

The BW of each cow was calculated based on the measurement of heart girth and body length using the following equation: BW (kg) = heart girth[2] (m)×body length (m)×96.475 (Heinrichs et al., 1992; Shen et al., 2010; Yu et al., 2014). Apparent nutrient digestibility was calculated by using the following equation: Apparent digestibility = 100– $[(N_f/N_d)\times(M_d/M_f)]\times100$, where N_f = concentration of the nutrient in the fecal, N_d = concentration of the nutrient in the consumed diet, M_d = concentration of the acid insoluble ash (AIA) in the consumed diet and M_f = concentration of the AIA in the fecal (Stojanovic et al., 2014).

Data on DMI, P intake, body measurements, nutrient apparent digestibility coefficient, and fecal and urine P were analyzed using GLM of SAS (SAS Institute, 2000). Blood biochemical parameters were analyzed using PROC MIXED of SAS (SAS Institute, 2000); treatment, time, treatment×time, and block were included as fixed effects in the model; Heifers were the random effect. Probability values of p<0.05 were used to define statistical significance and values of p<0.10 and p≥0.05 were accepted as statistical trends.

RESULTS AND DISCUSSION

Feed ingredients, experimental diets, dry matter intake and P intake

The experimental diets had a similar composition of ingredients but with different P concentrations of 0.26%, 0.36%, and 0.42% (DM basis, Table 1). The Ca:P is different, but it is still within the normal range. NRC (2001) recommended that Ca:P is not critical on absorption of P and Ca in ruminants, unless the ratio is >7:1 or <1:1. DMI were similar among the treatments, average 2.1% of BW, which result from restricted feeding. Kertz (1987) believed that Holstein heifers must gain greater than 800 g/d and less than 1,000 g/d that not only could make heifers reach a BW of 570 kg at 24 mo of age, but also could avoid heifers' excessive fattening. The amount of P intake was increased with the increase of dietary P concentration (Table 2).

Body measurements and serum parameters

The skeletal measurements and calculated BW of Holstein heifers fed diets containing 0.26%, 0.36%, and 0.42% P were presented in Table 3. Heifers fed 0.26% P were similar in every measure of frame growth compared to the other two groups, suggesting that low-P ration had no effect on skeletal development. These observations are supported by Bjelland et al. (2011) and Esser et al. (2009), who reported no differences in skeletal growth of heifers due to the dietary addition of supplemental P. Similar results were also obtained from Hill et al. (2007). Mammary development is one of the most important criteria used to estimate lactation performance, and high-producing heifers tended to have lengthier teats and larger distances between teats compared with poor-producing heifers (Lin et al., 1987). This study determined that front teat length, rear teat length and teat distance were not differ among the 3 treatments (Figure 1), which indicated that the mammary development of the heifer was not affected by LP. The serum ALP, K, Mg concentrations were similar among treatments (Table 4), although a numerical decrease in ALP was observed with increased dietary P (p = 0.09). The serum P decreased with decreasing dietary P (p<0.05) and the serum P concentration of LP was consistent with results

Table 2. DMI and P intake of 8- to 10-month-old Holstein heifers

Item	Dietary treatment			SEM	p-value
	HP	MP	LP		
DMI (kg)	5.98	6.00	5.98	0.31	0.99
P intake (g/d)	25.13	21.60	15.57	1.12	<0.01

DMI, dry matter intake; HP, high phosphorus (P); MP, medium P; LP, low P; SEM, standard error of the mean.

Table 3. Calculated BW, and body measurements of 8- to10-month-old Holstein heifers

Item	Dietary treatment			SEM	p-value
	HP	MP	LP		
Calculated BW (kg)	289	291	297	4.86	0.52
Heart girth					
Initial (cm)	145.07	145.73	146.53	2.84	0.88
Final (cm)	153.00	152.80	152.33	2.70	0.97
Change[1] (cm/d)	0.14	0.13	0.10	0.02	0.20
Body length					
Initial (cm)	128.93	128.83	132.33	2.51	0.30
Final (cm)	134.87	135.13	136.60	2.67	0.78
Change (cm/d)	0.11	0.11	0.08	0.03	0.51
Body height					
Initial (cm)	107.80	108.60	110.80	2.37	0.43
Final (cm)	115.40	115.20	116.40	1.81	0.78
Change (cm/d)	0.13	0.12	0.10	0.02	0.14
Cannon bone[2]					
Initial (cm)	16.31	15.93	15.98	0.31	0.42
Final (cm)	16.48	16.12	16.11	0.35	0.49
Change (cm)	0.17	0.18	0.13	0.24	0.97

BW, body weight; LP, low phosphorus (P); MP, medium P; HP, high P; SEM, standard error of the mean.
[1] Change in body measurements from the beginning until the end of the trial.
[2] Value indicate cannon bone circumference.

reported by Bjelland (2011), who noting that blood P was 2.65 mmol/L with 0.30% dietary P in 8-month-old heifers. Wu et al. (2000) observed that serum P appeared higher for cows fed a high P diet compared with those receiving a low P diet; redundant dietary P was not utilized and was simply excreted in manure. The serum Ca content of heifers in LP was higher than that in MP and HP groups (p<0.05). Breves et al. (1985) reported that serum Ca increased when the amount of dietary P fed to sheep was decreased. Moreover, Kichura et al. (1982) believed that when dietary calcium is high, low dietary phosphorus seems helpful to enhance the activation of intestinal calcium absorption. No significant

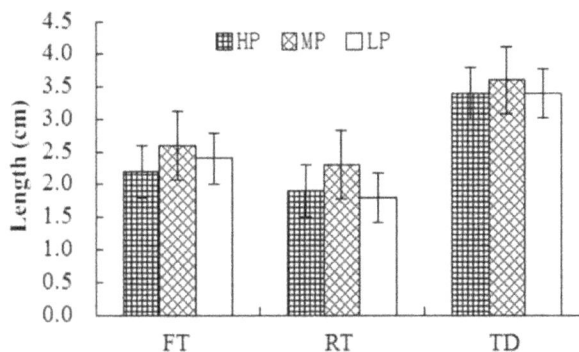

Figure 1. Udder measures of 8- to 10-month-old Holstein heifers fed varied dietary phosphorus (P). LP, low P; MP, medium P; HP, high P; FT, front teat length; RT, rear teat length; TD, teat distance, distance around gland front to rear.

Table 4. Effect of different dietary phosphorus (P) on serum biochemical parameters of 8- to 10-month-old Holstein heifers

Item	Dietary treatment			SEM	p-value		
	HP	MP	LP		Diet	Time	Diet×time
ALP (U/L)	120.7	125.3	135.9	4.85	0.09	<0.01	0.95
Ca (mmol/L)	2.46[a]	2.46[a]	2.54[b]	0.02	0.03	<0.01	0.49
P (mmol/L)	2.84[a]	2.71[ab]	2.64[b]	0.05	0.02	0.03	0.50
K (mmol/L)	5.14	5.19	5.06	0.07	0.45	<0.01	0.36
Mg (mmol/L)	0.80	0.84	0.80	0.02	0.31	<0.01	0.06

LP, low phosphorus (P); MP, medium P; HP, high P; SEM, standard error of the mean; ALP, alkalinephosphatase.
Values with different superscripts (a, b) are significantly different (p<0.05).

diet×time interaction was observed for serum P, Ca, K, Mg, and ALP.

Apparent digestibility of nutrients and manure P excretion

Even though the heifers in HP treatment consumed more P, the apparent P digestibility coefficients did not differ among the treatments (Table 5). Other research noted that the apparent digestibility of P in lactating cows less than 40% equates to an excessive of P intake (Wu et al., 2000), but it is still unknown for growing heifers. In regard to other nutrients, low dietary P had no effect on NDF, ADF, and the CP apparent digestibility coefficients, which is in agreement with previous results (Odongo et al., 2007; Xu et al., 2011). The excretion of P in manure was presented in Table 5. The total fecal P concentration decreased 35.62%, and the urine P concentration was reduced by 69.35% as dietary P decreased from 0.42% to 0.26% (p<0.05).

Table 5. Fecal phosphorus (P) and urine P and apparent digestibility coefficients for CP, P, NDF, and ADF of heifers fed different dietary P concentrations

Item	Dietary treatment			SEM	p-value
	HP	MP	LP		
CP (%)	54.3	58.5	58.3	0.51	0.13
P (%)	32.7	32.2	33.1	0.98	0.98
NDF (%)	54.7	56.6	55.4	0.58	0.76
ADF (%)	52.1	52.7	50.3	0.62	0.79
Fecal total P (% of DM)	0.73[a]	0.66[b]	0.47[c]	0.006	<0.01
Fecal water soluble P (% of DM)	0.38[a]	0.41[a]	0.26[b]	0.005	<0.01
Urine P (g/kg)	0.62[a]	0.22[b]	0.19[b]	0.025	<0.01

LP, low phosphorus (P); MP, medium P; HP, high P; SEM, standard error of the mean; CP, crude protein; NDF, neutral detergent fiber; ADF, acid detergent fiber; DM, dry matter.
Values with different superscripts (a, b, and c) are significantly different (p<0.05).

Nowadays, water soluble P has been seen as a viable index in environmental protection (Dou et al., 2002). This study determined that water soluble P account for more than 50% of total fecal P and similar with the results reported by Dou (2002) and Bernier (2014); and high dietary P not only led to higher total fecal P content, but it also increased the proportion of water soluble P (p<0.05). Previous studies demonstrated that fecal P excretion decreased by 23% as dietary P lessened by 0.1 percentage points (Wu et al., 2000); similar reductions in fecal P were also reported by Tallam (2005), and our results are also consistent with their findings.

CONCLUSIONS

In conclusion, this study revealed that reducing dietary P from 0.42% to 0.26% did not negatively affect heifers' growth performance, though it did significantly reduce manure P excretion into the environment. The dietary P content of 0.26% was sufficient for 8- to10-month-old Holstein heifers in China. Depending on the feed ingredients used, this concentration of P can be obtained without the addition of inorganic P supplement to the feed. However, further studies are required to identify the long-term effects of low dietary P on heifers.

ACKNOWLEDGMENTS

This work was supported by the National Key Technology R &D of China (no. 2012BAD12B02). The authors gratefully thank all staff at the Shanghai Lianjiang (Anhui) dairy farm for their inputs to this study.

REFERENCES

AOAC. 1990. Official Methods of Analysis. 15th edn. Association of Official Analytical Chemists, Arlington, VA, USA.

Bernier, J. N., M. Undi, K. H. Ominski, G. Donohoe, M. Tenuta, D. Flaten, J. C. Plaizier, and K. M. Wittenberg. 2014. Nitrogen and phosphorus utilization and excretion by beef cows fed a low quality forage diet supplemented with dried distillers grains with solubles under thermal neutral and prolonged cold conditions. Anim. Feed Sci. Technol. 193:9-20.

Bjelland, D. W., K. A. Weigel, P. C. Hoffman, N. M. Esser, and W. K. Coblentz. 2011. The effect of feeding dairy heifers diets with and without supplemental phosphorus on growth, reproductive efficiency, heath, and lactation performance. J. Dairy Sci. 94:6233-6242.

China Standard NY/T-34. 2004. Feeding Standard of Dairy Cattle. China Agric. Press, Beijing, China.

Correll, D. L. 1998. The role of phosphorus in the eutrophication of receiving waters: A review. J. Environ. Qual. 27:261-266.

Dou, Z., K. F. Konwlton, R. A. Kohn, Z. Wu, L. D. Satter, G.

Jhang, J. D. Toth, and J, D. Ferguson. 2002. Phosphorus characteristics of dairy feces affected by diets. J. Environ. Qual. 31:2058-2065.

Esser, N. M., P. C. Hoffman, W. K. Coblentz, M. W. Orth, and K. A. Weigel. 2009. The effect of dietary phosphorus on bone development in dairy heifers. J. Dairy Sci. 92:1741-1749.

Geisert, B. G., G. E. Erickson, T. J. Klopfenstein, C. N. Macken, M. K. Luebbe, and J. C. MacDonald. 2008. Phosphorus requirement and excretion of finishing beef cattle fed different concentrations of phosphorus. J. Anim. Sci. 88:2393-2402.

Guo, C. 2013. The Analysis of Diets Feed Nutrients and The Status of Nitrogen and Phosphorus Use in Scale Dairy Farms Which Are in The Eastern Region. BA Thesis. Zhejiang University, Hangzhou, China.

Heinrichs, A. J., G. W. Rogers, and J. B. Cooper. 1992. Predicting body weight and wither height in Holstein heifers using body measurements. J. Dairy Sci. 75:3576-3581

Hill, S. R., K. F. Knowlton, E. Kebreab, J. France, and M. D. Hanigan. 2008. A model of phosphorus digestion and metabolism in the lactating dairy cow. J. Dairy Sci. 91:2021-2032.

Hill, S. R., K. F. Knowlton, R. E. James, R. E. Pearson, G. L. Bethard, and K. J. Pence. 2007. Nitrogen and phosphorus retention and excretion in late-gestation dairy heifers. J. Dairy Sci. 90:5634-5642.

Imura, H. 2013. Environmental issues in China today: A view from Japan. In: Advances in Asian Human-Environmental Research. (Eds. M. Nüsser) Springer, Verlag, Japan. pp. 72-73.

Kertz, A. F., L. R. Prewitt, and J. M. Ballam. 1987. Increased weight gain and effects on growth parameters of Holstein heifer calves from 3 to 12 months of age. J. Dairy Sci. 70:1612-1622.

Kichura, T. S., R. L. Horst, D. C. Beitz, and E. T. Littledike. 1982. Relationships between prepartal dietary calcium and phosphorus, vitamin D metabolism, and parturient paresis in dairy cows. J. Nutr.112:480-487.

Li, R. G., Y. L. Xia, A. Z. Wu, and Y. Qian. 2000. Pollutants sources and their discharging amount in Taihu Lake Area of Jiangsu Province. J. Lake Sci. 8:147-153.

Lin, C. Y., A. J. Lee, A. J. McAllister, T. R. Batra, G. L. Roy, J. A. Vesely, J. M. Wauthy, and K. A. Winter. 1987. Intercorrelations among milk traits and body and udder measurements in Holstein heifers. J. Dairy Sci. 70:2385-2393.

NRC, 2001. Nutrient Requirements of Dairy Cattle. 7th edn. National Academy Press, Washington, DC. USA.

Odongo, N. E., D. McKnight, A. Koekkoek, J. W. fisher, P. Sharpe, E. Kebreab, J. France, and B. W. McBride 2007. Long-term effects of feeding diets without mineral phosphorus supplementation on the performance and phosphorus excretion high-yielding dairy cows. J. Anim. Sci. 87:639-646.

SAS Institute. 2000. SAS User's Guide: Statistics. Version 8.01. SAS Inst. Inc., Cary, NC, USA.

Shen, J. S., J. Q. Wang, H. Y. Wei, D. P. Bu, P. Sun, and L. Y. Zhou. 2010. Transfer efficiency of melamine from feed to milk in lactating dairy cows fed with different doses of melamine. J. Dairy Sci. 93:2060-2066.

Stojanovic, B., G. Grubic, N. Djordjevic, A. Bozickovic, A. Ivetic, and V. Davidovic. 2014. Effect of physical effectiveness on digestibility of ration for cows in early lactation. J. Anim. Physiol. Anim. Nutr. 98:714-721.

Tallam, S. K., A. D. Ealy, K. A. Bryan, and Z. Wu. 2005. Ovarian activity and reproductive performance of dairy cows fed different amount of phosphorus. J. Dairy Sci. 88:3609-3618.

Valk, H., L. B. J. Sebek, and A. C. Beynen. 2002. Influence of phosphorus intake on excretion and blood plasma and saliva concentrations phosphorus in dairy cows. J. Dairy Sci. 85:2642-2649.

Van-Soest, P. J., H. B. Robertson, and B. A. Lewis. 1991. Methods of dietary fiber, NDF and non-starch polysaccharides in relation to animal nutrition. J. Dairy Sci. 74:3583-3597.

Wu, Z. and L. D. Satter. 2000. Milk production and reproductive performance of dairy cows fed two concentrations of phosphorus for two years. J. Dairy Sci. 83:1052-1063.

Wu, Z., L. D. Satter, and R. Sojo. 2000. Milk production, reproductive performance, and fecal excretion of phosphorus by dairy cows fed three amount of phosphorus. J. Dairy Sci. 83:1028-1041.

Wang, C., Z. Liu, D. M. Wang, J. X. Liu, H. Y. Liu, and Z. G. Wu. 2014. Effect of dietary phosphorus content on milk production and phosphorus excretion in dairy cows. J. Anim. Sci. Biotechnol. 5:23.

Xu, J. H., W. Zhang, J. Huang, J. Jiang, C. M. Sun, and F. Mo. 2011. Effects of dietary phosphorus levels on apparent digestibility of nutrients in Simmental crossbreed replacement heifers. Chinese J. Anim. Nutr. 23:589-596.

Yu, Z., Y. X. Gao, Y. F. Cao, Q. F. Li, and J. G. Li. 2014. Study on growth and development pattern of Chinese Holstein calf and heifer. China Anim. Husb. Vet. Med. 41:121-125.

13

Performance and Metabolism of Calves Fed Starter Feed Containing Sugarcane Molasses or Glucose Syrup as a Replacement for Corn

C. E. Oltramari, G. G. O. Nápoles, M. R. De Paula, J. T. Silva, M. P. C. Gallo,
M. H. O. Pasetti, and C. M. M. Bittar*

Animal Science Department, University of São Paulo – USP/ESALQ, Piracicaba, SP 13418-900, Brazil

ABSTRACT: The aim of this study was to evaluate the effect of replacing corn grain for sugar cane molasses (MO) or glucose syrup (GS) in the starter concentrate on performance and metabolism of dairy calves. Thirty-six individually housed Holstein male calves were blocked according to weight and date of birth and assigned to one of the starter feed treatments, during an 8 week study: i) starter containing 65% corn with no MO or GS (0MO); ii) starter containing 60% corn and 5% MO (5MO); iii) starter containing 55% corn and 10% MO (10MO); and iv) starter containing 60% corn and 5% GS (5GS). Animals received 4 L of milk replacer daily (20 crude protein, 16 ether extract, 12.5% solids), divided in two meals (0700 and 1700 h). Starter and water were provided *ad libitum*. Starter intake and fecal score were monitored daily until animals were eight weeks old. Body weight and measurements (withers height, hip width and heart girth) were measured weekly before the morning feeding. From the second week of age, blood samples were collected weekly, 2 h after the morning feeding, for glucose, β-hydroxybutyrate and lactate determination. Ruminal fluid was collected at 4, 6, and 8 weeks of age using an oro-ruminal probe and a suction pump for determination of pH and short-chain fatty acids (SCFA). At the end of the eighth week, animals were harvested to evaluate development of the proximal digestive tract. The composition of the starter did not affect (p>0.05) concentrate intake, weight gain, fecal score, blood parameters, and rumen development. However, treatment 5MO showed higher (p<0.05) total concentration of SCFAs, acetate and propionate than 0MO, and these treatments did not differ from 10MO and 5GS (p>0.05). Thus, it can be concluded that the replacement of corn by 5% or 10% sugar cane molasses or 5% GS on starter concentrate did not impact performance, however it has some positive effects on rumen fermentation which may be beneficial for calves with a developing rumen. (**Key Words:** By-product, Sucrose, Butyrate, Ruminal Development)

INTRODUCTION

Corn is the main energy source used in the formulation of starters for calves, since it presents approximately 67% starch in its composition (Nocek and Tamminga, 1991). Starch is a non-structural carbohydrate with extensive and fast degradability generating high production of short chain fatty acid (SCFA) per unit time (Noziére et al., 2010). Fermentation on starch may result in increased lactic acid production, dramatically decreasing rumen pH, which decreases starter intake by dairy calves. However, the price of corn varies during the year, reaching the highest values in the offseason, burdening production costs. Because of this, alternatives for corn inclusion in starters may be interesting by the economic and production point of view.

Molasses is a by-product of the ethanol industry, consisting mainly of sugars quickly and extensively fermented in the rumen (Noziére et al., 2010), with approximately 93% of total digestible nutrients (TDN) of the corn (NRC, 2001). On the other hand, glucose syrup is a by-product of corn industry, composed of glucose and maltose, with 90.9% of TDN in dry matter (DM), higher than the corn, which lies close to 84.9% (NRC, 2001).

Molasses has been included in starter for milk-fed dairy calves not only for being a cheaper source of readily fermentable carbohydrate, but also by having positive impact on intake and assisting in particle agglutination in the concentrate (Hill et al., 2008). In addition, there are

* Corresponding Author: C. M. M. Bittar.
E-mail: carlabittar@usp.br

reports that molasses can increase the production of butyrate in the rumen (Kellogg and Owen, 1969a ; Martel et al., 2011) and thus accelerate the ruminal development, since that SCFA has an important role in the growth of rumen papillae (Tamate et al., 1962). The rapid and efficient transition from pre-ruminant (liquid diet) to functional ruminant (only solid diet) has substantial economic importance to producers.

Few studies have been conducted to elucidate the effects of partial replacement of corn by sugarcane molasses in the concentrate for milk-fed calves. On the other hand, there are no published studies using glucose syrup in the starter feed for this animal class. Thus, the objective of this work was to evaluate different levels of sugar cane molasses and glucose syrup to replace corn in the starter feed on performance and metabolism of calves.

MATERIAL AND METHODS

Thirty-six newborn Holstein calves were fed approximately 6 L of colostrum, during the first 12 hours after birth, receiving milk-replacer thereafter. After the colostrum feeding period, calves were individually housed in wood hutches (1.00×1.45 m) distributed in a grass field, at the experimental calf facility of the Department of Animal Science of the "Luiz de Queiroz" College of Agriculture, University of Sao Paulo. Study was conducted from February to June of 2011, with calves born from February to April. All calves were individually fed and received 4 L of milk replacer daily (Sprayfo Violeta, 20% crude protein (CP), 16% ether extract (EE), 12.5% solids, Sloten of Brazil Ltd., Santos, SP, Brazil). Calves were fed milk replacer in two equal feedings at 0700 and 1700 h and had free access to water and starter feed. Calves were blocked by weight and date of birth and assigned to one of the four starter feed compositions (9 calves/treatment), during an 8 week study: 0MO (0% MO and 65% corn), 5MO (5% of MO and 60% corn), 10MO (10% of MO and 55% corn) and 5GS (5% of GS and 60% corn).

Corn grain was ground to reach particle size close to 2 mm and blended with the other ingredients of the concentrate diets using a horizontal mixer (Lucato, Limeira, Brazil), resulting in a coarsely ground physical form. All starter concentrates were formulated according to NRC (2001) to have the same crude protein (CP) and minerals concentration (Table 1). The starter feed was supplied *ad libitum* and every morning remains were weighed to obtain the daily starter feed intake. Weaning was abruptly performed at the 8th week of age, when trial ended. Animals were weighed weekly, before the morning milk feeding, on a mechanical scale (ICS-300, Coimma Ltd. Piracicaba, SP, Brasil), and measurements of withers height, heart girth and hip width were taken. Every morning

Table 1. Ingredients and chemical composition of starter

Items	Treatments[1]			
	0MO	5MO	10MO	5GS
Ingredients (% DM)				
Corn	65	60	55	60
Soybean meal	24	24	23	24
Sugarcane molasses	0	5	10	0
Glucose syrup	0	0	0	5
Soybean hulls	10	10	11	10
Mineral/vitamin premix[2]	1	1	1	1
Chemical composition				
Dry matter (% fed basis)	88.0	87.4	87.8	87.9
Ash (% DM)	4.2	4.7	4.7	4.1
Crude protein (% DM)	19.6	19.9	19.3	19.5
Ether extract (% DM)	4.1	4.1	3.4	3.9
NDF (% DM)	16.4	18.8	18.7	16.9
ADF (% DM)	9.4	10.8	11.5	10.3
N-NDF (% total N)	12.2	8.6	10.1	7.9
N-ADF (% total N)	5.6	3.9	2.6	3.6
Lignin (% DM)	0.75	0.72	0.94	0.82
NFC (% DM)	55.7	52.5	53.9	55.6
ME (Mcal/kg DM)[3]	3.00	3.00	2.98	3.00
Net energy for gain (Mcal/kg DM)[3]	1.93	1.92	1.91	1.92

DM, dry matter; NDF, neutral detergent fiber; ADF, acid detergent fiber; N-NDF, nitrogen in NDF; N-ADF, nitrogen in ADF; NFC, non-fibrous carbohydrate; ME, metabolizable energy.
[1] 0MO, no molasses; 5MO, 5% molasses; 10MO, 10% molasses; 5GS, 5% glucose syrup.
[2] Mineral/vitamin premix composition: Ca 16.8%; P 4.2%; S 2.3%; Na 11.6%; Cl 8.0%; Mg 2.4; Co 38.2 ppm; Cu 343 ppm; I 30.2 ppm; Fe 578.2 ppm; Mn 1,146.4 ppm; Se 15.5 ppm; Zn 1,176.2 ppm; Vit. A 68,760 UI/kg; Vit. E 764 UI/kg; Vit. D 57,300 IU/kg.
[3] Value estimated by NRC (2001).

animal's feces were scored, as described by Larson et al. (1977), on a scale from 1 to 4 (1 = normal; 2 = loose; 3 = very loose, no watery separation; and 4 = very watery). To avoid variations a sole observer scored feces according to this scale, after being trained with the aid of pictures.

Ruminal fluid samples were collected at the 4th, 6th, and 8th week of age, using an oro-esophageal tube and a vacuum pump (TE-0581, Tecnal Ltd. Piracicaba, SP, Brazil), with the first fraction of fluid collected discarded to avoid saliva. Samples (50 mL) were filtered through appropriate cloth tissue (around 8 layers of surgical gauze) and pH was immediately measured (Tec-5, Tecnal Ltda., Brasil). Samples were frozen for latter analyses of SCFA. Samples were centrifuged at 15,000×g (Universal 320R, Hettich, Tuttlinger, Germany) for determination of SCFA, as described by Ferreira et al. (2009). Samples were analyzed by gas chromatograph (Hewlett Packard 5890 Series II GC, Wilmington, DE, USA) equipped with integrator (Hewlett Packard 3396 Series II Integrator, USA), and automatic injector (Hewlett Packard 6890 Series Injector, USA). A volume of 100 μL of the internal standard, 2-methylbutyric acid, 800 μL of sample and 200 μL sample of formic acid were pipette in a vial for gas chromatograph injection. A mixture of short-chain fatty acids of known concentration was used as external standard for calibration.

Blood samples were taken weekly, two hours after morning feeding, via jugular venipuncture by vacuum tubes containing sodium fluoride and potassium EDTA (Vacuette of Brazil, Campinas, SP, Brazil). Samples were centrifuged at 2,000×g, (20 min at 4°C) and plasma was stored until analysis. Specific enzymatic kits were used to analyzed plasma concentrations of glucose (Glicose HK Liquiform – Ref.: 85, Labtest Diagnóstica S.A., Lagoa Santa, MG, Brazil) and β-hydroxybutyrate (RANBUT – Ref.: RB1007, Randox Laboratories, Crumlin, UK) in an automatic biochemistry system (SBA-200, CELM, Barueri, SP, Brazil).

At weaning, with 8 weeks of age, animals were slaughtered by stunning and bleeding, to evaluate development of the upper digestive tract and rumen papillae. The compartments of the stomach (rumen, reticulum, omasum, and abomasum) were separated and weighed individually. Reticulum-rumen had its maximum volume measured by filling it with water. Samples from the ventral sac of the rumen were collected and preserved in 10% formaldehyde solution. Number, height and width of papillae (cm^2) were measured as by Lesmeister and Heinrichs (2004), through a stereoscopic microscope equipped with a scale.

Samples of starter feed were periodically sampled and ground through a 1-mm mesh for dry matter (DM), ashes, and EE determination according to AOAC (1990); nitrogen was determined by combustion, according to the Dumas method, using an N analyzer by LECO, model FP-528 (St. Joseph, MI, USA) and CP was calculated by multiplying results by 6.25; free-ash neutral detergent fiber (NDF), acid detergent fiber (ADF) and lignin were determined, according to the method of Van Soest et al. (1991), using sodium sulfite and thermo stable amylase when required; and nitrogen in NDF and ADF were determined according to Licitra et al. (1996). The TDN values were calculated according to the equation proposed by Weiss (1993) and non-fiber carbohydrates (NFC) by the equation: NFC = 100 – (crude protein+ether extract+neutral detergent fiber$_{cp}$+ashes), expressed as g/kg DM, being NDFcp the NDF free of crude protein.

Data concerning concentrate intake, body weight, average daily gain, body measurements, as well as ruminal and plasma parameters were analyzed as repeated measurements using the PROC MIXED from SAS software according to the model (1). The best covariance structure was identified from different covariance structures (arma (1, 1), ar (1); arh (1); Toep; Toeph; UN; CS) by comparing the AICC statistic (Akaike Information Criteria Corrected). Differences were considered significant at p<0.05 unless otherwise stated. Data for morphometric measurements of the proximal digestive tract and development of the rumen (papillae) were performed by the model (2), using the general linear model from SAS. Significance was adopted for values of p<0.05 for all parameters.

$$Y_{ijk} = \mu + T_i + B_j + W_k + T_i W_k + E_{ijk} \qquad (1)$$

$$Y_{ijk} = \mu + T_i + B_k + E_{ijk} \qquad (2)$$

Where, Y_{ijk} is the response variable, μ is the overall mean, T_i is the treatment effect, B_j is the block effect, W_k is the age effect, $T_i W_k$ is the interaction of treatment and age effects, and E_{ijk} is the residual effect.

RESULTS AND DISCUSSION

Starter intake was not affected (p>0.05) by the substitution of corn by molasses or glucose syrup, as well as by age or by the interaction of these factors (Table 2). All treatments resulted in concentrate intake higher than 700 g/d at weaning, considered suitable for Holstein calves subjected to early weaning (Quigley, 1996). It is recommended that calves present this level of intake at weaning, because it guarantees a minimum level of rumen development that allows calves to maintain weight gain

Table 2. Effect of sugarcane molasses and glucose syrup as a replacement for corn in the starter feed of dairy calves on average starter intake, body weight, average daily gain and gain of body measurements

	Treatments[1]				SEM	p[2]		
	0MO	5MO	10MO	5GS		T	A	T×A
Starter intake (g/d)	261.8	396.0	443.8	323.9	68.88	0.54	0.16	0.57
Body weight (kg)	39.6	40.8	42.8	40.3	1.207	0.30	<0.001	0.21
Average daily gain (g)	256.7	289.6	298.5	214.4	52.68	0.70	<0.001	0.11
Withers height gain (cm/wk)	0.74	0.84	0.54	0.75	0.146	0.50	<0.001	0.46
Heart girth gain (cm/wk)	1.18	1.38	1.56	1.24	0.213	0.58	<0.001	0.12
Hip width gain (cm/wk)	0.30	0.37	0.37	0.27	0.057	0.51	<0.001	0.67

SEM, standard error of the mean.
[1] 0MO, no molasses; 5MO, 5% molasses; 10MO, 10% molasses; 5GS, 5% glucose syrup.
[2] T, treatment effect; A, age (week) effect; T×A, treatment and age interaction effect.

after the interruption of liquid diet feeding. Concentrate intake is closely linked to production of SCFA, which are the main stimulators of development of rumen epithelium.

It was expected that the mean intake of starter containing sugar cane molasses treatments were higher than that observed for the control treatment, since this ingredient has been used to enhance palatability in the diet (Hill et al., 2008). Likewise, it was believed that the glucose syrup would also increase the starter intake, because it is basically composed of glucose and maltose, carbohydrates with fast ruminal fermentation and intestinal absorption. However, no significant differences for the starter intake were observed. Similar results to the present study were reported by Hill et al. (2008), who replaced the corn for molasses in the starter feed for Holstein calves weaned at 42 days and found no differences (p>0.05) in the consumption of calves receiving 5 (454 g/d) or 10% (381 g/d) of molasses. Lesmeister and Heinrichs (2005) evaluated the inclusion of molasses in the concentrate for calves from the first to the sixth week of age and observed a decrease in intake when inclusion was increased from 5% (509 g/d) to 12% (396 g/d). However, there was no difference (p>0.05) for weight gain and final body weight for all treatments (Table 2).

Starter composition had no effect (p>0.05) on daily gain at weaning and average daily gain. There was also no interaction between starter composition and age of calves for these variables (Table 2). However, there was an increase (p<0.01) in average daily gain as animals aged (Table 2). Weight loss was observed in all treatments between the first and second week, which may have been associated with low starter intake at this stage (Table 2) and the occurrence of diarrhea (Figure 1), associated with the low milk replacer feeding volume. According to Van Amburgh and Drackley (2005), calves fed 4 L of milk replacer requires the energy of 75% of total intake just for maintenance, which may affect body weight gain negatively, mainly at early ages as a result of the low starter intake

during this period.

Starter composition did not affect (p>0.05) initial, final or average weight (Table 2). However, there was an age effect (p<0.01), with increasing body weight over the course of weeks (Table 2). These results were expected, since the starter intake has increased considerably with the passage of weeks and did not differ among treatments during the experimental period.

Withers height, heart girth and hip width (Table 2) were not affected by the substitution of corn by molasses or glucose syrup, or by the interaction of this factor with age (p>0.05). However significant effects of animal age (p<0.001) for those body measurements were observed, increasing as animals aged. Withers height was lower than the standard for Holstein calves, which is 79.4 cm (±3.3 cm) (Heinrichs and Losinger, 1998). This may explain the lower withers height gain compared to that suggested by Hoffman (1997), which is between 1.3 and 1.4 cm per week until 2 months of age. However, other authors have reported similar values to those of the present study to assess the withers height of calves fed starter feed containing molasses (Hill et al., 2008) and animals with similar body frame (Ferreira and Bittar, 2010).

Fecal score was not affected (p>0.05) by the starter composition (Figure 1). Likewise, the interaction of starter composition and age was not significant (p>0.05). However, there was an age effect (p<0.001), with decreasing fecal score over the weeks. It was believed that the inclusion of molasses in the starter concentrate could increase the incidence of scores indicative of diarrhea, since there is no sucrase activity in calves up to 44 days (Huber et al., 1961), which could impair the use of sucrose with a consequent increase of osmotic pressure in the intestine.

Animals fed starter concentrate containing 5% or 10% molasses replacing corn tended (p<0.10) to have lower ruminal pH than others groups (Table 3). This result is probably due to the fast production of SCFA, since this

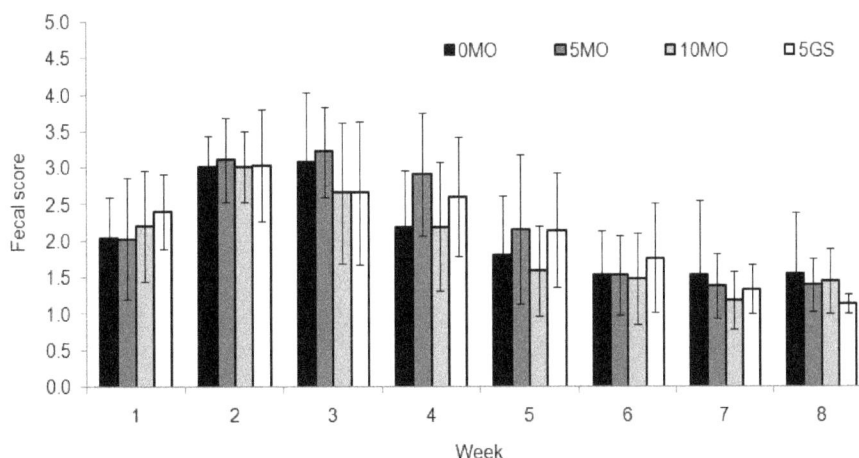

Figure 1. Fecal scores according to week of age, of calves with supplemented of sugarcane molasses or glucose syrup in replacing corn. 0MO, no molasses; 5MO, 5% molasses; 10MO, 10% molasses; 5GS, 5% glucose syrup.

ingredient has a high rumen fermentation rate (Oliveira et al., 2003). Thus, diets containing molasses can lead to a faster production of SCFA when compared to corn and, therefore, decreasing the pH of the rumen fluid. However, the same results were expected for calves consuming concentrate containing glucose syrup, since glucose and maltose, the main components of this ingredient, have high fermentation rate. The relatively low pH values found in this study may be related to the consumption of considerable amounts of concentrate. According to Quigley (1996), ruminal pH varies with the rate at which SCFA and ammonia are produced and absorbed by the rumen wall or other microorganisms. Since milk-fed calves do not have a fully developed rumen papillae, the absorption area is small and therefore, there may be an accumulation of SCFA in ruminal fluid, resulting in a drop in pH (NRC, 2001). Age and the interaction of age and starter composition presented a significant effect on rumen pH (Table 3). As animals aged and starter intake increased, rumen pH decreased because of higher fermentation end products.

Animals fed starter concentrate containing 5% molasses replacing corn had higher (p<0.05) molar total SCFA than those fed 0MO. However, due to the higher rate and extent of fermentation of sugarcane molasses and glucose syrup when compared to corn, higher values were expected for molar SCFA concentration in ruminal fluid of 10MO and 5GS fed animals. Lesmeister and Heinrichs (2005) reported higher plasma concentration of total SCFA in calves fed 12% of sugarcane molasses in the starter feed compared to those who received 5%. According to these authors, a higher plasma concentration of SCFA indicates higher metabolic activity of ruminal epithelium (absorptive rate) or increase in the production of these compounds in the rumen resulting from a more digestible diet.

Similar results were observed for individual SCFA molar concentrations (Table 3), with higher values (p<0.05) for calves fed 5MO than Control calves, with no differences

among other treatments. A higher concentration of propionate was expected for animals fed sugarcane molasses or glucose syrup, because of the faster and more extensive fermentation of these ingredients. However, there may have been greater lactic acid production (not measured), which would explain the reduction in ruminal pH with increasing age of the animals. Ruminal butyrate molar concentration tended to be higher (p = 0.07) in animals receiving 10% molasses replacing corn in the starter concentrate as compared to animals fed 0MO (Table 3). This result is of great importance in the nutrition of calves, since this SCFA is primarily responsible for the growth of rumen papillae (Quigley, 1996). Increases in the concentration of butyrate concentration by adding molasses or sucrose are described in the literature for both *in vitro* (Kellogg and Owen, 1969b) and *in vivo* (Martel et al., 2011) studies.

Even though there was an age effect for total and individual SCFA rumen molar concentration, there was no significant interaction of age and concentrate composition (p<0.05). Similarly as for the total SCFA concentration, these results are a response to the increasing starter concentrate intake from the fourth to the sixth and the eighth weeks of age. Similar results are usually found in the literature, since SCFA are generated by the fermentation of organic matter present in the rumen, which increases as animals increase the consumption of solid feed.

The concentration of ammonia-N in rumen fluid was not affected (p>0.05) by the inclusion of sugarcane molasses or glucose syrup replacing corn in the starter concentrate (Table 3). Even though there was an age effect (p<0.01), with decreasing values as animals aged, there was no interaction between age and starter composition. As the rumen develops and is colonized by microorganisms and there is a proper supply of nutrients to ferment, there is an increase in the ability to use and recycle ammonia-N, resulting in decreased values of rumen ammonia-N

Table 3. Effect of sugarcane molasses and glucose syrup as a replacement for corn in the starter feed of dairy calves on mean values of pH and short chain fatty acids

	Treatments[1]				SEM	p[2]		
	0MO	5MO	10MO	5GS		T	A	T×A
pH	5.59	5.14	5.14	5.26	0.138	0.09	0.01	0.01
Short chain fatty acids (mmol/L)								
Acetic	57.63[b]	68.00[a]	63.66[ab]	60.94[ab]	2.515	0.04	0.01	0.17
Propionic	33.65[b]	48.07[a]	44.69[ab]	38.20[ab]	3.577	0.03	0.01	0.40
Butyric	7.81	10.66	12.77	11.90	1.376	0.07	0.01	0.06
Total	104.2[b]	132.3[a]	127.6[ab]	116.8[ab]	7.177	0.04	0.01	0.08
Acetic:propionic	2.11	1.60	1.53	1.89	0.184	0.11	0.01	0.53
Ammonia-N	18.03	16.18	19.30	16.36	1.889	0.58	0.01	0.39

SEM, standard error of the mean.
[1] 0MO, no molasses; 5MO, 5% molasses; 10MO, 10% molasses; 5GS, 5% glucose syrup.
[2] T, treatment effect; A= age (week) effect; T×A = treatment and age interaction effect.
[a,b] Lower letters in the same row differ for p<0.05.

Table 4. Effect of sugarcane molasses and glucose syrup as a replacement for corn in the starter feed of dairy calves on mean values of plasma glucose and β-Hydroxybutyric acid concentration

| | Treatments[1] | | | | SEM | p[2] | | |
	0MO	5MO	10MO	5GS		T	A	T×A
Glucose (mg/dL)	96.5	101.8	95.9	92.3	3.078	0.16	0.48	0.17
βHBA (mmol/L)	0.14	0.14	0.13	0.13	0.012	0.97	0.86	0.80
Lactate (mg/dL)	9.16	9.65	9.47	10.06	0.804	0.88	0.66	0.63

SEM, standard error of the mean; βHBA, β-hydroxybutyrate.
[1] 0MO, no molasses; 5MO, 5% molasses; 10MO, 10% molasses; 5GS, 5% glucose syrup.
[2] T, treatment effect; A, age (week) effect; T×A, treatment and age interaction effect.

concentration. The concentration of ammonia-N remained above 5 mg/dL in all samples, suggesting that there was sufficient nitrogen for proper microbial growth. According to Leng and Nolan (1984), 5 mg/dL is the minimum concentration required to satisfactory microbial protein synthesis.

Plasma concentrations of glucose and β-hydroxybutyrate (βHBA) were not affected by inclusion of co-products replacing corn in the concentrate, nor by the age effect or the interaction between age and starter composition (Table 4). Glucose plasma concentration usually decreases as animal's age (Quigley et al., 1991; Haga et al., 2008) and rumen develops as a result of solid feed intake. According to Haga et al. (2008), glucose is the primary energy source for calves with an undeveloped rumen, with lactose from the liquid diet being the main supply. As animals increase starter intake and the rumen develops, there is a very low absorption of glucose from the diet, and most of plasma glucose has its origin on hepatic gluconeogenesis from propionate. From that point on, animals rely mostly on ketone bodies as energy source, mainly in peripheral tissues (Haga et al., 2008).

In addition, an increase in the concentration of βHBA with advancing age of the animals was expected, since there was an increase in the starter intake from the fourth week of age (Figure 1) and a consequent increase in the molar concentration of butyrate in ruminal fluid. The concentration of βHBA is highly correlated with the starter intake (Quigley et al., 1991) and has been used as a parameter for monitoring rumen development. β-hydroxybutyrate is the product of butyrate metabolism by rumen epithelial cells (ketogenesis), but also may undergo oxidative metabolism through β-oxidation and the citric acid cycle (Wiese et al., 2013).

According to Davis and Drackley (1998), propionic and butyric acids are the primary stimulators of growth of rumen tissue, in part because they are extensively metabolized by the rumen epithelium during absorption. This metabolism provides energy for growth of epithelial tissue and muscle contractions. However, even though there were some differences in SCFA concentrations (Table 3), no effects were observed for rumen development as a result of inclusion of sugarcane molasses replacing corn in the concentrate (Table 5). Higher molar concentration of total SCFA and propionic acid for animals fed 5MO (Table 3), as compared to animals fed 0MO, had no effect on rumen development. There was no effect on rumen development probably because the increase on total SCFA and propionic

Table 5. Effect of sugarcane molasses and glucose syrup as a replacement for corn in the starter feed of dairy calves on mean values of forestomach morphometrics

| | Treatments[1] | | | | SEM | p |
	0MO	5MO	10MO	5GS		
Total tract (g)	1,153.9	1,433.0	1,306.5	1,161.1	106.92	0.25
Rumen-reticulum (g)	764.9	973.4	871.7	752.8	84.33	0.26
% of total tract	65.77	66.34	65.65	64.23	1.977	0.91
Capacity (L)	8.31	9.39	8.50	7.85	0.888	0.68
Omasum (g)	167.2	207.0	184.8	144.9	21.32	0.28
% of total tract	14.40	14.42	13.59	12.55	1.109	0.65
Abomasum(g)	226.7	260.2	250.1	263.3	11.80	0.18
% of total tract	20.26	19.66	20.79	23.12	1.850	0.62
Papillae number (number/cm^2)	94.1	75.2	74.2	87.3	7.17	0.18
Papillae height (mm)	1.69	2.03	3.38	1.76	0.565	0.13
Papillae width (mm)	1.00	1.10	1.36	1.10	0.209	0.62
Papillae area (cm^2)	1.93	2.57	5.25	2.00	1.258	0.21

SEM, standard error of the mean.
[1] 0MO, no molasses; 5MO, 5% molasses; 10MO, 10% molasses; 5GS, 5% glucose syrup.

acid was accompanied by increased acetate molar concentration, with no effect on the acetate:propionate ratio (Table 3).

Reticulum-rumen weight and its proportion of the total tract weight as well as its volume were also not affected by starter composition (p>0.05). Those values are higher than those found by Nussio et al. (2003) and Ferreira et al. (2009), with similar animals and experimental design, most likely due to the higher starter intake observed in the present study. According to Quigley (1996), the reticulum-rumen of a four-week-old calf should correspond to 60% of the total weight of the forestomach, the omasum to 13% and the abomasum to 27%. Values observed are higher for the reticulum-rumen and lower for the abomasum proportion, probably due to the higher age at slaughter and higher concentrate intake as compared to a four-week-old calf. Papillae measurements were not affected by inclusion of sugarcane molasses or glucose syrup as a replacement for corn in the concentrate (Table 5). Other researchers reported a trend of greater height and width of papillae of calves fed 12% molasses in the concentrate compared to those receiving 5% (Lesmeister and Heinrichs, 2005). Values observed are lower than those suggested by Huber (1969) for calves eight weeks old (5 to 7 mm).

CONCLUSION

Replacement of corn by 5% or 10% of sugarcane molasses, or by 5% glucose syrup in the starter concentrate had no effect on performance, fecal score or rumen development. However, feeding molasses increased total SCFA and propionic acid concentration, which may be beneficial for calves with a developing rumen. Therefore, these energy sources may be included in the starter concentrate, in these substitution rates, for dairy calves during the liquid-feeding phase.

ACKNOWLEDGMENTS

The authors wish to express their appreciation for the financial support provided by the CNPq and FAPESP.

REFERENCES

AOAC. 1990. Official Methods of Analysis, 18th edn. AOAC, Arlington, VA, USA.

Davis, C. L. and J. K. Drackley. 1998. The Development Nutrition and Management of the Young Calf. Iowa State University Press, Ames, Iowa, USA.

Ferreira, L. S. and C. M. M. Bittar. 2010. Performance and plasma metabolites of dairy calves fed starter containing sodium butyrate, calcium propionate or sodium monensin. Animal 5:239-245.

Ferreira, L. S., C. M. M. Bittar, V. P. Santos, W. R. S. Mattos, and A. V. Pires. 2009. Effect of inclusion of sodium butyrate, calcium propionate or sodium monensin in the starter feed on ruminal parameters and forestomach development in dairy calves. Braz. J. Anim. Sci. 38:2238-2246.

Haga, S., S. Fujimoto, T. Yonezawa, K. Yoshioka, H. Shingu, Y. Kobayashi, T. Takahasshi, Y. Otani, K. Katoh, and Y. Obara. 2008. Changes in hepatic key enzymes of dairy calves in early weaning production systems. J. Dairy Sci. 91:3156-3164.

Heinrichs, A. J. and W. C. Losinger. 1998. Growth of Holstein dairy heifers in the United States. J. Anim. Sci. 76:1254-1260.

Hill, T. M., H. G. Baterman II, J. M. Aldrich, and R. L. Schlotherbeck. 2008. Effects of feeding different carbohydrates sources and amounts to young calves. J. Dairy Sci. 91:3128-3137.

Hoffman, P. C. 1997. Optimum body size of Holstein replacement heifers. J. Anim. Sci. 75:836-845.

Huber, J. T. 1969. Development of the digestive and metabolic apparatus of the calf. J. Dairy Sci. 52:1303-1315.

Huber, J. T., N. I. Jacobson, and R. S. Allen. 1961. Digestive enzyme activities in the young calf. J. Dairy Sci. 44:1494-1501.

Kellogg, D. W. and F. G. Owen. 1969a. Relation of ration sucrose level and grain content to lactation performance and rumen fermentation. J. Dairy Sci. 52:657-662.

Kellogg, D. W. and F. G. Owen. 1969b. Alterations of in vitro rumen fermentation patterns with various levels of sucrose and cellulose. J. Dairy Sci. 52:1458-1460.

Larson, L. L., F. G. Owen, J. L. Albright, R. D. Appleman, R. C. Lamb, and L. D. Muller. 1977. Guidelines toward more uniformity in measuring and reporting calf experimental data. J. Dairy Sci. 60:989-991.

Leng, R. A. and J. V. Nolan. 1984. Nitrogen metabolism in the rumen. J. Dairy Sci. 67:1072-1089.

Lesmeister, K. E. and A. J. Heinrichs. 2004. Effects of corn processing on growth characteristics, rumen development, and rumen parameters in neonatal dairy calves. J. Dairy Sci. 87:3439-3450.

Lesmeister, K. E. and A. J. Heinrichs. 2005. Effects of adding extra molasses to a texturized calf starter on rumen development, growth characteristics, and blood parameters in neonatal dairy calves. J. Dairy Sci. 88:411-418.

Licitra, G., T. M. Hernandez, and P. J. Van Soest. 1996. Standardization of procedures of nitrogen fractionation of ruminant feeds. Anim. Feed Sci. Technol. 57:347-358.

Martel, C. A., E. C. Titgemeyer, L. K. Mamedova, and B. J. Bradfort. 2011. Dietary molasses increases ruminal pH and enhances ruminal biohydrogenation during milk fat depression. J. Dairy Sci. 94:3995-4004.

National Research Council. 2001. Nutrient Requirement in Dairy Cattle. 7th edn. National Academy of Science, Washington, DC, USA.

Nocek, J. E. and S. Tamminga. 1991. Site of digestion of starch in the gastrointestinal tract of the dairy cows and its effects on milk yield and composition. J. Dairy Sci. 74:3598-3629.

Nozière, P., I. Ortigues-Marty, C. Loncke, and D. Sauvant. 2010. Carbohydrate quantitative digestion and absorption in ruminants: from feed starch and fiber to nutrients available for tissues. Animal 4:1057-1074.

Nussio, C. M. B., F. A. P. Santos, M. Zopollatto, A. V. Pires, J. B. Morais, and J. J. R. Fernandes. 2003. Ruminal fermentation parameters and metric measurements of the rumen of dairy calves fed processed corn (steam-rolled vs. steam-flaked) and monensin. Braz. J. Anim. Sci. 32:1021-1031.

Oliveira, M. V. M., F. M. Vargas Jr, L. M. B. Sanchez, W. Paris, A. Frizzo, I. P. Haygert, D. Montagner, A. Weber, and L. Cerdótes. 2003. Ruminal degradability and intestinal digestibility of feeds by means of associated technical *in situ* and mobile nylon bag. Braz. J. Anim. Sci. 32:2023-2031.

QuigleyIII, J. D., Z. P. Smith, and R. N. Heitmann. 1991. Changes in plasma volatile fatty acids in response to weaning and feed intake in young calves. J. Dairy Sci. 74:258-263.

Quigley III, J. D. 1996. Feeding prior to weaning. In: Proceedings of Calves, Heifers and Dairy Profitability National Conference. Harrisburg, PA, USA. pp. 245-255.

Tamate, H., A. D. McGilliard, N. L. Jacobson, and R. Getty. 1962. Effect of various dietaries on the anatomical development of the stomach in the calf. J. Dairy Sci. 45:408-420.

Van Amburgh, M. and J. K. Drackley. 2005. Current perspectives on the energy and protein requirements of the pre-weaned calf. In: Calf and Heifer Rearing: Principles of Rearing the Modern Dairy Heifer from Calf to Calving. (Ed. P. C. Gransworthy). Nottingham University Press, Nottingham, UK. pp. 67-82.

Van Soest, P. J., J. B. Robertson, and B. A. Lewis. 1991. Methods for dietary fiber neutral detergent fiber, and non-starch polysaccharides in relation to animal nutrition. J. Dairy Sci. 74:3583-3597.

Weiss, W. P. 1993. Predicting energy values of feeds. J. Dairy Sci. 76:1802-1811.

Wiese. B. I., P. Górka, T. Mutsvangwa, E. Okine, and G. B. Penner. 2013. Short communication: Interrelationship between butyrate and glucose supply on butyrate and glucose oxidation by ruminal epithelial preparations. J. Dairy Sci. 96:5914-5918.

14

Supplementation of Dried Mealworm (*Tenebrio molitor* larva) on Growth Performance, Nutrient Digestibility and Blood Profiles in Weaning Pigs

X. H. Jin, P. S. Heo, J. S. Hong, N. J. Kim[1], and Y. Y. Kim*

Department of Agricultural Biotechnology, College of Animal Life Sciences,
Seoul National University, Seoul 151-921, Korea

ABSTRACT: This experiment was conducted to investigate the effects of dried mealworm (*Tenebrio molitor* larva) on growth performance, nutrient digestibility and blood profiles in weaning pigs. A total of 120 weaning pigs (28±3 days and 8.04±0.08 kg of body weight) were allotted to one of five treatments, based on sex and body weight, in 6 replicates with 4 pigs per pen by a randomized complete block design. Supplementation level of dried mealworm was 0%, 1.5%, 3.0%, 4.5%, or 6.0% in experimental diet as treatment. Two phase feeding programs (phase I from 0 day to 14 day, phase II from 14 day to 35 day) were used in this experiment. All animals were allowed to access diet and water *ad libitum*. During phase I, increasing level of dried mealworm in diet linearly improved the body weight (p<0.01), average daily gain (ADG) (p<0.01) and average daily feed intake (ADFI) (p<0.01). During phase II, ADG also tended to increase linearly when pigs were fed higher level of dried mealworm (p = 0.08). In addition, increasing level of dried mealworm improved the ADG (p<0.01), ADFI (p<0.05) and tended to increase gain to feed ratio (p = 0.07) during the whole experimental period. As dried mealworm level was increased, nitrogen retention and digestibility of dry matter as well as crude protein were linearly increased (p = 0.05). In the results of blood profiles, decrease of blood urea nitrogen (linear, p = 0.05) and increase of insulin-like growth factor (linear, p = 0.03) were observed as dried mealworm was increased in diet during phase II. However, there were no significant differences in immunoglobulin A (IgA) and IgG concentration by addition of dried mealworm in the growth trial. Consequently, supplementation of dried mealworm up to 6% in weaning pigs' diet improves growth performance and nutrient digestibility without any detrimental effect on immune responses. (**Key Words:** Dried Mealworm, Growth Performance, Insect, *Tenebrio molitor* Larva, Weaning Pigs)

INTRODUCTION

More than 10 million of insect species have been identified, which account for a half of all creatures in earth. Among of them, approximately 1,500 species of insects are known to be an edible protein source for humans and animals (Ng, 2001). Edible insects contain higher protein content compared with plant protein, 18 kinds of amino acids including essential amino acids and vitamin and mineral properties that have been evaluated in detail (MacEvilly, 2000). Furthermore, insects have several useful physiological characteristics such as high efficiency reproductive ability, high feed conversion rate (or high conversion rate of organic matter), and easy rearing with low feed cost (Liu et al., 2010).

The Sánchez-Muros et al. (2014) also reported that edible insects were used as an alternative source of feed ingredients in the animal feed industry. From those reasons, insects have been proposed as a sustainable alternative protein source for animal feed.

The insect market for animal feed is continually increasing in the world, especially focused on *Tenebrio molitor* larva. *Tenebrio molitor* has been known to be an

* Corresponding Author: Y. Y. Kim.
E-mail: yooykim@snu.ac.kr
[1] National Academy of Agricultural Science, Wanju 565-851, Korea.

acceptable protein source in poultry diets (Ramos-Elorduy et al., 2002), and the nutritional value of *Tenebrio molitor* was also reported (Nergui et al., 2012).

Moreover, a few studies have been conducted to evaluate the application and large-scale production of *Tenebrio molitor* to supply the animal feed industry (Hernandez, 1987; Lagunes and Garcia, 1994). However, there is still lack of published data of the effects of dried mealworm on weaning pigs.

Therefore, the objective of this experiment was to investigate the effects of dried mealworm (*Tenebrio molitor* larva) supplementation as a protein source on growth performance, nutrient digestibility, and blood profiles in weaning pigs.

MATERIALS AND METHODS

Experimental animal, treatment and diet

A total of 120 crossbred ([Yorkshire×Landrace]× Duroc) weaning pigs (28±3 d of age, 8.04±0.08 kg body weight [BW]) were assigned to one of five treatments considering sex and body weight in a randomized complete block design. Four pigs were reared in each pen of a weaning pigs' house (concrete-slot floor, 0.90×2.15 m). The control treatment group was provided with corn-barley-soybean meal-based diet and the other groups were provided with diet containing 1.5%, 3.0%, 4.5%, or 6.0% of dried mealworm powder during 35 days after weaning. Feed and water were provided *ad libitum* through a feeder and a nipple during the whole experimental period. The temperature was kept at 30°C during the first 7 days and lowered 1°C every week. An experimental period consists of two phases (Phase I: from 0 day to 14 day, Phase II: from 14 day to 35 day), and body weight and feed intake were recorded at the end of each phase to calculate average daily gain (ADG), average daily feed intake (ADFI) and gain to feed ratio (G:F ratio). Corn-barley-soybean meal-based experimental diets were formulated to contain dried mealworm at levels of 0%, 1.5%, 3.0%, 4.5%, or 6.0%. Dried mealworm powder replaced soybean meal (SBM). and soy oil because of the high energy and protein content in dried *tenebrio molitor* larva powder (metabolizable energy 5,258 kcal/kg, crude protein [CP] 46.44%). All nutrients of experimental diets met or slightly exceeded the nutrient requirements as specified by NRC (2012). Mealworms were harvested until approximate 70 d of age and the air-dried *Tenebrio molitor* larvae were obtained from National Institute of Agricultural Sciences (Wanju, Korea). Dried mealworms were ground wholly by grinder for mealworm powder type before mixing the experimental feed. Dried mealworm (dry matter [DM] basis) contained approximately 5.84% of moisture, 43.27% of CP, 32.93% of crude fat, 4.86% of crude ash, respectively. Amino acid and

fatty acid composition of dried mealworm are showed in Table 1. Formulas and chemical compositions of experimental diets are presented in Table 2 and 3.

Apparent total tract digestibility

A total of 20 crossbred pigs (10.05±0.98 kg BW) were assigned to individual metabolic crates and allotted to one of five treatments with 4 replicates in completely randomized design. Each pig was fed 200 g of phase II diet twice per day at 7:00 and 19:00, minimum level of feeding was 2% per body weight which was over 2 times the maintenance energy requirement (NRC, 1998). After 5 days adaptation period, a 5 day collection period was started with the addition of 1% chromium oxide in experimental diets as an initial marker. As a finishing marker, 1% ferric oxide was added in each experimental diet at 6th day of collection period. Collection of feces was started when the chromium

Table 1. Composition of total amino acid[1] and fatty acid[2] in dried mealworm larvae (dry matter bases)

Items	%
Total amino acid composition	
Aspartic acid	3.07
Threonine	1.57
Serine	1.86
Glutamic acid	4.57
Glycine	2.04
Alanine	3.15
Valine	3.14
Isoleucine	1.39
Leucine	2.81
Tyrosine	2.63
Phenylalanine	1.36
Lysine	1.86
Histamine	1.07
Arginine	2.03
Proline	2.23
Methionine	0.54
Cysteine	0.35
Total	35.67
Fatty acid composition	
C14	2.85
C15	7.10
C16	9.33
C16:1	2.12
C18	2.40
C18 1n9	40.78
C18 2n6	33.58
C18 3n6	1.85
Total	100.00

[1] Analyzed by laboratory of animal nutrition and biochemistry, Seoul National University, Korea.
[2] Analyzed by laboratory of animal nutrition and biochemistry, Seoul National University, Korea.

Table 2. The formula and chemical composition of experimental diets (phase I, d 0 to 14); as fed basis

Items	Tenebrio molitor larva (%)				
	0	1.5	3.0	4.5	6.0
Ingredients (%)					
Corn	30.95	31.03	31.02	31.05	31.04
Soy bean meal (44%)	35.10	33.34	31.51	29.80	28.01
Mealworm larva	0.00	1.50	3.00	4.50	6.00
Whey powder	3.00	3.00	3.00	3.00	3.00
Lactose	12.00	12.00	12.00	12.00	12.00
Barley	12.42	13.32	14.37	15.26	16.29
Soy-oil	3.22	2.51	1.79	1.10	0.39
MCP	1.22	1.22	1.20	1.20	1.20
Limestone	0.94	0.94	0.94	0.96	0.96
L-lysine·HCl	0.40	0.39	0.39	0.39	0.38
DL-methionine	0.40	0.39	0.39	0.39	0.39
Vit. Mix[1]	0.12	0.12	0.12	0.12	0.12
Min. Mix[2]	0.12	0.12	0.12	0.12	0.12
Salt	0.20	0.20	0.20	0.20	0.20
Choline-Cl (25%)	0.10	0.10	0.10	0.10	0.10
Zinc oxide	0.10	0.10	0.10	0.10	0.10
Total	100.00	100.00	100.00	100.00	100.00
Chemical composition					
ME (kcal/kg)[3]	3,400	3,400	3,400	3,400	3,400
Moisture (%)[4]	6.55	6.57	6.78	6.83	6.83
Crude protein (%)[4]	21.44	21.53	21.56	21.66	21.34
Ether extract (%)[4]	5.78	5.79	5.45	5.50	5.69
Crude ash (%)[4]	5.00	5.05	4.96	4.84	4.90
Total lysine (%)[3]	1.35	1.35	1.35	1.35	1.35
Total methionine (%)[3]	0.39	0.39	0.39	0.39	0.39
Total calcium (%)[3]	0.80	0.80	0.80	0.80	0.80
Total phosphorus (%)[3]	0.65	0.65	0.65	0.65	0.65

MCP, monocalcium phosphate; ME, metabolizable energy.
[1] Provided the following per kilogram of diet: vitamin A, 8,000 IU; vitamin D₃,1,600 IU; vitamin E, 32 IU; d-biotin, 0.4 mg; riboflavin, 3.2 mg; calcium pantothenic acid, 8 mg; niacin, 16 mg; vitamin B₁₂, 12 mg; vitamin K, 2.4 mg.
[2] Provided the following per kilogram of diet: Mn, 24.8 mg; CuSO₄, 54.1 mg; Fe, 127.3 mg; Zn, 84.7 mg; Co, 0.3 mg; Se, 0.1 mg; I, 0.3 mg.
[3] Calculated value.
[4] Analyzed value.

Table 3. The formula and chemical composition of experimental diets (phase II, d 14 to 35); as fed basis

Items	Tenebrio molitor larva (%)				
	0	1.5	3.0	4.5	6.0
Ingredients (%)					
Corn	32.55	33.49	34.44	35.38	36.29
Soy bean meal (44%)	27.28	25.61	23.95	22.27	20.63
Mealworm larva	0.00	1.50	3.00	4.50	6.00
Lactose	4.00	4.00	4.00	4.00	4.00
Barley	30.00	30.00	30.00	30.00	30.00
Soy-oil	3.42	2.65	1.87	1.10	0.34
MCP	1.03	1.00	1.01	1.00	1.00
Limestone	0.66	0.69	0.68	0.70	0.71
L-Lysine·HCl	0.43	0.43	0.42	0.42	0.41
DL-methionine	0.09	0.09	0.09	0.09	0.08
Vit. Mix[1]	0.12	0.12	0.12	0.12	0.12
Min. Mix[2]	0.12	0.12	0.12	0.12	0.12
Salt	0.20	0.20	0.20	0.20	0.20
Choline-Cl(25%)	0.10	0.10	0.10	0.10	0.10
Total	100.00	100.00	100.00	100.00	100.00
Chemical composition					
ME (kcal/kg)[3]	3,350	3,350	3,350	3,350	3,350
Moisture (%)[4]	9.23	9.61	9.37	9.25	8.66
Crude protein (%)[4]	18.99	18.31	18.86	18.68	18.32
Ether extract (%)[4]	5.63	5.60	5.71	5.65	5.58
Crude ash (%)[4]	4.38	4.24	4.28	4.26	4.24
Total lysine (%)[3]	1.23	1.23	1.23	1.23	1.23
Total methionine (%)[3]	0.36	0.36	0.36	0.36	0.36
Total calcium (%)[3]	0.70	0.70	0.70	0.70	0.70
Total phosphorus (%)[3]	0.60	0.60	0.60	0.60	0.60

MCP, monocalcium phosphate; ME, metabolizable energy.
[1] Provided the following per kilogram of diet: vitamin A, 8,000 IU; vitamin D₃, 1,600 IU; vitamin E, 32 IU; d-biotin, 0.4 mg; riboflavin, 3.2 mg; calcium pantothenic acid, 8 mg; niacin, 16 mg; vitamin B₁₂, 12 mg; vitamin K, 2.4 mg.
[2] Provided the following per kilogram of diet: Mn, 24.8 mg; CuSO₄, 54.1 mg; Fe, 127.3 mg; Zn, 84.7 mg; Co, 0.3 mg; Se, 0.1 mg; I, 0.3 mg.
[3] Calculated value.
[4] Analyzed value.

oxide appeared in the feces and kept until the appearance of ferric oxide in the feces. Urine samples were collected during collection period in plastic containers containing 50 mL of 4 N H_2SO_4 to prevent evaporation of nitrogen prior to nitrogen retention analysis. Fecal and urinary samples were stored at –20°C until the end of collection period and the feces were dried in a drying oven at 60°C for 72 h and then ground to 1mm in a Wiley mill for chemical analysis including moisture, protein, fat, and ash contents by AOAC methods (1995).

Blood sampling and analysis

In each treatment, 6 pigs with average body weight were bled through the anterior vena cava to analyze blood urea nitrogen (BUN), insulin-like growth factor (IGF-1) and immune response (IgA, IgG) at initial day and the ends of phases (phase I and phase II). Blood samples were collected in disposable culture tubes and centrifuged for 15 min by 3,000 rpm at 4°C (Eppendorf centrifuge 5810R, Hamburg, Germany). The serum was carefully transferred to 1.5 mL micro tubes and stored at –20°C until analysis. Total BUN concentration was analyzed using an analyzer (Ciba-Corning model, Express Plus, Ciba Corning diagnostics Co., Basel, Switzerland). The immunoglobulin G (IgG) and immunoglobulin A (IgA) concentration were analyzed by enzyme-linked immunosorbent assay (ELISA) assay by the manufacture's protocols (ELISA Starter Accessory Package, Pig IgG ELISA Quantitation Kit, Pig IgA ELISA

Quantitation Kit; Bethyl, Montgomery, TX, USA). The concentration of IGF-1 was analyzed by hormone analyzer (Immulite 2000, DPC, Malvern, PA, USA).

Chemical analysis

Diets and feces were ground by a Cyclotec 1093 Sample Mill (Foss Tecator, Hillerod, Denmark) and then analyzed. The contents of DM (procedure 967.03; AOAC, 1995), ash (procedure 923.03; AOAC, 1995). The nitrogen content was analyzed by using the Kjeldahl procedure with Kjeltec (KjeltecTM 2200, Foss Tecator, Höganäs, Sweden) and calculating the CP content (Nitrogen×6.25; procedure 981.10; AOAC, 1995).

Statistical analysis

Statistical analysis was carried out by least squares mean comparisons using PDIFF option of general linear model procedure (SAS, 2002; SAS Inst. Inc., Cary, NC, USA). Each pen was considered as experimental unit in measuring growth performance, while individual pig was used as experimental unit for analyzing nutrient digestibility, nitrogen retention and blood characteristics. Orthogonal polynomial contrasts were performed to determine linear and quadratic effects of inclusion levels of dried mealworm. Statistical differences were considered highly significant differ at $p<0.01$, significant differ at $p<0.05$, tendency between $p \geq 0.05$ and $p \leq 0.10$.

RESULTS

The effects of supplementation of dried mealworm on growth performance in weaning pigs are presented in Table 4. Increasing dried mealworm level in weaning pig's diet significantly increased the body weight at end of phase I and phase II (linear, $p<0.01$). During phase I, ADG and ADFI were linearly increased as dried mealworm level increased in diet (linear, $p<0.01$). During Phase II (14 to 35th day), increasing dried mealworm in weaning pig diet tended to improve the ADG of pigs (linear, $p = 0.08$), but ADFI was not altered by addition of dried mealworm. Although, significant improvement was not observed in G:F ratio during phase I and phase II, increasing dried mealworm level in diet resulted in a tendency of improvement to G:F ratio during overall experimental period (linear, $p<0.07$).

Addition of dried mealworm in weaning pigs' diet showed improvements of nutrient digestibility and nitrogen retention (Table 5). There were linear improvements in DM (linear, $p = 0.05$) and CP (linear, $p = 0.05$) digestibility by increasing levels of dried mealworm, and crude ash digestibility showed tendency of improvement by dried mealworm as well (linear, $p = 0.06$). In nitrogen retention, pigs fed a higher amount of dried mealworm showed a reduction of nitrogen excretion through feces (linear, $p = 0.05$), which resulted in a linear increase in nitrogen retention (linear, $p = 0.05$).

Changes of BUN, serum IGF-1, and serum immunoglobulins concentration by the addition of dried mealworm during the feeding trial are presented in Table 6 and 7. Increasing level of dried mealworm in diet linearly decreased BUN concentration at 35 d after weaning (linear, $p<0.05$), and a similar decreasing tendency was observed at

Table 4. Effect of dried mealworm larvae supplementation on growth performance in weaning pigs[1,2]

Criteria	Tenebrio molitor larva (%)					SEM	p-value	
	0	1.5	3.0	4.5	6.0		Lin.	Quad.
Body weight (kg)								
Initial	8.05	8.04	8.05	8.04	8.04	-	-	-
2 week	9.63	9.80	10.01	10.51	10.66	0.310	0.01	0.83
5 week	17.78	18.34	19.10	19.77	20.22	0.531	0.01	0.90
Average daily gain (g)								
0-2 week	113	126	140	176	188	11.6	0.01	0.81
2-5 week	388	407	433	441	455	15.2	0.08	0.82
Overall	278	294	316	335	348	11.0	0.01	0.91
Average daily feed intake (g)								
0-2 week	250	270	283	336	349	13.3	0.01	0.77
2-5 week	721	736	754	753	773	25.4	0.44	0.96
Overall	532	550	566	586	604	17.3	0.05	0.96
Gain:feed ratio								
0-2 week	0.452	0.467	0.495	0.524	0.539	0.028	0.25	0.28
2-5 week	0.539	0.552	0.575	0.586	0.589	0.015	0.24	0.89
Overall	0.521	0.538	0.565	0.573	0.576	0.011	0.07	0.62

SEM, standard error of mean.
[1] A total of 120 crossbred pigs were fed from average initial body weight 8.04±0.08 kg.
[2] Each least squares mean for all treatments represents 6 observations respectively.

Table 5. Effect of dried mealworm larvae supplementation on nutrient digestibility in weaning pigs[1,2]

Criteria	Tenebrio molitor larva (%)					SEM	p-value	
	0	1.5	3.0	4.5	6.0		Lin.	Quad.
Nutrient digestibility (%)								
Dry matter	90.13	92.33	92.93	93.80	94.22	0.715	0.05	0.53
Crude protein	86.29	90.25	91.27	92.17	93.04	1.141	0.05	0.47
Crude ash	67.62	67.17	71.11	72.20	76.01	1.841	0.06	0.66
Crude fat	81.40	81.67	82.96	82.79	81.55	1.328	0.87	0.65
N-retention (g/d)								
N-intake	5.30	5.34	5.34	5.31	5.36	-	-	-
N-feces	0.73	0.52	0.47	0.42	0.37	0.060	0.05	0.48
N-urine	2.37	2.55	2.55	2.51	2.56	0.155	0.80	0.85
N-retention	2.20	2.27	2.33	2.38	2.42	0.156	0.05	0.98

SEM, standard error of mean.

[1] A total of 20 crossbred pigs were fed from average initial body weight 10.05±0.98 kg.

[2] Least squares means of 4 observations per treatment.

14 d after weaning (linear, p = 0.08). In contrast to BUN concentration, linear increase of serum IGF-1 concentration was observed at 35th day when dried mealworm was increased in diet (linear, p<0.05). However, increase of dried mealworm in diet did not alter serum immunoglobulins.

DISCUSSION

Studies using dietary insects as feed ingredient were mostly focused on poultry nutrition, and showed that using insect as a protein source in poultry diets had positive effects on growth performance (Téguia et al., 2002; Awoniyi et al., 2003; Shen et al., 2006). However, little information is available about using insect as a protein source in pig diets (Ni and Tang, 1993; Zhang and Zhou, 2002).

In general, pigs have to adapt rapid changes in feed with low enzyme activity after weaning (Jensen et al., 1997). Therefore, the composition of the starter diet after weaning requires highly digestible ingredients to avoid growth check or high mortality from weaning stress (Mahan and Newton, 1993). Although animal-origin protein sources' price is more expensive, it is widely utilized in weaning pigs' diet rather than soybean meal or other plant-derived ingredients (Evans and Leibholz, 1979; Stoner et al., 1990; Kats et al., 1994). Beneficial effects of dried mealworm as feed ingredients have rarely been explored, but using other insect larvae as protein sources in swine feed has been evaluated by several studies. Huang and Zhang (1984) reported that

Table 7. Effect of dried mealworm larvae supplementation on serum IgG and IgA concentration in weaning pigs[1]

Criteria	Tenebrio molitor larva (%)					SEM	p-value	
	0	1.5	3.0	4.5	6.0		Lin.	Quad.
IgG (mg/mL)								
Initial	2.09	2.09	2.09	2.09	2.09	-	-	-
2 week	1.89	2.04	1.93	2.18	2.11	0.102	0.20	0.91
5 week	4.56	4.43	4.54	4.50	4.45	0.309	0.90	0.25
IgA (mg/mL)								
Initial	1.29	1.29	1.29	1.29	1.29	-	-	-
2 week	1.47	1.60	1.55	1.63	1.56	0.069	0.15	0.96
5 week	2.48	2.34	2.41	2.39	2.52	0.154	0.95	0.93

IgG, immunoglobulin G; IgA, immunoglobulin A; SEM, standard error of mean.

[1] Least squares means of 6 observations per treatment.

Table 6. Effect of dried mealworm larvae supplementation on BUN and IGF-1 in weaning pigs[1]

Criteria	Tenebrio molitor larva (%)					SEM	p-value	
	0	1.5	3.0	4.5	6.0		Lin.	Quad.
BUN (mg/dL)								
Initial	9.40	9.40	9.40	9.40	9.40	-	-	-
2 week	14.87	14.27	12.15	12.03	12.08	0.617	0.08	0.52
5 week	12.02	11.12	10.85	10.32	10.28	0.658	0.05	0.85
IGF-1 (ng/mL)								
Initial	75.12	75.12	75.12	75.12	75.12	-	-	-
2 week	83.77	87.77	86.10	87.13	93.08	3.169	0.37	0.78
5 week	107.67	123.00	129.58	131.48	136.34	4.007	0.03	0.44

BUN, blood urea nitrogen; IGF-1, insulin-like growth factor-1; SEM, standard error of mean.

[1] Least squares means of 6 observations per treatment.

feeding diet containing maggot meal in weaning pigs' diet increased weight gain by 3.5% and reduced production cost by 13.2% compared to diet containing fish meal. Zhang and Zhou (2002) presented evidence that silkworm fed growth-finishing pigs could increase ADG by 23.6% and shorten the finishing period. In case of *Tenebrio molitor* studies, Chen et al. (2012) reported that increasing *Tenebrio molitor* protein concentrate up to 6% in weaning pig diets linearly improved body weight and body weight gain. Although current study used full-fat *Tenebrio molitor* larva, the results showed that dietary *Tenebrio molitor* have benefits as a protein source for weaning pigs. Moreover, improvement of ADFI by increasing the amount of *Tenebrio molitor* larvae in current study demonstrated that the flavor of *Tenebrio molitor* improved the palatability of diet and increased feed intake in weaning pigs.

In the present study, increasing level of dried mealworm in weaning pigs' diet improved DM digestibility. Improving growth performance with mealworm can be explained with current result of nutrients digestibility. Animal protein sources have better availability compared to plant-derived protein sources because of the balanced amino acid composition in animal protein (Cromwell, 1998). Nergui et al. (2012) represented that mealworm had various kinds of amino acids and their composition met the requirements of domestic animals. Newton et al. (1977) reported that addition of black soldier fly larvae in pig diet showed similar apparent CP digestibility as soybean meal, and Hwangbo and Hong (2009) demonstrated that broilers fed diet containing 30% housefly larvae meal had a higher apparent CP and amino acid digestibility than that of broilers fed basal diet. In addition, larvae accumulate lipids in their body which contain high levels of energy to meet their energy requirement in the pupal stage. Therefore, addition of *Tenebrio molitor* larvae in the diet caused a decrease in the amount of soy oil added in the present study. In contrast to animal protein sources, plant derived fat sources generally have better bioavailability than animal derived fats in young pigs because plant oil contains a high proportion of unsaturated fatty acid (Overland et al., 1996; Smith et al., 1996; Van Oeckel et al., 1996; Leskanich et al., 1997). Furthermore, Newton et al. (1977) demonstrated that black soldier fly larvae had higher fat digestibility than that of SBM and Finke (2002) represented that *Tenebrio molitor* larvae contained high levels of unsaturated fatty acid and had suitable unsaturated fatty acid:saturated fatty acid ratio which may have resulted in the maintenance of fat digestibility as did the replacement of soy oil with mealworm powder in current study.

Serum IGF-1 and BUN concentration are affected by the nutritional status of the animal, and have been used to predict the trends of growth and nutrient digestibility (Eggum, 1970; Etherton et al., 1987). The IGF-1 as growth hormone plays an important role in controlling the structure, function of cardiovascular system and skeletal maturation (Bayes-genis et al., 2000). Current study showed that supplemented dried mealworm stimulated the IGF-1 secretion and it had a positive influence on growth and feed efficiency. High level of BUN indicated that excessive amino acids were metabolized and circulated in the blood (Malmolf, 1988). Therefore, BUN concentration can be considered as an indicator for measurement of protein property and amino acid availability by animals (Eggum, 1970). In current study, BUN concentration had a tendency to decrease linearly as supplementation level of dried mealworm powder increased on 2 week and 5 week (linear, p = 0.08, 0.05). These results in the current study corresponded to the growth performance that improved BW and ADG with the improved nitrogen retention of nutrient digestibility.

Chitin may have a positive effect on the functioning of the immune system (Lee et al., 2008; Sánchez-Muros et al., 2014) and it could improve the immune status of the animals (Harikrishnan et al., 2012). Yuanqing et al. (2013) recommended 500 mg/kg of chitosan as an antibiotic substitute in weaning pigs. Huang et al. (2005) reported that supplementation 100 mg/kg of chitosan oligosaccharide improved ADG and ileal digestibility in broilers. In current study, *Tenebrio molitor* larva had 11.56 mg/g of chitosan (not suggested in tables) and larvae in the 6% treatment contained 0.07% of chitosan in experimental diet. On the other hand, some researchers demonstrated that using insects should increase edible safety because insects also contained chemical defense substances as toxin produced by exocrine gland (Wang et al., 2001). Mealworm has been known to contain benzoquinone as a toxin which is secreted by defensive gland of *Tenebrio moilor* (Attygalle et al., 1991) However, supplementation of dried mealworm did not have significant effects on blood IgG and IgA concentration as immune response.

CONCLUSION

Inclusion of dried mealworm (*Tenebrio molitor* larvae) up to 6% in weaning pig's diet is beneficial for weaning pigs by improvement of growth performance. Dried mealworm supplementation increased feed intake and nutrient digestibility without any detrimental effect on immune response. Dried mealworm powder is available as protein source for weaning pigs' diet.

ACKNOWLEDGMENTS

This research was supported by "Cooperative Research Program for Agriculture Science and Technology Development (Project No. PJ009226)" Rural Development Administration, Republic of Korea.

REFERENCES

AOAC. 1995. Official Methods of Analysis. 16th edn. Association of Official Analytical Chemists, Arlington, VA, USA.

Attygalle, A. B., C. L. Blankespoor, J. Meinwald, and T. Eisner. 1991. Defensive secretion of *Tenebrio molitor* (Coleoptera: Tenebrionidae). J. Chem. Ecol. 17:805-809.

Awoniyi, T. A. M., V. A. Aletor, and J. M. AIna. 2003. Performance of broiler-chickens fed on maggot meal in place of fishmeal. Int. J. Poult. Sci. 2:271-274.

Bayes-Genis, A., C. A. Conover, and R. S. Schwartz. 2000. The insulin-like growth factor axis: a review of atherosclerosis and restenosis. Circ. Res. 86:125-130.

Chen, Z. B. 2012. Analysis for nutritional value of four kinds of insects and use of Tenebrio molitor power in weaning pig production. China Knowledge Resource Integrated Database (CNKI), Shandong Agricultural University, Shandong, China.

Cromwell, G. L. 1998. Feeding swine. In: Livestock Feeds and Feeding. 4th ed. Prentice-Hall, Upper Saddle River, NJ, USA. 354 p.

Eggum, B. O. 1970. Blood urea measurement as a technique for assessing protein quality. Br. J. Nutr. 24:983-988.

Etherton, T. D., J. P. Wiggins, C. M. Evock, C. S. Chung, J. F. Rebhun, P. E. Walton, and N. C. Steele. 1987. Stimulation of pig growth performance by porcine growth hormone: determination of the dose-response relationship. J. Anim. Sci. 64:433-443.

Evans, D. F. and J. Leibholz. 1979. Meat meal in the diet of the early-weaned pig. I. A comparison of meat meal and soya bean meal. Anim. Feed Sci. Technol. 4:33-42.

Finke, M. D. 2002. Complete nutrient composition of commercially raised invertebrates used as food for insectivores. Zoo Biol. 21:269-285.

Harikrishnan, R., J. S. Kim, C. Balasundaram, and M. S. Heo. 2012. Dietary supplementation with chitin and chitosan on haematology and innate immune response in *Epinephelus bruneus* against *Philasterides dicentrarchi*. Exp. Parasitol. 131:116-124.

Hernandez, C. 1987. Elaboration of a sweet yellow mealworm T. molitor (Coleoptera: Tenebrionidae). Thesis. University of Quimica, Auton, Mexico.

Huang, Z. Z. and N. Z. Zhang. 1984. Development of new source of protein feed - rearing maggots. Feed Res. 1987:17-21.

Huang, R. L., Y. L. Yin, G. Y. Wu, Y. G. Zhang, T. J. Li, L. L. Li, M. X. Li, Z. R. Tang, J. Zhang, B. Wang, J. H. He, and X. Z. Nie. 2005. Effect of dietary oligochitosan supplementation on ileal digestibility of nutrients and performance in broilers. Poult. Sci. 84:1383-1388.

Hwangbo, J. and E. C. Hong. 2009. Utilization of house fly-maggots, a feed supplement in the production of broiler chickens. J. Environ. Biol. 30:609-614.

Jensen, M. S., S. K. Jensen, and K. Jakobsen. 1997. Development of digestive enzymes in pigs with emphasis on lipolytic activity in the stomach and pancreas. J. Anim. Sci. 75:437-445.

Kats, L. J., J. L. Nelssen, M. D. Tokach, R. D. Goodband, T. L. Weeden, S. S. Dritz, J. A. Hansen, and K. G. Friesen. 1994. The effects of spray-dried blood meal on growth performance of the early-weaned pig. J. Anim. Sci. 72:2860-2869.

Lagunes, L. A. and L. Garcia. 1994. Two Insects Productivity Obtained by Recycling of Organics Made of Animal and Vegetable. Ph. M. Thesis, University of Ciencia, Auton, Mexico.

Lee, C. G., C. A. Da Silva., J. Y. Lee., D. Hartl, and J. A. Elias. 2008. Chitin regulation of immune responses: an old molecule with new roles. Curr. Opin. Immunol. 20:684-689.

Leskanich, C. O., K. R. Matthews, C. C. Warkup, R. C. Noble, and M. Hazzledine. 1997. The effect of dietary oil containing (n-3) fatty acids on the fatty acid, physiochemical, and organoleptic characteristics of pig meat and fat. J. Anim. Sci. 75:673-683.

Liu, Y. S., F. B. Wang., J. X. Cui, and L. Zhang. 2010. Recent status and advances on study and utilization of Tenebrio molitor. J. Environ. Entomol. 32:106-114.

MacEvilly, C. 2000. Bugs in the system. Nutr. Bull. 25:267-268.

Mahan, D. C. and E. A. Newton. 1993. Evaluation of feed grains with dried skim milk and added carbohydrate sources on weanling pig performance. J. Anim. Sci. 71:3376-3382.

Malmolf, K. 1988. Amino acid in farm animal nutrition metabolism, partition and consequences of imbalance. J. Agric. Res. 18:191-193.

Newton, G. L., C. V. Booram, R. W. Barker, and O. M. Hale. 1977. Dried Hermetia illucens larvae meal as a supplement for swine. J. Anim. Sci. 44:395-400.

Ng, W. K., F. L. Liew, L. P. Ang, and K. W. Wong. 2001. Potential of mealworm (*Tenebrio molitor*) as an alternative protein source in practical diets for African catfish, *Clarias gariepinus*. Aquac. Res. 32:273-280.

Ni, X. J. and G. J. Tang. 1993. Evaluation of optimal silkworm supplementation in suckling piglets diet. ZheJiang J. Anim. Sci. Vet. Med. 3:49-49.

NRC. 1998. Nutrient Requirements of Swine. 10th Ed. National Academy Press, Washington, DC, USA.

NRC. 2012. Nutrient Requirements of Swine. 11th Ed. National Academy Press, Washington, DC, USA.

Overland, M., O. Taugbol, A. Haug, and E. Sundstol. 1996. Effect of fish oil on growth performance, carcass characteristics, sensory parameters, and fatty acid composition in pigs. Acta Agric. Scand. Anim. Sci. 46:11-17.

Ramos-Elorduy, J., E. A. Gonzalez., A. R. Hernandez, and J. M. Pino. 2002. Use of Tenebrio molitor (Coleoptera: Tenebrionidae) to recycle organic wastes and as feed for broiler chickens. J. Econ. Entomol. 95:214-220.

Ravzanaadii, N., S. H. Kim, W. H. Choi, S. J. Hong, and N. J. Kim. 2012. Nutritional Value of Mealworm, Tenebrio molitor as Food Source. Int. J. Indust. Entomol. 25:93-98.

Sánchez-Muros, M. J., F. G. Barroso, and F. Manzano-Agugliaro. 2014. Insect meal as renewable source of food for animal feeding: A review. J. Cleaner Prod. 65:16-27.

SAS Institute, 2002. SAS/STAT User's Guide: Version 9.1. SAS Institute, Cary, NC, USA.

Shen, H., X. L. Pan, and J. G. Wang. 2006. Effect of Tenebrio molitor L. supplementation on growth performance and protein deposition in broilers. Heilongjiang Anim. Vet. Sci. 2006(08):61-62.

Smith, D. R., D. A. Knabe, and S. B. Smith. 1996. Depression of lipogenesis in swine adipose tissue by specific dietary fatty acids. J. Anim. Sci. 74:975-983.

Stoner, G. R., G. L. Allee, J. L. Nelssen, M. E. Johnston, and R. D. Goodband. 1990. Effect of select menhaden fish meal in starter diets for pigs. J. Anim. Sci. 68:2729-2735.

Téguia, A., M. Mpoame, and J. O. Mba. 2002. The production performance of broiler birds as affected by the replacement of fish meal by maggot meal in the starter and finisher diets. Tropicultura 20:187-192.

Van Oeckel, M. J., M. Casteels, N. Warnants, L. Van Damme, and Ch. V. Boucque. 1996. Omega-3 fatty acids in pig nutrition: implications for the intrinsic and sensory quality of the meat. Meat Sci. 44:55-63.

Wang, X. P., C. L. Lei, and C. Y. Niu. 2001. The defensive secretion from insects. Institute of insect sources, Huazhong Agricultural University, Wuhan, China.

Yuanqing, X., S. Binlin, G. Yiwei, L. Tiyu, L. Junliang, Y. Ping, and G. Xiaoyu. 2013. Effects of chitosan on the development of immune organs and gastrointestinal tracts in weaned piglets. Feed Ind. 3:008.

Zhang, J. H. and E. F. Zhou. 2002. Feed Resource and Utilization. China agriculture press, BeiJing, China.

15

Effects of Dietary Energy Levels on the Physiological Parameters and Reproductive Performance of Gestating Gilts

S. S. Jin, S. W. Jung, J. C. Jang, W. L. Chung, J. H. Jeong, and Y. Y. Kim*

School of Agricultural Biotechnology, and Research Institute for Agriculture and Life Science,
Seoul National University, Seoul 151-921, Korea

ABSTRACT: This experiment was conducted to investigate the effects of dietary energy levels on the physiological parameters and reproductive performance of gestating first parity sows. A total of 52 F1 gilts (Yorkshire×Landrace) were allocated to 4 dietary treatments using a completely randomized design. Each treatment contained diets with 3,100, 3,200, 3,300, or 3,400 kcal of metabolizable energy (ME)/kg, and the daily energy intake of the gestating gilts in each treatment were 6,200, 6,400, 6,600, and 6,800 kcal of ME, respectively. During gestation, the body weight (p = 0.04) and weight gain (p = 0.01) of gilts linearly increased with increasing dietary energy levels. Backfat thickness was not affected at d110 of gestation by dietary treatments, but increased linearly (p = 0.05) from breeding to d 110 of gestation. There were no significant differences on the litter size or litter birth weight. During lactation, the voluntary feed intake of sows tended to decrease when the dietary energy levels increased (p = 0.08). No difference was observed in backfat thickness of the sows within treatments; increasing energy levels linearly decreased the body weight of sows (p<0.05) at d 21 of lactation and body weight gain during lactation (p<0.01). No significant differences were observed in the chemical compositions of colostrum and milk. Therefore, these results indicated that high-energy diets influenced the bodyweight and backfat thickness of sows during gestation and lactation. NRC (2012) suggested that the energy requirement of the gestation gilt should be between 6,678 and 7,932 kcal of ME/d. Similarly, our results suggested that 3,100 kcal of ME/kg is not enough to maintain the reproductive performance for gilts during gestation with 2 kg feed daily. Gilts in the treatment 3,400 kcal of ME/kg have a higher weaning number of piglets, but bodyweight and backfat loss were higher than other treatments during lactation. But bodyweight and backfat loss were higher than other treatments during lactation. Consequently, an adequate energy requirement of gestating gilts is 6,400 kcal of ME/d. (**Key Words:** Energy Level, Gilts, Body Weight, Backfat Thickness, Reproductive Performance)

INTRODUCTION

Sow productivity has developed extensively in recent decades due to the improvement of the high genetic potential in modern sows. According to the Canadian Centre for Swine Improvement (2015) the average litter size is 14.00 in modern sows, mainly due to the increased sow productivity. Efforts to meet the nutrient requirements of high-producing sows have been undertaken by supplementing nutrients to support both the normal reproductive cycle and body maturation (Boyd et al., 2000).

Primiparous sows are particularly sensitive to adequate energy during gestation, as they are still growing and utilizing ingested nutrients to support body maturation, growth of the fetus and body maintenance (Jang et al., 2014).

However, prior research in energy supplementation in the diet is still controversial. Almeida (2000) reported that high dietary energy levels during the ovulation period can increase the ovulation rate and promote progesterone secretion to increase fetal survival. In addition, Kongsted (2005) demonstrated that higher energy intake in gestation period may reduce the risk of being culled due to pregnancy failure because of insufficient body weight (BW) gain and backfat deposition. Recently, NRC (2012) suggested that higher energy intake is good for the development and

* Corresponding Author: Y. Y. Kim.
E-mail: yooykim@snu.ac.kr

growth of the fetus, corresponding tissues (placenta, uterus, and mammary tissue) and deposition of maternal lipids and proteins. The energy requirement of gestating gilts and sows should be between 6,678 and 8,182 kcal of metabolizable energy (ME)/d, which is 1,650 kcal higher than previous recommendations (NRC, 1998). However, more research is needed to understand the precise energy requirements of gestating gilts.

Therefore, the objective of the current study is to determine the optimal dietary energy level for reproductive performance in high-producing modern sows.

MATERIALS AND METHODS

Animal preparation

A total of 52 gilts (Large White×Landrace) that were, on average, 150 d old and weighed approximately 85 kg were selected and housed in an 11×14 m barn. Sows were provided feed and water *ad libitum* until reaching 120 kg of BW and then moved to individual gestation stall cages with concrete slatted floors (0.64×2.40 m). Diets were fed individually, twice daily with 800 g each time for an ADG of 750 g/d. Gilts were mated at an average BW of 136 kg after three or four estrus cycles. Semen (Darby AI center, Chung Ju, Korea) collected from 88 boars (Duroc) in the same batch was provided for the artificial insemination (AI) of gilts.

Experimental design and animal management

A total of 52 crossbred gilts (large White×Landrace) averaging 240 days of age with a BW of 135.82±0.85 kg were allotted to 4 dietary treatments by BW and backfat thickness (BFT) in a completely random design with 13 replicates. Experimental diets for gestating gilts were formulated to contain 13.08% crude protein, 0.86% lysine, 0.90% calcium and 0.70% phosphorus, with energy contents of 3,100, 3,200, 3,300, or 3,400 kcal of ME/kg, respectively. Feed was provided at 2.0 kg/d for all treatments. Lactating diets contained 3,265 kcal ME/kg, 17.07% CP, 1.26% lysine, 0.90% calcium, and 0.70% phosphorus, respectively (Table 1). All other nutrients were formulated to meet or exceed the NRC requirements (2012). Gilts were housed in temperature-controlled rooms and placed in an individual crate (2.4×0.65 m) with a concrete floor until d 110 of gestation. After d 110 of gestation, pregnant gilts were washed and moved into farrowing crates (2.4×1.8 m). During the lactation period, all sows were fed the same commercial lactation diet. After farrowing, the lactation diet was increased gradually from 1.0 kg/d until 5 d postpartum and then provided *ad libitum* during the lactation period. Weaning was performed at approximately 21 d.

Table 1. Formula and chemical composition of gestating and lactating diets (%)

Items	Gestating diets (ME, kcal/kg)				Lactating diets
	3,100	3,200	3,300	3,400	
Ingredients (%)					
Corn	56.59	54.56	52.53	50.50	67.51
Soybean meal (46% CP)	10.09	10.44	10.78	11.12	25.57
Sugar molasse					1.00
Tallow	0.45	2.13	3.82	5.50	-
Soy oil	-	-	-	-	1.30
Barley	25.00	25.00	25.00	25.00	-
Rapeseed meal	3.60	3.60	3.60	3.60	-
L-lysine·Hcl	0.41	0.40	0.40	0.40	0.60
DL-methionine	0.04	0.04	0.04	0.04	-
Dicalciumphosphate	2.36	2.39	2.41	2.43	2.30
Limestone	0.86	0.84	0.82	0.81	0.85
Vit. Mix[1]	0.10	0.10	0.10	0.10	0.20
Min. Mix[2]	0.10	0.10	0.10	0.10	0.10
Salt	0.25	0.25	0.25	0.25	0.42
Choline chloride-50	0.15	0.15	0.15	0.15	0.15
Chemical compositions[3] (%)					
ME (kcal/kg)	3,100	3,200	3,300	3,400	3,265
CP	13.08	13.08	13.08	13.08	17.07
Lys	0.86	0.86	0.86	0.86	1.26
Met	0.23	0.23	0.23	0.23	0.25
Ca	0.90	0.90	0.90	0.90	0.90
Total P	0.70	0.70	0.70	0.70	0.70
Available P	0.42	0.42	0.42	0.42	0.41

ME, metabolizable energy; CP, crude protein.
[1] Provided per kg of diet: Vit. A, 10,000 IU; Vit. D$_3$, 1,500 IU; Vit. E, 35 IU; Vit. K$_3$, 3 mg; Vit. B$_2$, 4 mg; Vit. B$_6$, 3 mg; Vit. B$_{12}$, 15 μg; pantothenic acid, 10 mg; biotin, 50 μg; niacin, 20 mg; folic acid 500 μg.
[2] Provided per kg of diet: Fe, 75 mg; Mn, 20 mg; Zn, 30 mg; Cu, 55 mg; Se 100 μg; I, 250 μg; Co, 250 μg.
[3] Calculated value.

Measurements and analysis

The BW and BFT at the P2 position of sows were measured. Body length was measured from the center of both ears to the tail with a measuring tape. Blood samples were collected at breeding, 110 days of gestation, 24 h post farrowing, and 21 days of lactation. The number of total born, piglets born alive, still born, and mummified fetuses as well as the piglet BW were recorded. The fat and protein mass of primiparous sows were calculated using the equations of Dourmad et al. (1997).

$$EBW \text{ (kg)} = \text{sow empty live weight estimated from the live weight } (= 0.905 \times BW^{1.013})$$

$$Fat \text{ (kg)} = -26.4 + 0.221 \times (EBW, \text{kg}) + 1.331 \times (\text{Backfat, mm})$$

Protein (kg) = 2.28+0.178×(EBW, kg)
   +0.333×(Backfat, mm)

Blood samples were collected from the jugular vein of sows and piglets with heparinized tubes and centrifuged immediately at 3,000 rpm at 4°C, and then, plasma was separated and stored at –20°C until later analysis. Colostrum and milk were collected from the first and second teats at 24 h and 21 d postpartum after an intravascular injection of 5 IU oxytocin (Komi oxytocin inj. Komipharm International Co., Ltd., Siheung, Korea) in the ear. All samples were stored at –20°C until analysis. Proximate analysis of colostrum and milk samples was conducted using a Milkoscan FT 120 (FOSS Electric, Sungnam, Korea). Plasma free fatty acid (FFA) concentrations were determined according to the colorimetric Acyl-CoA synthetase Acyl-CoA oxidase (ACS-ACOD) method (Shimizu et al., 1979) using a commercial kit (Wako FFA c Kit; Wako chemical, Osaka, Japan). The serum glucose and blood urea nitrogen (BUN) concentrations were analyzed ug a kinetic UV assay (Glucose Hexokinase Kit; UREA/BUN Kit, Roche, Mannheim, Germany). The fatty acid content in colostrum was analyzed on a Agilent 7890 Gas Liquid Chromatograph (Agilent Technologies, Palo Alto, CA, USA) equipped with a flame ionization detector and a SP-2560 (i.d. 100 m×0.25 mm×0.20 μm) film column. Nitrogen was used as carrier gas, injector core temperature was 250°C, detector temperature was 260°C and column temperature was programmed to begin at 170°C, then increase to 250°C and remain at 240°C for 40 min. Chromatography was calibrated with a mixture of 37 different fatty acids (FAME 37; Supelco Inc., Bellefonte, PA, USA), this standard contained fatty acids ranging from C4:0 to C24:1n9, and samples were added 250 μL of internal standard spike solution (Pentadecanoic acid; Sigma, Saint Louis, MO, USA) by the method of AOAC (1990). Colostrum samples were centrifuged at 105,000×g at 4°C for 1 hour and the supernatant was separated and kept frozen until colostrum immunoglobulin G (IgG) and immunoglobulin A (IgA) analysis. Colostrum IgG and IgA concentrations were determined using ELISA according to the manufacture's protocols (Elisa Starter Accessory Package, pig IgG ELISA Quantification Kit, pig IgA ELISA Quantification kit; Bethyl, Montgomery, TX, USA).

Statistical analysis

Data were analyzed by analysis of variance with a completely randomized design using the general linear model procedure implemented in SAS (SAS Institute, 2004). The least squares means were calculated for each independent variable. Orthogonal polynomial contrasts were used to determine the linear and quadratic effects by increasing the dietary energy levels in gestation for all measurements of sows and piglets. Individual sows and their litters were used as the experimental unit. The alpha level used for the determination of significance for all analyses was 0.05 and for the determination of trends was p>0.05 and p<0.10.

RESULTS

The effects of the energy level on BW and BFT are presented in Table 2. Body weight and BW gain in gestation increased as the dietary energy level increased (linear response, p<0.05). Back fat thickness was not affected by dietary treatments during the gestation period, but back fat gain from breeding to 110 d of gestation increased linearly (p<0.05) as the dietary energy level increased. Body length was not affected by dietary treatments. During the lactation period, the BW and BW gain of sows decreased as energy levels increased (linear response, p<0.05), but back fat thickness was not affected by treatments. Consequently, the BW change of gilts from mating to 3 weeks of lactation tended to decrease as energy levels increased (linear response, p = 0.06). During the lactation period, the voluntary feed intake of sows tended to decrease (linear response, p = 0.08) when dietary energy levels increased (Table 2). The weaning to estrus interval (WEI) was not significantly affected by treatments, but the culling rates of sows were highest in the 3,100 kcal/kg ME treatment (Table 2). The estimated protein and fat masses of gilts were calculated based on the BW and BFT (Dourmad et al., 1997). The protein mass and fat mass increased as energy levels increased (linear response, p<0.01, and p<0.05, respectively) during the gestation period (Table 3). During the lactation period, the protein mass and fat mass decreased (linear and quadratic response, p<0.01, and p<0.05, respectively), while the dietary energy level increased (Table 3). There were no significant differences in the total number of pigs born per litter or litter birth weight (Table 4). There were also no significant differences in the fat, protein, lactose, or solid-not-fat (SNF) content of colostrum and milk (Table 5). There was no significant difference in the fatty acid content of colostrum due to dietary energy levels (Table 6). The BUN concentration in the serum of gilts tended to increase with dietary energy level at d 110 of gestation and 24 h postpartum (linear response, p = 0.06, and p = 0.07, respectively) (Table 7). The glucose concentrations tended to increase with dietary energy level (linear response, p = 0.07) at 24 h postpartum, but the FFA concentration was not affected (Table 7). There were no significant differences in the concentration of IgG or IgA in colostrum (Table 8).

DISCUSSION

In the current study, increasing energy intake led to a

Table 2. Effects of the dietary energy levels on the body weight, back-fat, body length, feed intake and WEI of primiparous gestating and lactating sows

Criteria	Treatment				SEM	p-value	
	3,100[1]	3,200	3,300	3,400		Linear	Quadratic
Gestation							
No. of sows	13	13	13	13			
Body weight (kg)							
Breeding (d)	136.00	135.95	135.68	135.35	0.85	0.64	0.96
D 110	178.50	182.18	182.59	185.81	1.42	0.04	0.74
Breeding-110 d (change)	42.50[b]	46.22[ab]	46.91[ab]	50.46[a]	1.08	0.01	0.69
Backfat thickness (mm)							
Breeding (d)	19.32	19.18	18.55	19.08	0.56	0.90	0.55
D 110	20.72	21.27	20.45	23.08	0.70	0.22	0.26
Breeding-110 d (change)	1.40	2.09	1.91	4.00	0.43	0.05	0.32
Body length (cm)							
Breeding (d)	113.90	115.00	114.17	113.62	0.56	0.31	0.40
D 110	121.55	122.18	120.96	121.96	0.58	0.83	0.80
Breeding-110 d (change)	7.65	7.18	6.79	8.35	0.53	0.51	0.33
Lactation							
Body weight (kg)							
24 h post farrowing	164.72	165.68	162.08	169.00	1.28	0.53	0.26
Weaning	174.11[a]	173.45[a]	168.88[b]	167.27[b]	1.55	0.03	0.70
Post farrowing-21 d (gain)	9.39[a]	7.77[a]	6.79[a]	−1.73[b]	1.24	0.01	0.10
Breeding-21 d postpartum (gain)	39.72	37.50	33.79	31.92	1.31	0.06	0.90
Backfat thickness (mm)							
24 h post farrowing	21.06	20.32	20.83	21.77	0.64	0.57	0.49
Weaning	18.94	18.73	19.37	18.27	0.57	0.87	0.77
0-21 d (gain)	−2.12	−1.59	−1.46	−3.5	0.35	0.19	0.08
Breeding -21 d (gain)	0.56	−0.45	0.66	−0.81	0.47	0.91	0.72
Average feed intake (kg/d)	5.99	5.77	5.79	5.32	0.11	0.08	0.16
Weaning to estrus interval	5.29	5.27	5.17	5.67	0.18	0.99	0.20
Culling rate (%)	38	7	7	15	-	-	-

WEI, weaning to estrus interval; SEM, standard error of mean; ME, metabolizable energy.
[1] Energy intake ME kcal/kg.
[a, b] Means with different superscripts indicate significant differences (p<0.05).

higher BW and BW gain from 42.5 to 50.46 kg, which agreed with the results of Noblet et al. (1990), who observed an increased BW of gilts during gestation with higher energy intake. This observation may be attributed to the higher BFT due to the high dietary energy during gestation (Long et al., 2010). In the present study, although gilts consumed an equal amount of amino acids, fat tissue and protein tissue was increased during gestation, but decreased during lactation, which implies that the energy supply is important to maintain adequate BW and BFT for subsequent reproductive cycles. These results are in agreement with previous studies, which demonstrated that increased feed intake during gestation increased sow weight loss during subsequent lactation (Piao et al., 2010). Moreover, Long (2010) demonstrated that providing high-energy feed during gestation increased BW and BFT loss in the lactation period. The BFT loss during lactation was observed in the highest energy level treatment, which may

be attributed to reduced feed intake during lactation due to higher dietary energy levels in gestation.

In this study, WEI was not affected by treatments and the culling rate was highest in the 3,100 kcal/kg ME treatments. Previous studies indicated that increasing body fatness at farrowing is likely to decrease feed intake during lactation (Xue et al., 1997). Low feed intake during lactation could result in higher BW loss and subsequently lead to several common reproductive problems, such as an increased WEI interval (Baidoo et al., 1992), increased anestrus incidence after weaning, and decreased conception rate (Kirkwood and Thacker, 1988). Although feed intake decreased and BW loss increased during lactation with increasing energy levels during gestation, high dietary energy levels did not increase the culling rate in the first parity. However, treatment with 3,100 kcal of ME/kg showed the highest culling rate (38%) due to pregnancy failure and post-weaning anestrus, which suggests that low

Table 3. Effects of the dietary energy level on the estimated protein, fat mass and its gain of primiparous gestating and lactating sows

Criteria	Treatment				SEM	p-value	
	3,100[1]	3,200	3,300	3,400		Linear	Quadratic
Estimated protein mass[2] in gestation (kg)							
Breeding (d)	31.57	31.90	31.59	31.76	0.27	0.93	0.95
d 110	39.80[b]	40.50[ab]	40.54[ab]	41.72[a]	0.39	0.04	0.41
Breeeding-110 d (gain)	8.23[b]	8.60[ab]	8.95[ab]	9.95[a]	0.30	0.01	0.31
Estimated protein mass in lactation (kg)							
24 h postpartum	37.44	37.36	36.91	38.40	0.36	0.47	0.28
d 21	38.34	38.16	37.59	36.95	0.40	0.16	0.70
0-21 d (gain)	0.90[b]	0.80[b]	0.67[b]	−1.46[a]	0.28	0.01	0.03
Breeeding-21 d (gain)	6.76	6.26	5.99	5.19	0.41	0.23	0.71
Estimated fat mass[3] in gestation (kg)							
Breeding (d)	26.94	27.98	27.16	27.71	0.82	0.88	0.94
110 d	39.45	40.57	40.71	43.74	1.08	0.09	0.39
Breeeding-110 d (gain)	12.51	12.59	13.56	16.03	0.80	0.03	0.26
Estimated fat mass in lactation (kg)							
24 h postpartum	36.58	35.79	35.72	38.43	0.99	0.51	0.38
d 21	35.75	35.32	35.21	33.40	0.97	0.44	0.73
0-21 d (gain)	−0.83[b]	−0.47[b]	−0.51[b]	−5.03[a]	0.62	0.02	0.03
Breeding-21 d (gain)	8.81	7.35	8.05	5.70	1.12	0.48	0.75

SEM, standard error of mean; ME, metabolizable energy.
[1] Energy intake ME kcal/kg.
[2] Prediction equation from Dourmad et al. (1997): $2.28+0.178\times(EBW, kg)+0.333\times(Backfat, mm)$.
[3] Prediction equation from Dourmad et al. (1997): $-26.4+0.221\times(EBW, kg)+1.331\times(Backfat, mm)$
[a, b] Means with different superscripts indicate significant differences (p<0.05).

energy intake during gestation may increase the risks of being culled due to pregnancy failure (Kongsted, 2005). The pregnancy rate and post-weaning anestrus may be correlated with some hormones, and the dietary energy levels differentially affect the release of some reproductive hormones (Kemp et al., 1995). Whitley (2002) reported that insulin targets during reproduction had little connection to nutrition, the body condition, and other management factors. These results indicated that dietary energy levels during pregnancy for optimal longevity are not easy to establish. Further study is needed to clearly demonstrate a possible correlation between dietary energy levels during gestation and reproductive hormones. WEI is delayed by low plasma insulin concentrations during lactation, and the

Table 4. Effects of the dietary energy level on the reproductive performance and growth of progeny of primiparous sows

Criteria	Treatment				SEM	p-value	
	3,100[1]	3,200	3,300	3,400		Linear	Quadratic
Reproductive performance							
Total born	12.11	13.00	12.33	12.00	0.33	0.36	0.66
Born alive	11.44	12.27	11.75	11.54	0.33	0.42	0.76
Still birth	0.67	0.73	0.58	0.46	0.08	0.61	0.56
After cross-fostering	11.33	11.18	11.25	11.62	0.24	0.65	0.31
Weaning pigs	11.00[ab]	10.27[b]	10.58[ab]	11.31[a]	0.24	0.66	0.02
Litter weight on lactation (kg)							
Litter birth weight	14.47	16.03	15.42	15.29	0.44	1.00	0.62
After cross-fostering	14.29	14.30	14.40	14.59	0.43	0.85	0.77
Weaning litter weight	58.16	56.82	57.49	63.55	1.36	0.31	0.19
Piglet weight on lactation (kg)							
Piglet birth weight	1.19	1.25	1.25	1.30	0.03	0.23	0.96
After cross-fostering	1.26	1.28	1.27	1.26	0.03	0.94	0.82
Weaning piglet weight	5.35	5.56	5.47	5.64	0.09	0.10	0.60

SEM, standard error of mean; ME, metabolizable energy.
[1] Energy intake ME kcal/kg.
[a,b] Means with different superscripts indicate significant differences (p<0.05).

Table 5. Effects of the dietary energy level in gestating sows on the components of colostrum and milk in primiparous lactating sows.

Items	Treatment				SEM	p-value	
	3,100[1]	3,200	3,300	3,400		Linear	Quadratic
Chemical composition of colostrum at 24 h postpartum (%)							
Fat	6.44	7.40	7.33	7.67	0.36	0.69	0.24
Protein	8.26	8.15	7.62	7.30	0.34	0.10	0.47
Lactose	4.15	4.04	4.21	4.31	0.07	0.12	0.22
Solids-not-fat	12.79	12.55	12.25	11.96	0.27	0.11	0.56
Chemical composition of sow milk at 21 d postpartum (%)							
Fat	5.56	7.01	6.58	6.86	0.21	0.10	0.06
Protein	4.19	4.58	4.66	4.52	0.07	0.20	0.08
Lactose	6.08	5.93	5.85	5.81	0.06	0.12	0.41
Solids-not-fat	10.50	10.64	10.72	10.50	0.05	0.91	0.13

SEM, standard error of mean; ME, metabolizable energy.

[1] Energy intake ME kcal/kg.

concentration is increased by glucose but lowered by FFA concentrations in the plasma of weaning sows (Armstrong et al., 1986). In this study, the plasma glucose and FFA concentrations of sows at weaning were not affected by treatments, indicating that dietary energy levels did not affect WEI.

NRC (2012) suggested dietary energy levels for 140 kg BW gilts during gestation between 6,678 kcal of ME/kg to 7,932 kcal of ME/kg daily. However, our results suggested that litter size was not affected by increasing dietary energy, which may be considered to be a balance between gestation preparation and early gestation. High energy levels had

Table 6. Effects of dietary energy level in gestating sows on the fatty acids composition of colostrum in primiparous lactating sows (mg/g)

Items	Treatment				SEM[2]	p-value	
	3,100[1]	3,200	3,300	3,400		Linear	Quadratic
Fatty acid composition of colostrum at 24h postpartum (mg/g)							
Saturated							
Myristic (C14:0)	0.53	0.76	0.80	0.47	0.076	0.76	0.04
Palmitic (C16:0)	7.94	9.51	9.80	6.47	0.983	0.63	0.21
Heptadecanioic (C17:0)	0.16	0.18	0.17	0.13	0.014	0.44	0.35
Stearic (C18:0)	2.27	2.59	2.45	1.90	0.274	0.61	0.43
Arachidic (C20:0)	0.06	0.07	0.06	0.05	0.007	0.57	0.38
Henicosanoic (C21:0)	0.08	0.09	0.12	0.06	0.010	0.75	0.13
Behenic (C22:0)	0.05	0.05	0.05	0.04	0.004	0.49	0.35
Monounsaturated							
Palmitoleic (C16:1)	1.19	1.77	2.24	1.17	0.218	0.80	0.05
Nervonic (C24:1)	0.07	0.09	0.09	0.06	0.007	0.52	0.18
Oleic (C18:1n9c)	13.63	16.34	18.40	12.38	1.929	0.92	0.30
Eicosanoic (C20:1)	0.13	0.16	0.16	0.11	0.021	0.78	0.37
Erucic (C22:1n9)	0.03	0.04	0.03	0.03	0.004	0.78	0.48
Polyunsaturated							
Linoleic (C18:2n6c)	5.88	6.73	7.52	4.64	0.681	0.63	0.19
Cis-11,14-Eicosadienoic (C20:2)	0.17	0.19	0.21	0.14	0.021	0.73	0.29
γ-linolenic (C18:3n6)	0.13	0.15	0.17	0.08	0.019	0.58	0.20
α-linolenic (C18:3n3)	0.27	0.32	0.38	0.23	0.033	0.87	0.15
Arachidonic (C20:4n6)	0.40	0.37	0.51	0.29	0.046	0.67	0.36
Cis-11,14,17-Eicosatrienoic (C20:3n3)	0.03	0.03	0.04	0.03	0.004	0.99	0.31
Cis-5,8,11,14,17Ecicosapentaenoic (C20:5n3)	0.03	0.03	0.04	0.02	0.003	0.99	0.29
Cis-4,7,10,13,16,19-Docosahexaenoic (C22:6n3)	0.02	0.02	0.03	0.02	0.003	0.86	0.55

SEM, standard error of mean; ME, metabolizable energy.

[1] Energy intake ME kcal/kg.

Table 7. Effects of the dietary energy level on the BUN, glucose and free fatty acid concentrations in the plasma of primiparous sows

Criteria	Treatment				SEM	p-value	
	3,100[1]	3,200	3,300	3,400		Linear	Quadratic
BUN (mg/dL)							
Breeding (d)	10.60	10.60	10.60	10.60			
110 d	9.06	8.72	7.08	7.82	0.37	0.07	0.39
24 h postpartum	11.68	10.48	9.06	9.50	0.59	0.06	0.36
Weaning	20.76	19.70	17.12	18.68	0.83	0.24	0.44
Glucose (mg/dL)							
Breeding (d)	76.20	76.20	76.20	76.20			
110 d	74.00	81.60	69.80	73.00	2.56	0.39	0.57
24 h postpartum	87.50	88.40	89.25	95.20	1.77	0.07	0.37
Weaning	82.50	82.00	81.40	82.80	3.37	0.99	0.90
Free fatty acid (μEq/L)							
Breeding (d)	140.60	140.60	140.60	140.60			
110 d	145.20	146.00	189.40	183.20	12.10	0.22	0.90
24 h postpartum	257.60	211.00	232.20	292.60	21.03	0.56	0.27
Weaning	182.20	158.00	171.40	170.20	12.81	0.86	0.69

BUN, blood urea nitrogen; SEM, standard error of mean; ME, metabolizable energy.
[1] Energy intake ME kcal/kg.

positive effects on the ovulatory rate, but high energy levels might decrease embryonic survival after breeding (Jindal et al., 1996), while low energy supplies from day 3 after mating until day 15 do not affect embryo survival.

There are many of scientific studies assessing the effects of increased feed or energy intake in gestating sows on piglet birth weight (NRC, 1998). Daily energy requirements during gestation include maintenance for the sow maternal gain as well as uterine growth. Increased energy intake during late gestation can positively affect fetal growth (NRC, 2012). In this experiment, there was no treatment effect on litter birth weight and individual piglet birth weight, which is consistent with the results from Long (2010), who demonstrated that the average piglet BW at farrowing was not affected by energy levels in gestation diets. Increased feed intake during gestation also did not increase the litter weight or individual piglet weight (Piao et al., 2010). This result demonstrates that increased feed intake during gestation could increase BW gain and BFT in gestating gilts. However, increased BW and backfat did not increase litter weight or individual piglet weight.

Yang (2008) reported that there was no change in colostrum composition when the energy levels increased from 13.7 to 14.2 MJ of ME/kg during gestation. Our study suggests that there were no significant differences in the chemical composition (fat, protein, lactose and SNF) of sow colostrum and milk, which is consistent with the results from Willams (1985), who confirmed that the chemical composition of colostrum and milk was not affected by dietary energy levels during gestation because sows mobilized their internal reserves to compensate for deficient nutrients. In the present study, the composition of fatty acids in colostrum did not show any changes due to the dietary energy level of the gestation diets. The fatty acid composition of colostrum is affected by dietary fat levels (Christon et al., 1999). Several studies have investigated how the composition of fatty acids in colostrum is affected by energy levels in gestation diets, and the mechanism is not yet clearly understood. In this experiment, the composition of fatty acids did not show any changes due to the different treatments, but the differences in energy levels among treatments simply may not have been high enough to affect the composition of fatty acids in colostrum.

It is well known that BUN is directly related to protein intake and inversely related to protein quality (Hahn et al., 1995) and the retention of dietary nitrogen in the body (Whang and Easter, 2000). In the present study, the BUN concentration in serum tended to decrease with the increasing energy level at 110 days of gestation and 24 hours post-farrowing. Ruiz (1971) reported that the BUN

Table 8. Effects of the dietary energy level in gestating sows on immune parameters of colostrum[1] in primiparous lactating sows

Items	Treatment				SEM	p-value	
	3,100[2]	3,200	3,300	3,400		Linear	Quadratic
IgG (mg/mL)	5.44	5.58	5.36	4.95	0.60	0.81	0.86
IgA (mg/mL)	5.58	4.03	5.46	5.25	0.67	0.94	0.58

SEM, standard error of mean; ME, metabolizable energy.
[1] 24 hours postpartum. [2] Energy intake ME kcal/kg.

concentration was higher in swine fed low-energy diets compared to pigs fed high-energy diets, which may suggest that energy intake in sows is affected by protein metabolism during gestation. The glucose concentration influences insulin secretion (Quesnel and Prunier, 1998), and insulin can reduce the oxidation of FFA for the production of energy (Gamble and Cook, 1985). The glucose concentration increased with higher energy levels 24 hours post-farrowing in the current study, whereas FFAs were not affected by treatments. Xue (1997) demonstrated that increased energy levels during gestation could decrease glucose utilization and subsequently decrease feed intake during lactation. Piao (2010) also suggested that increased feed intake in gestating gilts may cause sows to become insensitive to insulin, thereby exhibiting a smaller response in glucose clearance and decreased feed intake during lactation. Therefore, it can be assumed that the effect of insulin on feed intake in lactation might depend on the body condition of sows and metabolites of glucose, which may explain our result that BW and BFT decreased with energy levels during lactation.

A greater intake of immunoglobulins in colostrum may increase immune function in nursing pigs. Some previous research reported that polyunsaturated fatty acids (PUFA) can influence production of immunoglobulins in mammals. Mitre et al. (2005) reported that supply of shark-liver oil to sows from d 80 of gestation to farrowing could be increased IgG concentrations in the colostrum, but not IgA. Mateo et al. (2009) reported increased IgG concentrations in the colostrum and milk of sows fed a fish product. Clearly, dietary n-3 and n-6 FA's are involved in immune and inflammatory processes.

In the present study, the concentrations IgG and IgA in colostrum was not affected by the tallow in gilt diets, which has lower PUFA than other energy source.

IMPLICATIONS

NRC (2012) suggested that the energy requirement of the gestation gilt should be between 6,678 and 7,932 kcal of ME/d. Similarly, our results suggested that 3,100 kcal of ME/kg is not enough to maintain the reproductive performance for gilts during gestation with 2 kg feed daily. Gilts in the treatment 3,400 kcal of ME/kg tended to have a higher weaning number of piglets, but bodyweight and backfat loss were higher than other treatments during lactation. Consequently, an adequate energy requirement of gestating gilts is 6,400 kcal of ME/d.

ACKNOWLEDGMENTS

This work was carried out with the support of "Cooperative Research Program for Agriculture Science and Technology Development (Project No. PJ009226012013)" Rural Development Administration, Republic of Korea.

REFERENCES

Almeida, F. R. C. L., R. N. Kirkwood, F. X. Aherne, and G. R. Foxcroft. 2000. Consequences of different patterns of feed intake during estrous cycle in gilts on subsequent fertility. J. Anim. Sci. 78:1566-1563.

AOAC. 1990. Official Methods of Analysis. 15th ed. National Research Council, Association of Official Analytical Chemists, Arlington, VA, USA.

Armstrong, J. D., J. H. Britt, and R. R. Kraeling. 1986. Effect of restriction of energy during lactation on body condition, energy metabolism, endocrine changes and reproductive performance in primiparous sows. J. Anim. Sci. 63:1915-1925.

Baidoo, S. K., F. X. Aherne, R. N. Kirkwood, and G. R. Foxcroft. 1992. Effect of feed intake during lactation and after weaning on sow reproductive performance. Can. J. Anim. Sci. 72:911-917.

Boyd, R. D., K. J. Touchette, G. C. Castro, M. E. Johnston, K. U. Lee, and In K. Han. 2000. Recent advances in amino acid and energy nutrition of prolific sows. Asian Australas. J. Anim. Sci. 13:1638-1652.

Christon, R., G. Saminadin, H. Lionet, and B. Racon. 1999. Dietary fat and climate alter food intake, performance of lactating sows and their litters and fatty acid composition of milk. Anim. Sci. 69:353-365.

Dourmad, J. Y., M. Etienne, J. Noblet, and D. Causeur. 1997. Prediction of the chemical composition of the reproductive sows from their bodyweight and backfat depth- Utilization for determining the energy rcordance. J. Rech. Porc. France. 29:255-262.

Gamble, M. S. and G. A. Cook. 1985. Alteration of the apparent Ki of carnitine palmitoyltransferase for malonyl-CoA by the diabetic state and reversal by insulin. J. Biol. Chem. 260:9516-9519.

Hahn, J. D., R. R. Biehl, and D. H. Baker. 1995. Ideal digestible lysine for early- and late-finishing swine. J. Anim. Sci. 73:773-784.

Jang, Y. D., S. K. Jang, D. H. Kim, H. K. Oh, and Y. Y. Kim. 2014. Effects of dietary protein levels for gestating gilts on reproductive performance, blood metabolites and milk composition. Asian Australas. J. Anim. Sci. 27:83-92.

Jindal, R., J. R. Cosgrove, F. X. Aherne, and G. R. Foxcroft. 1996. Effect of nutrition on embryonal mortality in gilts: Association with progesterone. J. Anim. Sci. 74:620-624.

Kemp, B., N. M. Soede, F. A. Helmond, and M. W. Bosch. 1995. Effects of energy source in the diet on reproductive hormones and insulin during lactation and subsequent estrus in multiparous sows. J. Ahim. Sci. 73:3022-3029.

Kongsted, A. G. 2005. A review of the effect of energy intake on pregnancy rate and litter size–discussed in relation to group-housed non-lactating sows. Livest. Prod. Sci. 97:13-26.

Kirkwood, R. N. and P. A. Thacker. 1988. Nutritional factors affecting embryonic mortality in pigs. Pig News Info. 9:15-21.

Long, H. F., W. S. Ju, L. G. Piao, and Y. Y. Kim. 2010. Effect of dietary energy levels of gestating sows on physiological parameters and reproductive performance. Asian Australas. J. Anim. Sci. 23:1080-1088.

Mitre, R., M. Etienne, S. Martinais, H. Salmon, P. Allaume, P. Legrand, and A. B. Legrand. 2005. Humoral defence improvement and haematopoiesis stimulation in sows and offspring by oral supply of shark-liver oil to mothers during gestation and lactation. Br. J. Nutr. 94:753-762.

Mateo, R. D., J. A. Carroll, Y. Hyun, S. Smith, and S. W. Kim. 2009. Effect of dietary supplementation of n-3 fatty acids and elevated concentrations of dietary protein on the performance of sows. J. Anim. Sci. 87:948-959.

Noblet, J., J. Y. Dourmad, and M. Etienne. 1990. Energy utilization in pregnant and lactating sows: Modeling of energy requirements. J. Anim. Sci. 68:562-572.

NRC. 1998. Nutrient Requirements of Swine, 10th ed. Natl. Acad. Press, Washington, DC, USA.

NRC. 2012. Nutrient Requirements of Swine, 11th ed. Natl. Acad. Press, Washington, DC, USA.

Piao, L. G., W. S. Ju, H. F. Long, and Y. Y. Kim. 2010. Effects of various feeding methods for gestating gilts on reproductive performance and growth of their progeny. Asian Australas. J. Anim. Sci. 23:1354-1363.

Quesnel, H. and A. Prunier. 1998. Effect of insulin administration before weaning on reproductive performance in feed restricted primiparous sows. Anim. Reprod. Sci. 51:119-129.

Ruiz, M. E., R. C. Ewan, and V. C. Speer. 1971. Serum metabolites of pregnant and hysterectomized gilts fed two levels of energy. J. Anim. Sci. 32:1153-1159.

SAS Institute Inc. 2004. SAS/STAT User's Guide, SAS Institute Inc., Cary, NC, USA.

Shimizu, S., K. Inoue, Y. Tani, and H. Yamada. 1979. Enzymatic microdetermination of serum free fatty acids. Analytical biochemistry. 98:341-345.

Whang, K. Y. and R. A. Easter. 2000. Blood urea nitrogen as an index of feed efficiency and lean growth potential in growing finishing-swine. Asian Australas. J. Anim. Sci. 13:811-816.

Whitley, N. C., M. Thomas, J. L. Ramirez, A. B. Moore, and N. M. Cox. 2002. Influences of parity and level of feed intake on reproductive response to insulin administration after weaning in sows. J. Anim. Sci. 80:1038-1043.

Williams, I. H., W. H. Close, and D. J. A. Cole. 1985. Strategies for sow nutrition: Predicting the response of pregnant animals to protein and energy intake. In: Recent Advances in Animal Nutrition (Eds. W. Haresign and D. J. A. Cole). Butterworth, London, UK. pp. 133-147.

Xue, J. L., Y. Koketsu, G. D. Dial, J. E. Pettigrew, and A. Sower. 1997. Glucose tolerance, luteinizing hormone release, and reproductive performance of first-litter sows fed two levels of energy during gestation. J. Anim. Sci. 75:1845-1852.

Yang, Y., S. Heo, Z. Jin, J. Yun, P. Shinde, J. Choi, B. Yan, and B. Chae. 2008. Effects of dietary energy and lysine intake during late gestation and lactation on blood metabolites, hormones, milk composition and reproductive performance in multiparous sows. Arch. Anim. Nutr. 62:10-21.

16

Nutritional Performance of Cattle Grazing during Rainy Season with Nitrogen and Starch Supplementation

Ísis Lazzarini, Edenio Detmann*, Sebastião de Campos Valadares Filho, Mário Fonseca Paulino,
Erick Darlisson Batista, Luana Marta de Almeida Rufino,
William Lima Santiago dos Reis, and Marcia de Oliveira Franco

Department of Animal Science, Universidade Federal de Viçosa, Viçosa, MG CEP 36570-000, Brazil

ABSTRACT: The objective of this work was to evaluate the effects of supplementation with nitrogen and starch on the nutritional performance of grazing cattle during the rainy season. Five rumen cannulated Nellore steers, averaging 211 kg of body weight (BW), were used. Animals grazed on five signal grass paddocks. Five treatments were evaluated: control (forage only), ruminal supplementation with nitrogen at 1 g of crude protein (CP)/kg BW, ruminal supplementation with starch at 2.5 g/kg BW, supplementation with nitrogen (1 g CP/kg BW) and starch (2.5 g/kg BW), and supplementation with nitrogen (1 g CP/kg BW) and a mixture of corn starch and nitrogenous compounds (2.5 g/kg BW), thereby resulting in an energy part of the supplement with 150 g CP/kg of dry matter (DM). This last treatment was considered an additional treatment. The experiment was carried out according to a 5 ×5 Latin square design following a 2×2+1 factorial arrangement (with or without nitrogen, with or without starch, and the additional treatment). Nitrogen supplementation did not affect (p>0.10) forage intake. Starch supplementation increased (p<0.10) total intake but did not affect (p<0.10) forage intake. There was an interaction between nitrogen and starch (p<0.10) for organic matter digestibility. Organic matter digestibility was increased only by supplying starch and nitrogen together. Nitrogen balance (NB) was increased (p<0.10) by the nitrogen supplementation as well as by starch supplementation. Despite this, even though a significant interaction was not observed (p>0.10), NB obtained with nitrogen plus starch supplementation was greater than NB obtained with either nitrogen or starch exclusive supplementation. Supplementation with starch and nitrogen to beef cattle grazing during the rainy season can possibly improve digestion and nitrogen retention in the animal.. (**Key Words:** Beef Cattle, Digestibility, Intake, Metabolism, Nellore Steers, Nitrogen Retention)

INTRODUCTION

In the rainy season, tropical forages are not considered deficient in crude protein (CP). However, benefits of supplementation of grazing cattle with nitrogenous compounds in the rainy season have been reported (Zervoudakis et al., 2008; Costa et al., 2011a; Figueiras et al., 2015). However, the effects of supplemental nitrogen in the rainy season have been considered to be mainly metabolic when contrasted to the dry season, when the nitrogen has the main role of improving the microbial activity in the rumen (Detmann et al., 2014a).

Recent studies conducted in the tropics have showed that nitrogen supplementation can improve the concentration of ruminal ammonia nitrogen (RAN) (Costa et al., 2011a) and decrease the participation of recycled nitrogen on total nitrogen assimilated by ruminal microorganisms (Batista et al., 2016). Additionally, nitrogen supplementation can improve the total availability of nitrogen for anabolic purposes by direct supply or by decreasing the muscle protein breakdown (Detmann et al., 2014a). Thus, the use of metabolizable energy from the forage could be increased by supplementation with nitrogen.

However, the inclusion of additional energy resources, mainly in the form of non-fibrous carbohydrates (NFC), could adversely affect the performance of ruminant animals

* Corresponding Author: Edenio Detmann.
E-mail: detmann@ufv.br

in the rainy season because it could negatively affect the energy-protein balance in the diet, generating dietary and metabolic constraints that would lead to a decrease in nitrogen retention in the animal (Costa et al., 2011b; Detmann et al., 2014a).

Nevertheless, information related to the study of the inclusion of additional nitrogen and energy resources in the use of basal forage for beef cattle grazing during the rainy season is still scarce in tropical conditions. Thus, the objective of this work was to evaluate the effects of supplementation with nitrogen and starch on the nutritional performance of grazing cattle during the rainy season.

MATERIAL AND METHODS

The experiment was performed from January to March 2009 during the rainy season in Viçosa, Minas Gerais, Brazil (20°45′ S, 42°52′ W). The experimental period presented a total rainfall and average temperature of 677 mm and 22.5°C, respectively. Five Nellore steers, averaging 211±17 kg of body weight (BW), were surgically fitted with ruminal cannulae and kept under continuous grazing in individual signal grass (*Brachiaria decumbens*) paddocks of approximately 0.34 ha. Put-and-take animals were not used in this experiment. All surgical and animal care procedures were approved by the University Animal Care Committee. Ruminal fistulae and their surrounding areas were cleaned routinely during the experiment. Water and a mineral mixture were available to the steers at all times. A corral annexed to the paddocks was used for management of animals and sampling. The distribution of animals to the paddocks was randomly performed at the beginning of each experimental period.

Five treatments were evaluated: control (without supplementation), supplementation with nitrogen at 1 g of crude protein (CP)/kg BW, supplementation with 2.5 g of starch (Amisol 3408, CornProducts Co., Santana do Parnaíba, SP, Brazil)/kg BW, supplementation with nitrogen (1 g CP/kg BW) and starch (2.5 g/kg BW), and supplementation with nitrogen (1 g CP/kg BW) and a mixture of corn starch and nitrogenous compounds (2.5 g/kg BW), thereby resulting in an energy part of the supplement with 150 g CP/kg of dry matter (DM). This last treatment was considered an additional treatment where it was objectified to evaluate if an additional nitrogen supply could be case any benefit on animal performance.

A mixture of urea, ammonium sulfate, and albumin (Maximus, Arve Alimentos Co., Viçosa, MG, Brazil) was used as a source of nitrogenous compounds at a ratio of 4.5:0.5:1.0. The supplement amount was calculated based on BW at the beginning of each experimental period and placed in two portions of equal weight in the rumen of the animals daily at 0600 h and 1800 h during the experimental period.

The nitrogen supplement lacked carbohydrate content, thereby allowing the supplementation effects with nitrogen to be evaluated without any supplementary source of fiber or energy interfering with the measurements. Albumin was included in the supplement to meet the microbial requirements for true degradable protein and to supply essential substrates, such as branched chain volatile fatty acids.

The experiment consisted of five 15-day experimental periods. The first five days of each experimental period were used to adapt the animals to the supplements. In order to minimize the possible effects of paddocks on experimental treatments, the animals were rotated among the five paddocks every experimental period.

Available forage was estimated by cutting five square areas (0.5×0.5 m) in each paddock that were randomly chosen on the first day of each experimental period. Samples were oven-dried (60°C), processed in a knife mill (1-mm), and analyzed for DM content (method INCT-CA G-003-1; Detmann et al., 2012). Average forage availability during the experiment was 14.8±1.39 ton DM/ha.

Evaluation of the consumed forage was performed by hand-plucked sampling on the first, fourth, and seventh days of each experimental period. Samples were oven-dried (60°C) and processed in a knife mill (1- and 2-mm). Pooled samples were produced for each paddock and experimental period.

Fecal excretion was estimated using titanium dioxide as an external marker. The marker was infused (20 g/d) into the rumen of each animal at 1200 h from the first to the eighth day of each experimental period. Fecal sampling started on sixth day (Titgemeyer et al., 2001) and samples were taken from the rectum of each animal according to the following schedule: sixth day—0800 h and 1400 h, seventh day—1000 h and 1600 h, and eighth day—1200 h and 1800 h. The fecal samples were oven-dried (60°C) and processed in a knife mill (1- and 2-mm). Next, pooled samples were produced for each animal and experimental period.

To evaluate the RAN concentration and rumen pH, samples of ruminal fluid were taken on the ninth day of each experimental period at 0600 h, 1200 h, 1800 h, and 2400 h. Samples were collected manually from the liquid-solid interface of the rumen mat, filtered through a triple layer of cheesecloth, and submitted for a pH assessment using a digital potentiometer (TEC-3P-MP, Tecnal, Piracicaba, SP, Brazil). An aliquot of 40-mL was subsequently separated, fixed with 1 mL of H_2SO_4 (1:1), and frozen (−20°C) for a posterior RAN analysis.

On the 15th day of each experimental period, urine spot samples were obtained before (0600 h) and approximately six hours (1200 h) after the morning supplementation. Samples were filtered through cheesecloth, and a 10-mL aliquot was separated, diluted with 40 mL H_2SO_4 (0.036 N), and frozen ($-20°C$). Concomitantly to the urine samples, blood was collected from the jugular vein of each animal using test tubes containing separator gel and a coagulation accelerator (BD Vacutainer SST II Advance, São Paulo, SP, Brazil). Samples were centrifuged at $2,700×g$ for 20 minutes to obtain the serum, which was frozen ($-20°C$).

Forage and fecal samples (processed to pass through a 1-mm screen sieve) were analyzed for DM, (method INCT-CA G-003-1) organic matter (OM; method INCT-CA M-001/1), CP (method INCT-CA N-001/1), and neutral detergent fiber corrected for ash and protein (NDFap; using thermostable α-amylase and omitting the use of sodium sulfite; methods INCT-CA F-002/1, INCT-CA M-002/1 and INCT-CA N-004/1) contents, according to the standard methods for feed analysis of the Brazilian National Institute of Science and Technology in Animal Science (INCT-CA; Detmann et al., 2012). Supplement samples were analyzed for DM, OM , and CP contents (Table 1).

The fecal samples were evaluated for titanium dioxide content according to a colorimetric method (method INCT-CA M-007/1; Detmann et al., 2012). The fecal excretion of DM was obtained as the ratio of the daily dose to the fecal content of the marker. The estimates of forage intake were obtained using indigestible NDF (iNDF) as an internal marker. The iNDF contents in feces and forage were estimated in the samples processed to pass a 2-mm screen sieve using a 288-hours *in situ* incubation procedure (method INCT-CA F-008/1; Detmann et al., 2012).

The RAN content in the ruminal fluid samples was evaluated using an indophenol colorimetric method (method INCT-CA N-006/1; Detmann et al., 2012). The concentrations obtained at the different sampling times were pooled by animal and period to obtain a single value that represented the average daily RAN concentration. Ruminal pH values were combined in a similar manner.

Blood serum samples, after thawing, were analyzed for

urea concentration (enzyme-colorimetric method, Bioclin K047, Belo Horizonte, MG, Brazil).

After thawing, urine samples were pooled by animal and experimental period and then analyzed for creatinine contents estimated by the modified Jaffé method (Bioclin K016-1), total nitrogen content (method INCT-CA N-001/1; Detmann et al., 2012), uric acid (LCF enzymatic-colorimetric method, Human 10687, Itabira, MG, Brazil), and allantoin (colorimetric method; Chen and Gomes, 1992).

Total urinary volume was estimated using the ratio of creatinine excretion per unit of BW to its concentration in the urine (Chizzotti et al., 2006). Excretion of purine derivatives was calculated from the sum of the amounts of allantoin and uric acid excreted in the urine. From this finding, the absorbed purines were calculated by the following equation (Verbic et al., 1990):

$$AP = \frac{PD - 0.385 \times BW^{0.75}}{0.85} \qquad (1)$$

where AP is the amount of absorbed purines (mmol/d), PD is the amount of excreted purine derivatives (mmol/d), 0.85 is the recovery of absorbed purines as purine derivatives in the urine (mmol/mmol), and 0.385 is the excretion of endogenous purine derivatives in the urine per unit of metabolic size (mmol).

The microbial synthesis of nitrogenous compounds in the rumen (NMIC, g/d) was estimated according to Chen and Gomes (1992) as follows:

$$NMIC = \frac{70 \times AP}{0.83 \times R \times 1,000} \qquad (2)$$

where R is the $N_{RNA}:N_{TOTAL}$ ratio in the microorganisms (mg/mg), 70 is the nitrogen content in purines (mg/mol), and 0.83 is the intestinal digestibility of the microbial purines (mg/mg). It was adopted R = 0.176 (Valadares Filho, 1995).

The experiment was analyzed according to a 5×5 Latin square design following a 2×2+1 factorial arrangement (with or without nitrogenous compounds and with or without starch, plus an additional treatment). After the analysis of variance, the treatments were compared using contrasts (Table 2). All statistical procedures were performed using the GLM procedure of SAS (SAS Institute Inc., Cary, NC, USA, version 9.2) (α = 0.10). One animal developed problems unrelated to the experimental treatments during one experimental period; therefore, there was a loss of one experimental unit with respect to the variables associated with the voluntary intake and digestibility.

Table 1. Chemical composition of forage and supplements

Item	Supplements				Forage[1]
	NIT	STA	NIT+STA	ADI	
DM (% as fed)	96.0	88.3	92.1	92.2	19.9±0.6
OM (% of DM)	98.7	99.9	98.7	98.9	89.7±1.3
CP (% of DM)	204.3	0.0	58.2	65.8	13.5±1.4
NDFap (% of DM)	-	-	-	-	55.0±2.7
NDIP % of CP	-	-	-	-	17.2±4.2
iNDF (% of DM)	-	-	-	-	17.7±1.8

NIT, nitrogen; STA, starch; CP, crude protein; ADI, nitrogen+starch mixture with 15% of CP; DM, dry matter; OM, organic matter; NDFap, neutral detergent fiber corrected for ash and protein; NDIP, neutral detergent insoluble protein; iNDF, indigestible NDF.

[1] Mean±standard error (hand-plucked samples).

Table 2. Distribution of coefficients employed in the contrasts among treatments

Contrast[2]	Treatments[1]				
	CON	NIT	STA	NIT+STA	ADI
	---------------------- Orthogonal set ----------------------				
N	+1	−1	+1	−1	0
S	+1	+1	−1	−1	0
N×S	+1	−1	−1	+1	0
	---------------------- Additional contrast ----------------------				
A	0	0	0	+1	−1

[1] CON, control (without supplementation); NIT, supplementation with nitrogen; STA, supplementation with starch; NIT+STA, supplementation with nitrogen and starch; ADI, supplementation with nitrogen and a mixture of starch and nitrogen, which presented 150 CP/kg DM (additional treatment).

[2] N, effect of supplementation with nitrogen; S, effect of supplementation with starch; N×S, interaction between nitrogen and starch; A, additional comparison between treatments NIT+STA and ADI.

RESULTS

There was no ($p>0.10$) interaction between nitrogen and starch or the effect of the additional nitrogen supply to animals supplemented with nitrogen and starch in any of the variables associated with the voluntary intake (Table 3). Nitrogen supplementation increased ($p<0.01$) CP intake but did not affect ($p>0.10$) forage intake. Starch supplementation increased ($p<0.02$) total DM intake but did not affect ($p>0.10$) forage intake. In addition, starch supplementation positively affected ($p<0.02$) CP and digestible OM (DOM) intake (Table 3).

There were interactions of supplementation with nitrogen and starch on OM ($p<0.02$) and NDFap ($p<0.07$) digestibilities (Table 4). The study of this effect indicated that OM digestibility was only increased ($p<0.10$) when nitrogen and starch were provided together. Evaluation of the interaction on NDFap digestibility indicated that solely starch supplementation depressed ($p<0.10$) fiber digestibility. However, this effect was avoided ($p<0.10$) when nitrogen was provided along with starch (Table 5).

Total CP digestibility increased ($p<0.01$) when nitrogen compounds were supplied; however, it was not affected ($p>0.10$) by starch supplementation (Table 5). Similar to voluntary intake, none of the variables associated with total digestibility demonstrated an effect ($p>0.10$) of the additional nitrogen supply to animals receiving nitrogen and starch supplementation (Table 4).

There was no effect of treatment ($p>0.10$) on ruminal pH, which had an average value of 6.53 (Table 6). Additionally, there was an interaction ($p<0.02$) between nitrogen and starch on the RAN concentration (Table 6). In this sense, nitrogen supplementation increased ($p<0.10$) RAN concentration either with or without starch. However,

Table 3. Least squares means of intake (kg/d) of dry matter (DM), DM from forage (DMF), organic matter (OM), crude protein (CP), neutral detergent fiber corrected for ash and protein (NDFap), digested OM (DOM), digested NDFap (DNDF), and indigestible neutral detergent fiber (iNDF) according to the treatments

Item	Treatments[1]					SEM	p-value[2]			
	CON	NIT	STA	NIT+STA	ADI		N	S	N×S	A
DM	4.39	4.02	5.20	5.34	5.38	0.35	0.750	0.013	0.507	0.930
DMF	4.39	3.93	4.69	4.61	4.66	0.35	0.479	0.207	0.611	0.924
OM	3.98	3.59	4.69	4.87	4.87	0.32	0.747	0.012	0.414	0.984
CP	0.56	0.72	0.65	1.03	1.13	0.06	0.001	0.010	0.140	0.266
NDFap	2.46	2.13	2.59	2.60	2.48	0.20	0.477	0.178	0.444	0.664
DOM	2.15	1.88	2.41	2.96	2.99	0.22	0.556	0.015	0.109	0.919
DNDF	1.55	1.30	1.47	1.68	1.52	0.16	0.896	0.387	0.200	0.485
iNDF	0.74	0.71	0.82	0.77	0.75	0.06	0.507	0.284	0.921	0.760

[1] CON, control; NIT, nitrogen; STA, starch; NIT+STA, nitrogen+starch; ADI, nitrogen+starch mixture with 15% of CP.

[2] N, effect of nitrogen supplementation; S, effect of starch supplementation; N×S, interaction between nitrogen and starch supplementation; A, additional contrast to verify the effect of additional supply of nitrogen in the supplementation of nitrogen plus starch.

Table 4. Least squares means (%) of total digestibility of organic matter (OM), crude protein (CP), and neutral detergent fiber corrected for ash and protein (NDFap) according to the treatments

Item	Treatments[1]					SEM	p-value[2]			
	CON	NIT	STA	NIT+STA	ADI		N	S	N×S	A
OM	53.5	52.4	51.4	60.4	61.5	1.60	0.040	0.109	0.013	0.640
CP	52.6	67.6	48.0	72.1	75.6	3.11	<0.001	0.984	0.196	0.445
NDFap	63.0	61.1	56.9	63.7	61.2	2.02	0.267	0.416	0.065	0.396

[1] CON, control; NIT, nitrogen; STA, starch; NIT+STA, nitrogen+starch; ADI, nitrogen+starch mixture with 15% of CP.

[2] N, effect of nitrogen supplementation; S, effect of starch supplementation; N×S, interaction between nitrogen and starch supplementation; A, additional contrast to verify the effect of additional supply of nitrogen in the supplementation of nitrogen plus starch.

Table 5. Interaction between nitrogen and starch supplementation on total digestibilities of organic matter (OM) and neutral detergent fiber corrected for ash and protein (NDFap), and the concentration of ruminal ammonia nitrogen (RAN)

Nitrogen	Starch	
	Without	With
OM (%)		
Without	53.5[Aa]	51.4[Ba]
With	52.4[Ab]	60.4[Aa]
NDFap (%)		
Without	63.0[Aa]	56.9[Bb]
With	61.1[Aa]	63.7[Aa]
RAN (mg/dL)		
Without	5.57[Ba]	5.61[Ba]
With	17.20[Aa]	11.14[Ab]

[A,B,a,b] Means followed by different capital letters within a column or different lowercase letters within a row are different at $p < 0.1$.

starch decreased RAN concentration ($p < 0.10$) when it was provided along with nitrogen compared to the nitrogen supply only (Table 5). For any of the other variables shown in Table 6, there was no interaction ($p > 0.10$) between nitrogen compounds and starch.

Fecal nitrogen excretion was increased ($p < 0.01$) with starch supplementation; however, it was not affected ($p > 0.10$) by nitrogen supplementation. Conversely, urinary nitrogen excretion was increased ($p < 0.01$) by supplementation with nitrogen, but it was not affected ($p > 0.10$) by supplementation with starch (Table 6). Nitrogen balance (NB; g/d) increased with nitrogen supplementation ($p < 0.01$) as well as with starch supplementation ($p < 0.05$). However, the apparent efficiency of nitrogen utilization in the animals' body (ENU; g of nitrogen retained/g of nitrogen intake) was increased only ($p < 0.01$) when nitrogen was provided (Table 6).

Concentrations of serum urea nitrogen, both before ($p < 0.01$) and after ($p < 0.05$) supplementation, were significantly increased by the nitrogen supplementation, but they were not affected ($p > 0.10$) by starch supplementation (Table 6).

Production of NMIC was not affected ($p > 0.10$) by supplementation with nitrogen, but it was increased ($p < 0.02$) by starch supplementation (Table 6).

Effects of the additional nitrogen supply to animals supplemented with both starch and nitrogen were only observed on RAN concentration ($p < 0.05$) and NB ($p < 0.05$). In both cases, the additional supply of nitrogen increased the estimates obtained (Table 6).

DISCUSSION

In this study, no effects of nitrogen supplementation were observed on forage voluntary intake (Table 3). This behavior confirms the results obtained by other authors in tropical conditions when protein supplements were provided to grazing cattle during the rainy season (Zervoudakis et al., 2008; Costa et al., 2011c).

Results obtained in Brazilian studies indicated that the inclusion of supplemental nitrogen in the diet of cattle that are fed tropical grasses can improve the forage intake up to protein levels near to 10% (Figueiras et al., 2010; Sampaio et al., 2010). From this dietary CP level, the microbial nitrogen requirements would be met and stimulation on forage digestion would no longer be observed (Detmann et al., 2014b). These assumptions would be applied to the conditions of this study, considering the actual CP content of the basal forage (Table 1) and the absence of effects of supplementation with nitrogen on the DOM and digested NDF intakes (Table 3).

Table 6. Least squares means of ruminal pH, concentration of ruminal ammonia nitrogen (RAN; mg/dL), nitrogen intake (NI; g/d), fecal nitrogen (FN; g/d), urinary nitrogen (UN g/d), apparent balance of nitrogen compounds (NB; g/d), efficiency of nitrogen utilization (ENU; g/g nitrogen intake), serum concentrations of nitrogen urea before (NUSb; mg/dL) and after supplementation (NUSa; mg/dL) and production of microbial nitrogen in the rumen (NMIC g/d) according to the treatments

Item	Treatments[1]					SEM	p-value[2]			
	COM	NIT	STA	NIT+STA	ADI		N	S	N×S	A
pH	6.53	6.54	6.53	6.53	6.52	0.10	0.967	>0.999	0.967	0.942
RAN	5.57	17.20	5.61	11.14	14.61	1.09	<0.001	0.017	0.016	0.043
NI	89.2	115.6	104.7	164.2	180.6	9.83	0.001	0.010	0.140	0.266
FN	40.8	37.4	52.8	45.9	43.8	2.86	0.112	0.005	0.564	0.594
UN	36.9	44.2	31.3	47.1	34.9	5.09	0.054	0.799	0.442	0.117
NB	11.5	34.0	20.6	71.2	102.0	9.20	0.003	0.040	0.187	0.042
ENU	0.11	0.29	0.16	0.43	0.56	0.06	0.006	0.163	0.514	0.161
SUNb	10.4	16.2	8.9	16.2	16.6	0.95	<0.001	0.472	0.406	0.805
SUNa	13.3	15.5	9.4	20.2	22.6	2.36	0.047	0.874	0.134	0.529
NMIC	57.5	57.2	66.8	76.7	73.4	5.13	0.367	0.015	0.332	0.651

[1] CON, control; NIT, nitrogen; STA, starch; NIT+STA, nitrogen+starch; ADI, nitrogen+starch mixture with 15% of CP.
[2] N, effect of nitrogen supplementation; S, effect of starch supplementation; N×S, interaction between nitrogen and starch supplementation; A, additional contrast to verify the effect of additional supply of nitrogen in the supplementation of nitrogen plus starch.

Starch supplementation did not change forage intake. However, indirectly, small stimuli on forage intake were observed because the starch supplementation increased CP intake (p<0.02; Table 3). Starch has no nitrogen in its composition (Table 1), thus this small stimulus could only result from the increase in forage intake, which was observed, although without significant effects (p>0.10; Table 3).

This pattern apparently contradicts some results obtained in the tropics in which it was reported that NFC supplementation for grazing animals during the rainy season could lead to a high substitutive effect on forage intake (Costa et al., 2011b). Additionally, starch supplementation could decrease NDF utilization in the rumen, as seen through the reduction in NDF digestibility (Table 4). Theoretically, this decreased fiber degradation would increase the rumen fill effect of NDF and reduce the forage voluntary intake (Detmann et al., 2014b). However, at least one characteristic appears to be associated with the absence of negative effects of starch supplementation on forage intake. The forage grazed during the rainy season should be understood as a diet, in which one of the main nutritional characteristics is its capacity to supply energy and protein in accordance with the animal's requirements. Accordingly, dietary energy-to-protein ratio could be unbalanced by energy supplements as they may cause a relative excess of energy when the only source of dietary nitrogen is the forage CP (Detmann et al., 2014b). Such a dietary pattern would increase animal discomfort (Illius and Jessop, 1996) and hence decrease forage intake (Detmann et al., 2014a; b). However, the CP content of the forage in this study was above the levels usually seen in tropical pastures during the rainy season (average of 9.42% reported by Detmann et al., 2014b). Thus, under the conditions of this study, starch supplementation would cause less interference in the energy-to-protein ratio in the diet, leading to the absence of negative effects on forage intake. This is reinforced by the similarity in the CP levels in the diet for the control and animals supplemented only with starch (12.7% and 12.6%, on a DM basis, respectively).

In general, the effects of supplementation on the digestibility were characterized by an interaction between nitrogen and starch (Tables 4 and 5). Considering NDFap digestibility, it was found that nitrogen supplementation did not cause a positive effect, which reflects the fact that the basal forage has no deficiency of nitrogen compounds for growth of fibrolytic microorganisms. Moreover, starch supplementation depressed the NDFap digestibility (Table 5). This reflects the fact that the inclusion of readily fermentable carbohydrates favors the growth of NFC fermenting bacteria rather than fibrolytic bacteria. This phenomenon, known as the "carbohydrate effect", increases the competition for essential nutrients between groups of microbial species, decreasing the utilization of insoluble fiber by fibrolytic microorganisms that have lower competitive capacity (El-Shazly et al., 1961; Arroquy et al., 2005; Carvalho et al., 2011). However, when starch was combined with nitrogen compounds, its deleterious effects were avoided, indicating that the nitrogen supply minimized the competition events between microbial species in the rumen, making NDFap digestibility similar to that found in animals without supplementation. Such a result agrees with Heldt et al. (1999) and Arroquy et al. (2004), who found that an adequate supplying of rumen degradable protein for cattle fed low-quality forage was able to overcome the negative effect of supplemental NFC on fiber digestion.

On the other hand, the increase of OM digestibility when starch and nitrogen supplementation was combined (Table 5) seems to indicate that both compounds interact with each other with respect to their ruminal utilization. Accordingly, starch would imply a better microbial assimilation of additional nitrogen, which is indirectly perceived by the decreased RAN concentration compared with animals supplemented only with nitrogen (Table 5). Additionally, although NMIC has been increased only by starch supplementation, it can be seen that the observed value for starch plus nitrogen supplementation was higher than those observed with the isolated supplementation of nitrogen or starch (Table 6). This reflects the fact of supplementation with nitrogen has allowed better use of energy from the NFC in the rumen, which has been reported by other authors in tropical conditions (Souza et al., 2010). Even without significant interaction (p>0.10), DOM intake was higher when nitrogen and starch were offered together, which shows a positive effect on the total intake of digestible energy.

The NB has been improved with nitrogen supplementation as well as with starch supplementation. In the latter case, the increase of nitrogen retention in the animal agrees, at least partially, with the increased NMIC production obtained with starch supplementation (Table 6). Such increased NMIC is supposed to improve the metabolizable protein (MP) supply. However, the positive effects on NB were more prominent considering the supplementation with nitrogen, without effects on NMIC (Table 6).

Detmann et al. (2015) used a meta-analytical approach to quantify the impacts of the increase in microbial protein caused by supplemental nitrogen and the improvement in NB in the tropics. These authors found that, even using supplements based on non-protein nitrogen, the improvement in microbial nitrogen production responds only for 21% of the improvement on NB. Such a pattern indicates the occurrence of post-digestive and metabolic effects associated with supplemental CP as observed by other authors in the tropics (Costa et al., 2011a; Rufino,

2011; Detmann et al., 2014a).

Increases in nutritional performance with the use of nitrogen supplements have been attributed to improvements in nitrogen status in the animal (Egan, 1965a; Egan and Moir, 1965). In general, the nitrogen status defines the quantitative and qualitative availability of nitrogen compounds for all metabolic and physiological functions, including functions associated with the metabolism of other compounds (e.g., energy). Based on this concept, it can be established that the nitrogen compounds available for animal metabolism would be used for different metabolic functions followed by an order of priority, namely: survival, maintenance, and production (Detmann et al., 2014a). Several reports in the tropics has confirmed that supplemental nitrogen impacts negatively on breakdown rate of myofibrillar protein and positively on the blood concentration of anabolic hormones (e.g., IGF1), even without a concurrent source of supplemental energy (Rufino, 2011; 2015; Franco, 2015; Batista et al., 2016). These both impacts implies in a net increment in nitrogen accretion in the animal.

Some values of NB were apparently too high and seemed above biological limits of protein deposition in the lean tissues (Table 6). Therefore, the absolute values of NB should be evaluated with caution because they likely overestimate protein accretion. Gerrits et al. (1996) compared nitrogen retention obtained in digestion trials with protein accretion obtained by serial slaughter, and they reported that nitrogen retention overestimated protein accretion of growing cattle. This seems to occur mainly due the underestimation of urine nitrogen (e.g., volatile nitrogen losses from containers), fecal nitrogen (e.g., incomplete collection, volatile losses during either collection or drying), or both (Spanghero and Kowalski, 1997). However, when experimental procedures are standardized among treatments and experimental periods within an experiment, as performed in this study, in spite of bias in the absolute values, the relative comparisons between treatments should be valid.

Moreover, one of the metabolic functions of higher priority is nitrogen recycling to the gastrointestinal tract because a continuous supply of nitrogen for microbial growth in the rumen should be seen as a survival strategy (Egan, 1965b; Van Soest, 1994). Considering a dietetic situation in which there is no prominent deficiency of nitrogen compounds, the amount of nitrogen recycled to the rumen remains relatively constant (Marini and Van Ambourgh, 2003). Therefore, there is less nitrogen for tissue deposition under low nitrogen status because a greater percentage of nitrogen intake is directed towards recycling and, as a result, a lower percentage of nitrogen will be available for anabolic purposes.

Therefore, considering that efficiency of nitrogen utilization is more strongly associated with nitrogen supply than with energy supply (Detmann et al., 2014a), nitrogen retention in the body will then be improved by increasing the nitrogen status with the use of protein supplements (Table 6).

Moreover, even without an interaction effect (p>0.10), it was found that NB and ENU estimates were optimized by the combined supplementation of nitrogen and starch (Table 6). In ruminants, protein deposition efficiency depends on energy availability, and energy utilization efficiency depends on amino acids availability (Schroeder and Titgemeyer, 2008). Thus, even with improvements in nitrogen status, MP would not be retained in the body due to a probable relative deficiency of metabolizable energy. Thus, the relative MP excess (which in turn is determined by the metabolizable energy availability) would be eliminated, which would increase urinary nitrogen excretion, as found in this study (Table 6). Such a pattern was also reported by Lazzarini et al. (2013). In those circumstances, the additional inclusion of starch would provide energy for better MP retention, increasing the NB (Table 6). This confirms that tissue deposition is defined by an interactive process in which the energy and MP utilization efficiencies are interrelated (Schroeder and Titgemeyer, 2008; Lazzarini et al., 2013).

The purpose of including an additional treatment was to determine whether an additional supply of nitrogen compounds could have beneficial nutritional effects on animals supplemented with nitrogen and starch (Table 2). In this context, an increase in NB was observed (Table 6). According to Detmann et al. (2014a), nitrogen supplementation during the rainy season would provide positive effects on ENU up to RAN concentrations of approximately 13 mg/dL, since RAN is assumed to be an indicator of nitrogen availability. This would result in a better balance of nitrogen compounds between rumen and the bloodstream, improving the availability of nitrogen compounds for anabolic purposes. The additional supply of nitrogen compounds for animals receiving nitrogen and starch supplementation increased RAN concentration to levels greater than that described above (11.14 to 14.61 mg/dL), justifying the results obtained in NB.

CONCLUSION

The supplementation with starch and nitrogen compounds for grazing cattle during the rainy season results in interactive effects by improving digestion and nitrogen retention in the animal.

ACKNOWLEDGMENTS

The authors wish to thank the Conselho Nacional de Desenvolvimento Científico e Tecnológico (CNPq, Brazil), Fundação de Apoio à Pesquisa de Minas Gerais (FAPEMIG, Brazil), and Instituto Nacional de Ciência e Tecnologia – Ciência Animal (INCT – Ciência Animal) for providing financial support.

REFERENCES

Arroquy, J. I., R. C. Cochran, T. G. Nagaraja, E. C. Titgemeyer, and D. E. Johnson. 2005. Effect of types of non-fiber carbohydrate on *in vitro* forage fiber digestion of low-quality grass hay. Anim. Feed Sci. Technol. 120:93-106.

Arroquy, J. I., R. C. Cochran, M. Villareal, T. A. Wickersham, D. A. Llewellyn, E. C. Titgemeyer, T. G. Nagaraja, D. E. Johnson, and D. Gnad. 2004. Effect of level of rumen degradable protein and type of supplemental non-fiber carbohydrate on intake and digestion of low-quality grass hay by beef cattle. Anim. Feed Sci. Technol. 115:83-99.

Batista, E. D., E. Detmann, E. C. Titgemeyer, S. C. Valadares Filho, R. F. D. Valadares, L. L. Prates, L. N. Rennó, and M. F. Paulino. 2016. Effects of varying ruminally undegradable protein supplementation on forage digestion, nitrogen metabolism, and urea kinetics in Nellore cattle fed low-quality tropical forage. J. Anim. Sci. 94:201-216.

Carvalho, I. P. C., E. Detmann, H. C. Mantovani, M. F. Paulino, S. C. Valadares Filho, V. A. C. Costa, and D. I. Gomes. 2011. Growth and antimicrobial activity of lactic acid bacteria from rumen fluid according to energy or nitrogen source. Rev. Bras. Zootec. 40:1260-1265.

Chen, X. B. and M. J. Gomes. 1992. Estimation of Microbial Protein Supply to Sheep and Cattle Based on Urinary Excretion of Purine Derivatives - An Overview of the Technical Details. Rowett Research Institute, Buchsburnd Aberdeen, UK. 21 p.

Chizzotti, M. L., S. C. Valadares Filho, R. F. D. Valadares, F. H. M. Chizzotti, J. M. S. Campos, M. I. Marcondes, and M. A. Fonseca. 2006. Intake, digestibility and urinary excretion of urea and purine derivatives in heifers with different body weights. Rev. Bras. Zootec. 35:1813-1821.

Costa, V. A. C., E. Detmann, M. F. Paulino, S. C. V. Valadares Filho, L. T. Henriques, and I. P. C. Carvalho. 2011a. Total and partial digestibility and nitrogen balance in grazing cattle supplemented with non-protein and, or true protein nitrogen during the rainy season. Rev. Bras. Zootec. 40:2815-2826.

Costa, V. A. C., E. Detmann, M. F. Paulino, S. C. Valadares Filho, I. P. C. Carvalho, and L. P. Monteiro. 2011b. Intake and digestibility in cattle under grazing during rainy season and supplemented with different sources of nitrogenous compounds and carbohydrates. Rev. Bras. Zootec. 40:1788-1798.

Costa, V. A. C., E. Detmann, M. F. Paulino, S. C. Valadares Filho, L. T. Henriques, I. P. C. Carvalho, and T. N. P. Valente. 2011c. Intake and rumen dynamics of neutral detergent fiber in grazing cattle supplemented with non-protein nitrogen and, or true protein during the rainy season. Rev. Bras. Zootec. 40:2805-2814.

Detmann, E., M. A. Souza, S. C. Valadares Filho, A. C. Queiroz, T. T. Berchielle, E. O. S. Saliba, L. S. Cabral, D. S. Pina, M. M. Ladeira, and J. A. G. Azevedo. 2012. Métodos para análise de alimentos. 1st Ed. Suprema, Visconde do Rio Branco, MG, Brazil.

Detmann, E., E. E. L. Valente, E. D. Batista, and P. Huhtanen. 2014a. An evaluation of the performance and efficiency of nitrogen utilization in cattle fed tropical grass pastures with supplementation. Livest. Sci. 162:141-153.

Detmann, E., M. F. Paulino, S. C. Valadares Filho, and P. Huhtanen. 2014b. Nutritional aspects applied to grazing cattle in the tropics: A review based on Brazilian results. Semin-Cienc. Agrar. 35:2829-2854.

Detmann, E., E. D. Batista, M. O. Franco, L. M. A. Rufino, W. L. S. Reis, M. F. Paulino, S. C. Valadares Filho, and C. B. Sampaio. 2015. Contribution of the rumen microbial nitrogen obtained using supplementation to the body accretion of nitrogen in cattle fed tropical forages. In: III Simpósio Matogrossense de Bovinocultura de Corte, Cuiabá, MT, Brazil. pp. 1-3.

Egan, A. R. 1965a. Nutritional status and intake regulation in sheep. III. The relationship between improvement of nitrogen status and increase in voluntary intake of low-protein roughages by sheep. Aust. J. Agric. Res. 16:463-472.

Egan, A. R. 1965b. The fate and effects of duodenally infused casein and urea nitrogen in sheep fed on a low-protein roughage. Aust. J. Agric. Res. 16:169-177.

Egan, A. R. and R. J. Moir. 1965. Nutritional status and intake regulation in sheep. I. Effects of duodenally infused single doses of casein, urea, and propionate upon voluntary intake of a low-protein roughage by sheep. Aust. J. Agric. Res. 16:437-449.

El-Shazly, K., B. A. Dehority, and R. R. Johnson. 1961. Effect of starch on the digestion of cellulose *in vitro* and *in vivo* by rumen microorganisms. J. Anim. Sci. 20:268-273.

Figueiras, J. F., E. Detmann, S. C. Valadares Filho, M. F. Paulino, E. D. Batista, L. M. A. Rufino, T. N. P. Valente, W. L. S. Reis, and M. O. Franco. 2015. Nutritional performance of grazing cattle during dry-to-rainy transition season with protein supplementation. Arch. Zootec. 64:269-276.

Figueiras, J. F., E. Detmann, M. F. Paulino, T. N. P. Valente, S. C. Valadares Filho, and I. Lazzarini. 2010. Intake and digestibility in cattle under grazing supplemented with nitrogenous compounds during dry season. Rev. Bras. Zootec. 39:1303-1312.

Franco, M. O. 2015. Nutritional Performance and Metabolic Characteristics in Cattle Fed Tropical Forage and Supplemented with Nitrogenous Compounds and Energy. Ph.D. Thesis, Universidade Federal de Viçosa, Viçosa, MG, Brazil.

Gerrits, W. J. J., G. H. Tolman, J. W. Schrama, S. Tamminga, M. W. Bosch, and M. W. A. Verstegen. 1996. Effect of protein and protein-free energy intake on protein and fat deposition rates in preruminant calves of 80 to 240 kg live weight. J. Anim. Sci. 74:2129-2139.

Heldt, J. S., R. C. Cochran, C. P. Mathis, B. C. Woods, K. C. Olson, E. C. Titgemeyer, T. G. Nagaraja, E. S. Vanzant, and D.

E. Johnson. 1999. Effects of level and source of carbohydrate and level of degradable intake protein on intake and digestion of low-quality tallgrass-prairie hay by beef steers. J. Anim. Sci. 77:2846-2854.

Illius, A. W. and N. S. Jessop. 1996. Metabolic constraints on voluntary intake in ruminants. J. Anim. Sci. 74:3052-3062.

Lazzarini, I., E. Detmann, M. F. Paulino, S. C. Valadares Filho, R. F. D. Valadares, F. A. Oliveira, P. T. Silva, and W. L. S. Reis. 2013. Nutritional performance of cattle grazing on low-quality tropical forage supplemented with nitrogenous compounds and/or starch. Rev. Bras. Zootec. 42:664-674.

Marini, J. C. and M. E. Van Amburgh. 2003. Nitrogen metabolism and recycling in Holstein heifers. J. Anim. Sci. 81:545-552.

Rufino, L. M. A. 2011. Ruminal and/or Abomasal Nitrogenous Supplementation in Cattle Fed Tropical Forage. M.Sc. Thesis, Universidade Federal de Viçosa, Viçosa, MG, Brazil.

Rufino, L. M. A. 2015. Nutritional Performance and Metabolic Characteristics in Cattle Fed Tropical Forages in Response to Infrequent Supplementation with Nitrogenous Compounds. Ph.D. Thesis, Universidade Federal de Viçosa, Viçosa, MG, Brazil.

Sampaio, C. B., E. Detmann, M. F. Paulino, S. C. Valadares Filho, M. A. Souza, I. Lazzarini, I., P. V. Paulino, and A. C. Queiroz. 2010. Intake and digestibility in cattle fed low-quality tropical forage and supplemented with nitrogenous compounds. Trop. Anim. Health Prod. 42:1471-1479.

Schroeder, G. F. and E. C. Titgemeyer. 2008. Interaction between protein and energy supply on protein utilization in growing cattle: A review. Livest. Sci. 114:1-10.

Souza, M. A., E. Detmann, M. F. Paulino, C. B. Sampaio, I. Lazzarini, and S. C. Valadares Filho. 2010. Intake, digestibility and rumen dynamics of neutral detergent fiber in cattle fed low-quality tropical forage and supplemented with nitrogen and/or starch. Trop. Anim. Health Prod. 42:1299-1310.

Spanghero, M. and Z. M. Kowalski. 1997. Critical analysis of N balance experiments with lactating cows. Livest. Prod. Sci. 52:113-122.

Titgemeyer, E. C., C. K. Armendariz, D. J. Bindel, R. H. Greenwood, and C. A. Löest. 2001. Evaluation of titanium dioxide as a digestibility marker for cattle. J. Anim. Sci. 79:1059-1063.

Valadares Filho, S. C. 1995. Microbial protein synthesis, crude protein ruminal degradation and intestinal digestibility in cattle. In: Proceedings of International Symposium on the Nutritional Requirements of Ruminants José Carlos Pereira. (Org.). JARD Viçosa MG, Brazil. pp. 201-234.

Van Soest, P. J. 1994. Nutritional Ecology of the Ruminant. 2nd Ed. Ithaca: Cornell University Press, Ithaca, NY, USA. 476 p.

Verbic, J., X. B., N. A. MacLeod, and E. R. Ørskov. 1990. Excretion of purine derivatives by ruminants. Effect of microbial nucleic acid infusion on purine derivative excretion by steers. J. Agric. Sci. 114:243-248.

Zervoudakis, J. T., M. F. Paulino, L. S. Cabral, E. Detmann, S. C. Valadares Filho, and E. H. B. K. Moraes. 2008. Multiple supplements of self controlled intake for steers during the growing phase in the rainy season. Ciênc. Agrotec. 32:1968-1973.

17

Relationship between Molecular Structure Characteristics of Feed Proteins and Protein *In vitro* Digestibility and Solubility

Mingmei Bai[1,2], Guixin Qin[1,2,]*, Zewei Sun[1,2,]*, and Guohui Long[3]

[1] Animal Production and Product Quality and Security Key Lab, Ministry of Education,
Jilin Agricultural University, Changchun 130118, China

ABSTRACT: The nutritional value of feed proteins and their utilization by livestock are related not only to the chemical composition but also to the structure of feed proteins, but few studies thus far have investigated the relationship between the structure of feed proteins and their solubility as well as digestibility in monogastric animals. To address this question we analyzed soybean meal, fish meal, corn distiller's dried grains with solubles, corn gluten meal, and feather meal by Fourier transform infrared (FTIR) spectroscopy to determine the protein molecular spectral band characteristics for amides I and II as well as α-helices and β-sheets and their ratios. Protein solubility and *in vitro* digestibility were measured with the Kjeldahl method using 0.2% KOH solution and the pepsin-pancreatin two-step enzymatic method, respectively. We found that all measured spectral band intensities (height and area) of feed proteins were correlated with their the *in vitro* digestibility and solubility ($p \leq 0.003$); moreover, the relatively quantitative amounts of α-helices, random coils, and α-helix to β-sheet ratio in protein secondary structures were positively correlated with protein *in vitro* digestibility and solubility ($p \leq 0.004$). On the other hand, the percentage of β-sheet structures was negatively correlated with protein *in vitro* digestibility ($p < 0.001$) and solubility ($p = 0.002$). These results demonstrate that the molecular structure characteristics of feed proteins are closely related to their *in vitro* digestibility at 28 h and solubility. Furthermore, the α-helix-to-β-sheet ratio can be used to predict the nutritional value of feed proteins. (**Key Words:** Feed, Protein, Protein *In vitro* Digestibility, Fourier Transform Infrared Spectroscopy, Molecular Structure Characteristics, Protein Solubility)

INTRODUCTION

Nutritional value of protein in feedstuff not only depends on its amino acid content and composition proportion, but also on its molecular structure characteristics (Peng et al., 2014). Recent studies have highlighted the importance of determining the relationship between molecular spectral band characteristics of proteins and their intestinal absorption so as to clarify the mechanism of nitrogen supplementation (Qin et al., 2014).

The traditional method of determining the nutritional value of feeds involves destroying the three-dimensional structure of constituent proteins by acid or alkali treatment or by mechanical polishing, and then using so-called wet chemical methods to determine chemical composition. However, this approach provides no information at a molecular level on the nutritional value of feeds or on their capacity for digestion by livestock. Fourier transform infrared (FTIR) technique is a powerful tool for analyzing the structure of feed proteins (Xin et al., 2013a), which allows investigators to predict how they are processed and absorbed within the animals' digestive tract. However, there is certainly difficulty to use FTIR spectrum for quantifying the tertiary structure having higher portion of side chains of the conventional feedstuffs protein.

Recent studies have shown a close correlation between the molecular structure characteristics of feed proteins and their nutritional value (Yu and Nuez-Ortín, 2010; Samadi et

* Corresponding Authors: G. X. Qin.
E-mail: qgx@jlau.edu.cn / Zewei Sun. Tel:
E-mail: sunzewei@jlau.edu.cn
[2] College of Animal Science and Technology, Jilin Agricultural University, Changchun 130118, China.
[3] College of Life Science, Jilin Agricultural University, Changchun 130118, China.

al., 2013). Specifically, α-helices and β-sheets in the protein secondary structures can affect digestibility of feed proteins (Dyson and Wright, 1993). Certain features (amide I and II bands) of protein molecular structures of corn distiller's dried grains with solubles (corn DDGS) from various sources were correlated with intestinal protein digestibility, which can be used to predict the absorption of essential nutrients by the small intestine (Chen et al., 2014). Another study found that spectral band intensity characteristics of proteins affect protein quality, nutrient utilization, and digestion in dairy cows (Doiron et al., 2009). Therefore, establishing a relationship between the molecular structure characteristics of conventional feed proteins and their nutritional value will strengthen the current knowledge on protein nutritional value and broaden the scope of molecular spectroscopy as a bioanalytical tool for rapidly assessing the nutritive value of feed proteins.

To our knowledge, the relationship between molecular structure characteristics of conventional feed proteins and protein solubility (PS) and *in vitro* digestibility in monogastric animals has not been previously investigated. The objectives of this study were i) to assess protein molecular structure characteristics of conventional feed proteins by FTIR; ii) to examine the PS and *in vitro* digestibility of these proteins; and iii) to quantify the relationship between protein molecular structure characteristics and PS as well as *in vitro* digestibility.

MATERIALS AND METHODS

Sample preparation and treatment

Five conventional feed ingredients (soybean meal, fish meal, corn DDGS, corn gluten meal, and feather meal) were provided by Changchun Wellhope Animal Husbandry Co. (Changchun, China). All feed samples used in this experiment were collected from raw commodities and stored according to commodity storage standard, one month of storage period. One half of each sample was ground in the laboratory mill through a 1 mm screen for laboratory analysis and sub-samples (n = 3 per type of feed) was collected and stored at 4°C. The other half was stored as a backup at –20°C. Chemical profiles of the five feeds are presented in Table 1.

Determination of protein solubility

PS was determined according to the method as described by Dale (1987). Feedstuff sample with 1.5 g was placed in a 100 mL conical flask, and then added 75 mL of 0.2% (w/v) potassium hydroxide solution; the mixture was stirred on a magnetic stirrer for 20 min and then transferred to a 50 mL tube and centrifuged at 2,700 rpm for 10 min. 15 mL volume of supernatant was measured by the chemical method (AOAC, 1990), yielding an amount equivalent to 0.3 g of original sample ($15/75 = x/1.5$ g, where $x = 0.3$ g). PS (%) was calculated as crude protein (CP) content in 15 mL supernatant/CP content of the original sample×100%.

Determination of protein *in vitro* digestibility

Protein *in vitro* digestibility was evaluated according to a method as previously described by Boisen and Fernandez (1995). Feedstuff sample with 1.0 g (accurate to 0.001 g) was placed in a 100 mL conical flask, and then added 10 mL of 0.01 M hydrochloric acid (pH 2.0) and pepsin solution containing 1.0 mg porcine pepsin (product no. P7-000; Sigma, St. Louis, MO, USA). To prevent bacterial growth, 0.5 mL of chloramphenicol solution consisting of 0.5 g chloramphenicol (product no. 0230; Amresco, Solon, OH, USA) in 100 mL ethanol was added to the mixture, followed by incubation for 4 h at 37°C. After neutralization with 0.2 M sodium hydroxide, 10 mL of 0.2 M phosphate buffer (pH 6.8) was added. The pH was adjusted to 6.8 with 1 M HCl or 1 M NaOH, and the solution was mixed with 1 mL of a freshly prepared pancreatin solution containing 50 mg porcine pancreatin (product no.P7-545-100G; Sigma, USA). The flask was closed with a rubber stopper and incubated with continuous magnetic stirring at 39°C for 4, 8, 12, 16, 20, 24, and 28 h, respectively. After adding 5 mL of 20% sulfosalicylic acid, the sample was centrifuged at 15,000 rpm for 15 min; the supernatant was discarded and the precipitate was heated at 80°C for 24 h. CP digestibility (in %) was calculated as CP content of the original sample –CP content of precipitate/CP content of the original sample×100%.

Table 1. Chemical profiles of five conventional feed ingredients (dry matter basic)

Item	Soybean meal	Fish meal	Corn DDGS	Corn gluten meal	Feather meal
Dry matter (%)	98.12±0.05[a]	98.42±0.03[b]	97.84±0.11[c]	98.55±0.02[b]	98.01±0.03[ac]
Ash (%)	6.39±0.16[a]	16.85±0.03[b]	5.49±0.47[c]	1.72±0.05[d]	2.82±0.01[e]
Ether extract (%)	1.32±0.02[a]	7.14±0.18[b]	4.79±0.21[c]	3.68±0.04[d]	2.42±0.04[e]
Crude protein (%)	43.35±0.02[a]	66.86±0.02[b]	28.36±0.03[c]	63.18±0.01[d]	85.49±0.03[e]
Total carbohydrate (%)	48.94±0.20[a]	9.14±0.13[b]	61.35±0.72[c]	31.43±0.04[d]	9.26±0.00[b]

Different lowercase letters in the same line indicate significant differences (n = 3, p<0.05).
Dried samples were analyzed for dry matter (Association of Official Analytical Chemists [AOAC], 1990; method 930.15), ash (AOAC, 1990; method 942.05), ether extract (AOAC, 1990; method 920.35), and crude protein (AOAC, 1990; method 984.13). Total carbohydrate content was calculated as (100%–crude protein–ether extract–ash) (NRC, 2001).

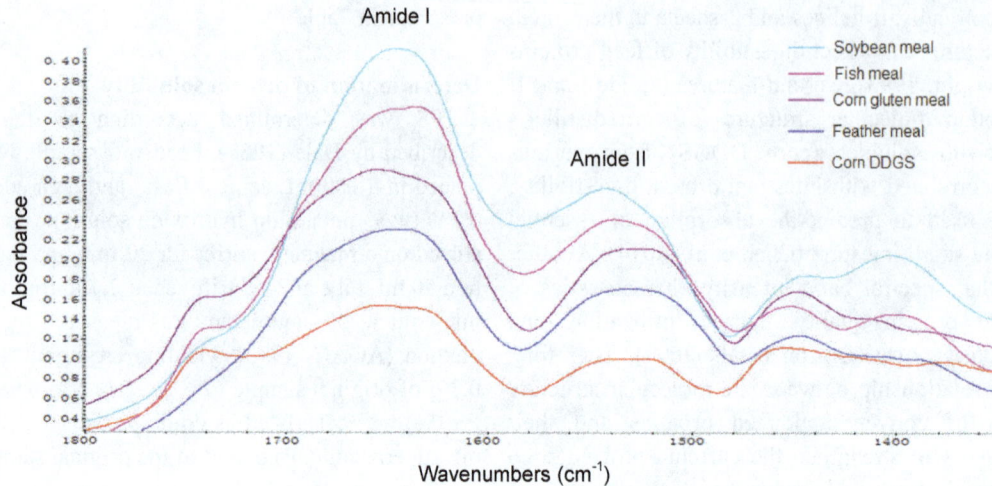

Figure 1. Fourier transform infrared (FTIR) spectra of five conventional feed ingredient proteins (baseline, ~1,720 to 1,479 cm^{-1}). Wavenumbers are as follows: amide I height, 1,646 cm^{-1}; amide II height, 1,542 cm^{-1}; amide I area, 1,720 to 1,583 cm^{-1}; and amide II area, 1,583 to 1,479 cm^{-1}.

Fourier transform infrared spectroscopy

FTIR spectroscopy was performed according to the method as described by Long et al (2015). Briefly, 2 mg of sample was mixed with 200 mg KBr in a vacuum dryer for 24 h; the mixture was uniformly ground and pressed into a tablet (tablet thickness = 0.25 mm; translucent shape) using an infrared (IR) tablet press (FW-4; Thermo Fisher Scientific, Waltham, MA, USA). FTIR spectra (FTIR-8400s; Shimadzu, Kyoto, Japan) were acquired in the mid-IR range (4,000 to 400 cm^{-1}) at a resolution of 4 cm^{-1} with 64 co-added scans. Each sample was run six times. As a control, a KBr pellet without protein was recorded under identical conditions. IR spectra were processed using OMNIC 8.0 software (Nicolet Analytical Instruments, Madison, WI, USA), and after baseline correction of the original map the sample and control spectra were compared. Amide I band narrowing was achieved with a full height of 38.2 and a resolution enhancement factor of 4.4. Fourier self-deconvolution using Origin 7.5 software (Origin Lab, Northampton, MA, USA) was used for Gaussian curve fitting in the region of the amide I band to separate overlapping bands. The secondary structure content of samples was detected from IR second-derivative amide I spectra by manually computing the relative peak areas under the bands assigned to a particular substructure.

Statistical analysis

Data were evaluated by analysis of variance with SPSS v.19.0.1 software (SPSS Inc., Chicago, IL, USA). The model used for protein molecular spectral band characteristics, PS and *in vitro* digestibility analysis was Y_{ij} = $\mu_i + e_{ij}$, where Y_{ij} is an observation on the dependent variable ij (i = 1, 2,. . .,5; j = 1, 2, 3), μ_i is the overall mean (i = 1, 2,. . .,5), and e_{ij} is the random error associated with the observation ij. Multiple comparisons of group means

were carried out using Duncan's method. Results are expressed as mean±standard error, with α = 0.05 indicating a statistically significant difference. Correlation analysis was performed using Pearson's correlation coefficient with 95% confidence limits (p<0.05).

RESULTS

Protein spectral band intensities

The protein IR spectrum showed two salient features, i.e., the amide I (1,700 to 1,600 cm^{-1}) and amide II (1,600 to 1,500 cm^{-1}) bands (Figure 1). Results of the protein spectral band intensities analyses are presented in Table 2. Significant differences were observed among feeds in terms of amide II heights and areas and amide I-to-II height and area ratios (p<0.001). Soybean meal had the highest values for amide I and II heights, amide I and II areas, and amide I-to-II height and area ratios, whereas corn DDGS had the lowest values for all of these features.

Protein secondary structures and solubility

The Gaussian curve fitting analysis of the amide I band peak was carried out based on the preliminary study of Meng-Xia and Yuan (2002), showed characteristic peaks at 1,611 to 1,639 cm^{-1} that were attributed to β-sheets. Other peaks were observed that corresponded to the following structures: random coils, 1,640 to 1,649 cm^{-1}; α-helices, 1,650 to 1,658 cm^{-1}; and β-turns, 1,660 to 1,700 cm^{-1} (Figure 2). Results of the protein secondary structures and solubility analyses are shown in Table 3. Soybean meal had the highest percentage of α-helices and random coil structures and α-helix-to-β-sheet ratios; feather meal had the highest percentage of β-sheet structures; and corn DDGS had the highest percentage of β-turn structures, while corn gluten meal had the lowest percentage of β-turn

Table 2. Molecular spectral band intensity characteristics of five conventional feed ingredient proteins

Item	Soybean meal	Fish meal	Corn DDGS	Corn gluten meal	Feather meal	SEM	p value
Amide I height	0.240[a]	0.183[b]	0.117[c]	0.138[d]	0.119[c]	0.005	<0.001
Amide II height	0.092[a]	0.076[b]	0.057[c]	0.084[d]	0.065[e]	0.002	<0.001
Ratio of amide I to II height	2.619[a]	2.413[b]	2.057[c]	1.645[d]	1.842[e]	0.021	<0.001
Amide I area	0.463[a]	0.353[b]	0.226[c]	0.266[d]	0.229[c]	0.009	<0.001
Amide II area	0.177[a]	0.146[b]	0.110[c]	0.161[d]	0.124[e]	0.003	<0.001
Ratio of amide I to II area	2.618[a]	2.420[b]	2.064[c]	1.646[d]	1.848[e]	0.021	<0.001

Corn DDGS, corn distiller's dried grains with solubles; SEM, standard error of the mean.
Different lowercase letters in the same line indicate significant differences (n = 3, p<0.05).

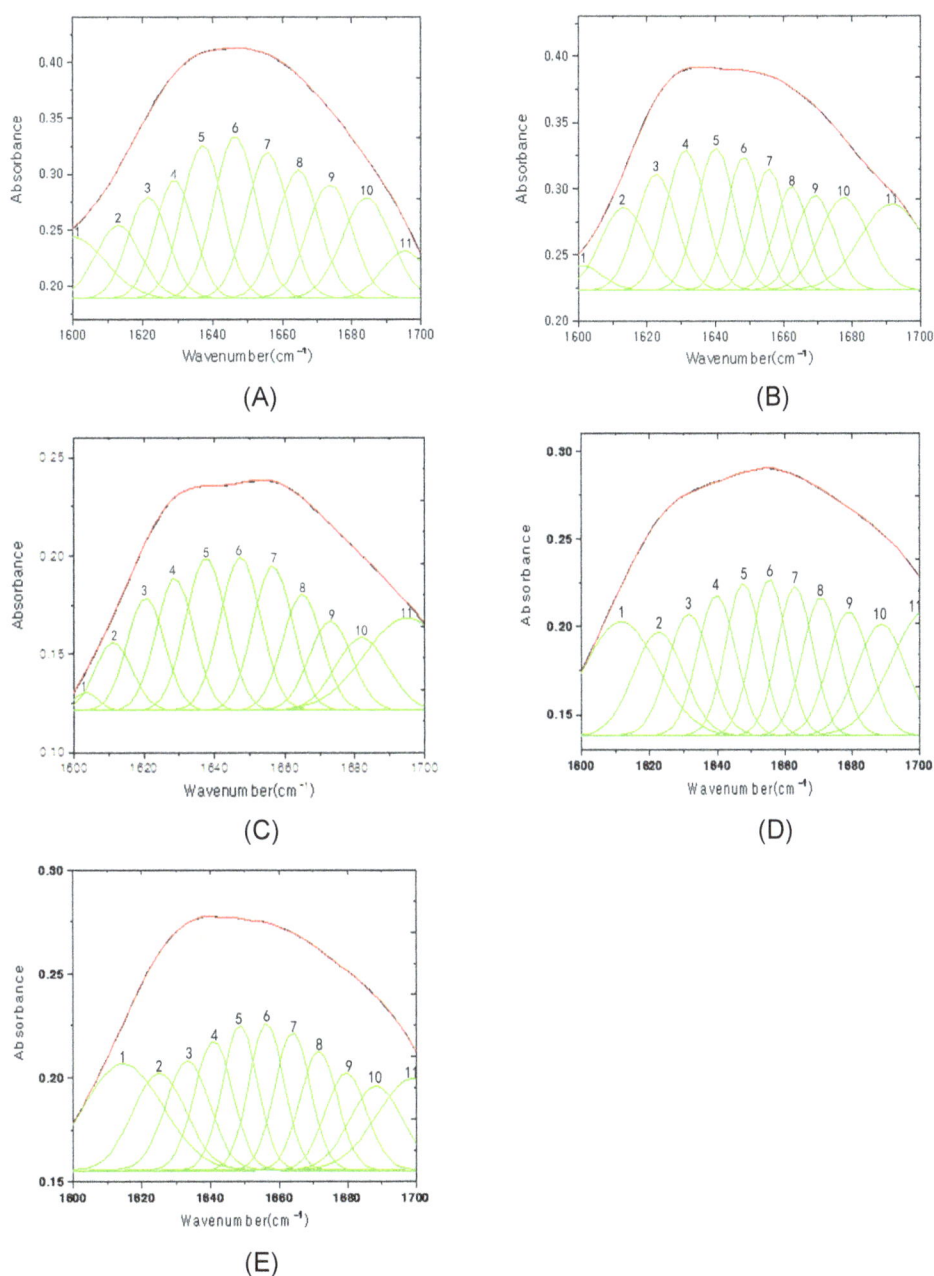

Figure 2. Curve-fitted individual component bands (A to E) of the amide I region in Fourier transform infrared (FTIR) spectra (1,700 to 1,600 cm^{-1}). (A) soybean meal; (B) fish meal; (C) corn distiller's dried grains with solubles (corn DDGS); (D) corn gluten meal; (E) feather meal. Using Fourier self-deconvolution, the positions of 11 sub-peaks were subjected to Gaussian curve-fitting in the amide I band.

Table 3. Protein solubility and secondary structures of five conventional feed ingredients

Item	Soybean meal	Fish meal	Corn DDGS	Corn gluten meal	Feather meal
Protein solubility (% of CP)	84.19 ± 0.42^a	33.77 ± 0.13^b	14.76 ± 0.02^c	12.31 ± 0.00^c	8.05 ± 0.00^c
Protein secondary structures (%)					
α-Helix	12.16 ± 0.50^a	10.82 ± 0.14^b	9.49 ± 0.33^c	8.66 ± 0.17^c	8.59 ± 0.12^c
β-Sheet	33.25 ± 1.44^a	36.45 ± 0.86^b	37.05 ± 0.10^b	36.70 ± 0.08^b	41.29 ± 0.84^c
β-Turn	35.95 ± 1.53^a	39.78 ± 0.33^b	43.20 ± 0.71^c	32.20 ± 0.38^d	39.67 ± 1.49^b
Random coil	12.34 ± 0.61^a	12.21 ± 0.07^a	10.68 ± 0.43^b	8.22 ± 0.05^c	9.64 ± 0.45^b
Ratio of α-helix to β-sheet	36.57 ± 0.05^a	29.70 ± 0.33^b	25.62 ± 0.82^c	23.58 ± 0.44^d	20.83 ± 0.70^e

Different lowercase letters in the same line indicate significant differences (n = 3, p<0.05).

and random coil structures. The rank order of PS was soybean meal>fish meal>corn DDGS>corn gluten meal>feather meal (p<0.05).

Protein *in vitro* digestibility

Evaluation of protein *in vitro* digestibility revealed significant differences among five conventional feed ingredients at each time point (p<0.05; Figure 3). The *in vitro* digestibility of proteins by pepsin-trysin increased over time. Soybean and feather meal had the highest and lowest *in vitro* digestibility values, respectively, at 28 h. The rank order of protein *in vitro* digestibility at this time point was soybean meal>fish meal>corn DDGS>corn gluten meal>feather meal (p<0.05). The lower *in vitro* digestibility of corn DDGS proteins at 28 h as compared to that at 24 h may be explained by the inhibition of enzymatic reactions due to accumulation of digestion products.

Correlation between protein spectral band intensities and protein *in vitro* digestibility and solubility

Coefficients of correlation between protein spectral band intensities (height and area) and protein *in vitro* digestibility at 28 h as well as solubility are shown in Table 4. Amide I height was positively correlated with PS (r = 0.958, p<0.001) and *in vitro* digestibility (r = 0.848, p< 0.001), as were amide II height (r = 0.707, p = 0.003 and r = 0.518, p = 0.048, respectively), the height ratio of amide I to II (r = 0.857, p<0.001 and r = 0.894, p<0.001, respectively), amide I area (r = 0.959, p<0.001 and r = 0.959, p<0.001, respectively), amide II area (r = 0.709, p = 0.003 and r = 0.519, p = 0.047, respectively), and area ratio of amide I to II (r = 0.854, p<0.001 and r = 0.893, p<0.001, respectively).

Correlation between protein secondary structures and protein *in vitro* digestibility and solubility

Coefficients of correlation between protein secondary structures and protein *in vitro* digestibility at 28 h as well as solubility are shown in Table 5. The percentage of α-helix structures was positively correlated with PS (r = 0.903, p<0.001) and *in vitro* digestibility (r = 0.916, p<0.001), as were the percentage of random coil structures (r = 0.694, p

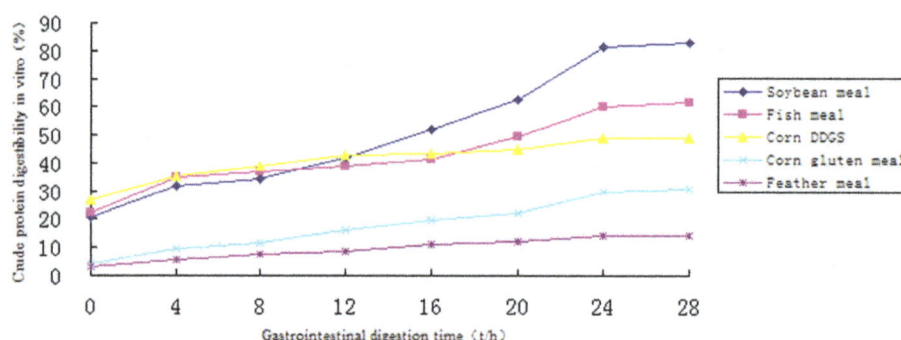

Figure 3. *In vitro* digestibility of five conventional feed ingredient proteins.

Table 4. Correlations between feed protein spectral band intensity characteristics and protein *in vitro* digestibility and solubility

Item	Peak height						Peak area					
	Amide I		Amide II		Height ratio amide I:II		Amide I		Amide II		Area ratio amide I:II	
	R	p value	R	p value	R	p value	R	p value	R	p value	R	p value
Digestible nutrients												
Protein *in vitro* digestibility	0.848**	<0.001	0.518*	0.048	0.894**	<0.001	0.849**	<0.001	0.519*	0.047	0.893**	<0.001
Protein solubility	0.958**	<0.001	0.707**	0.003	0.857**	<0.001	0.959**	<0.001	0.709**	0.003	0.854**	<0.001

R, correlation coefficient; * p<0.05, ** p<0.01.
Correlation coefficient obtained using the Pearson method.

= 0.004 and r = 0.789, p<0.001, respectively) and α-helix-to-β-sheet ratios (r = 0.955, p<0.001 and r = 0.966, p<0.001, respectively). However, the percentage of β-sheet structures was negatively correlated with both protein *in vitro* digestibility (r = –0.826, p<0.001) and solubility (r = –0.740, p = 0.002).

DISCUSSION

There is an increasing demand to improve the nutritional value of feed proteins for monogastric animals by optimizing feed protein sources and increasing the efficiency of protein utilization, which requires a fast and accurate estimation of the intestinal digestibility of feed proteins. Earlier findings suggest that the pepsin-pancreatin two-step enzymatic method is potentially a useful for evaluating *in vivo* digestibility (Boisen et al., 1995). However, previous studies have been unable to explain the variability in the digestion of different feed proteins, which are likely due to differences in protein structure.

The present study investigated the molecular spectral band characteristics of proteins in various types of feedstuff in relation to their digestibility and solubility. PS and protein spectral band intensities (height and area) in the feedstuff examined in this study varied widely. Protein secondary structures such as α-helices, β-sheets, and random coils as well as α-helix-to-β-sheet ratios in soybean meal and corn DDGS, and the relative quantitative amounts of α-helices and β-sheets in feather meal were inconsistent with earlier findings (Yu et al., 2004; Jiao et al., 2012). Even the reason for these differences is unclear, they may be related to variations inherent in feed sources, processing technology, or storage conditions (Peng et al., 2014).

We found that *in vitro* digestibility of soybean meal protein was lower than previously reported values (Boisen et al., 1995), which may have been due to the presence of anti-nutritional factors (Tan-Wilson et al., 1985) that can inhibit enzymatic reactions. On the other hand, fish meal protein *in vitro* digestibility was higher than that reported in another study (Huang et al., 2000) whereas feather meal protein *in vitro* digestibility was similar to results obtained by other investigators (Yu et al., 2004). This may be due to the presence of keratin—a scleroprotein that makes up 85% to 90% of feather meal—which has highly stable S-S and S-H linkages that are not readily broken down by animals

without processing (Papadopoulos et al., 1986). One possible reason for which corn DDGS and gluten meal had lower protein *in vivo* digestibility and PS is their low solubility in water due to a high content of alcohol-soluble proteins containing hydrophobic amino acids (Liu et al., 2013); alternatively, the S-S or O-H bonds in their structures may promote the formation of α-helices (Zhang and Yu, 2012), which are structurally stable and therefore not easily digested.

The amide I band mainly arises from stretching vibrations of the amide C=O group, while the amide II band is attributable to N-H bending (60%) and C-N stretching (40%) vibrations; these have also been used to determine protein structure or conformation (Yan et al., 2014), and the protein two primary spectral band intensities in terms of peak height and peak area indicate quantitative differences in protein functional groups (Peng et al., 2014). Corn DDGS and feather meal degraded slowly as compared to the other types of feed due in part to their low basic amino acid content, which is indirectly reflected by amide height and area. Trypsin is the main protease in the intestine and targets the peptide bond between lysine and arginine; therefore, the lack of basic amino acids would diminish the probability of contact between the CP in feed and digestive enzymes in the animal, thereby decreasing the amount and variety of oligopeptides and free amino acids that are released (Adler-Nissen, 1979).

The vibrational frequency of the amide I band is particularly sensitive to and can be used to predict protein secondary structure (Yu et al., 2004). Our findings confirmed that the percentage of β-sheet structures is closely related to the nutritional value of feed, with higher percentage being associated with lower PS and CP digestibility in the small intestine, since β-sheets have a large number of hydrogen bonds that can hinder protease activity. We also confirmed that the nutritional value of proteins feed differed according to α-helix-to-β-sheet ratio; previous studies have also reported that this ratio is a good predictor of nutritional value or digestibility of proteins feed (Theodoridou and Yu, 2013; Xin et al., 2013b). We found that the percentage of random coil structures was positively correlated with PS and digestibility, possibly because this structure is strong and flexible and has fewer hydrogen bonds that can impede enzyme access, which allows proteins to be readily degraded.

Table 5. Correlations between feed protein secondary structures and protein *in vitro* digestibility and solubility

Digestible nutrients	α-Helix		β-Sheet		Random coil		Ratio of α-helix to β-sheet	
	R	p value	R	p value	R	p value	R	p value
Protein *in vitro* digestibility	0.916**	<0.001	–0.826**	<0.001	0.789**	<0.001	0.966**	<0.001
Protein solubility	0.903**	<0.001	–0.740**	0.002	0.694**	0.004	0.955**	<0.001

R, correlation coefficient; * p<0.05, ** p<0.01.
Correlation coefficient obtained using the Pearson method.

ACKNOWLEDGMENTS

This study was supported by a grant from the National Key Basic Research Program Fund (973 Program; no. 2013CB127306).

REFERENCES

Adler-Nissen, J. 1979. Determination of the degree of hydrolysis of food protein hydrolysates by trinitro benzene sulfonic acid. J. Agric. Food Chem. 6:1256-1262.

AOAC. 1990. Official Methods of Analysis. 15th edn. Association of Official Analytical Chemists, Arlington, VA, USA.

Boisen, S. and J. A. Fernandez. 1995. Prediction of the apparent ileal digestibility of protein and amino acids in feedstuffs and feed mixtures for pigs by *in vitro* analyses. Anim. Feed Sci. Technol. 51:29-43.

Chen, L. M., X. W. Zhang, and P. Q. Yu. 2014. Molecular Spectroscopy Basis to Explore Molecular Structure in Relation to Nutritional Values and Metabolic Characteristics in Dairy Cattle of Chinese DDGS from Different Sources. Ph.D. Tianjin Agricultural University, Tianjin, China.

Dale, N. M., M. Araba, and E. Whittle. 1987. Protein solubility as an indicator of optimum processing of soybean meal. In: Proceedings of 1987 Georgia Nutrition Conference for the Feed Industry, Georgia Nutrition Society, Atlanta, GA, USA. pp. 88-95.

Doiron, K., P. Yu, J. J. McKinnon, and D. A. Christensen. 2009. Heat-induced protein structure and subfractions in relation to protein degradation kinetics and intestinal availability in dairy cattle. J. Dairy Sci. 92:3319-3330.

Dyson, H. J. and P. E. Wright. 1993. Peptide conformation and protein folding. Curr. Opin. Struct. Biol. 3:60-65.

Huang, R. L., Z. L. Tan, T. X. Xing, Y. F. Pan, and T. J. Li. 2000. An *in vitro* method for the estimation of ileal crude protein and amino acids digestibility using the dialysis tubing for pig feedstuffs. Anim. Feed Sci. Technol. 88:79-89.

Jiao, P. X., D. S. Liu, S. Zheng, and C. D. Chen. 2012. Study on protein secondary structures on the protein degradability of feeds in rumen of dairy cows. Feed Ind. 33:48-51.

Liu, B., P. Thacker, J. McKinnon, and P. Yu. 2013. In-depth study of the protein molecular structures of different types of dried distillers grains with solubles and their relationship to digestive characteristics. J. Sci. Food Agric. 93:1438-1448.

Long, G. H., J. Yuan, H. Pan, Z. Sun, Y. Li, and G. X. Qin. 2015. Characterization of thermal denaturation structure and morphology of soy glycinin by FTIR and SEM. Int. J. Food Prop. 18:763-774.

Meng-Xia, X. and L. Yuan. 2002. Studies on the hydrogen bonding of aniline's derivatives by FT-IR. Spectrochim. Acta A Mol. Biomol. Spectrosc. 58:2817-2826.

Papadopoulos, M. C., A. R. El Boushy, A. E. Roodbeen, and E. H. Ketelaars. 1986. Effects of processing time and moisture content on amino acid composition and nitrogen characteristics of feather meal. Anim. Feed Sci. Technol. 14:279-290.

Peng, Q. H., N. A. Khan, Z. Wang, and P. Q. Yu. 2014. Relationship of feeds protein structural makeup in common Prairie feeds with protein solubility, *in situ* ruminal degradation and intestinal digestibility. Anim. Feed Sci. Technol. 194:58-70.

Qin, G. X., Z. W. Sun, G. H. Long, T. Wang, and M. M. Bai. 2014. The physicochemical property of feedstuff proteins and its effects on the nutritional value. Chin. J. Anim. Nutr. 26:1-7.

Samadi, K. Theodoridou, and P. Yu. 2013. Detect the sensitivity and response of protein molecular structure of whole canola seed (yellow and brown) to different heat processing methods and relation to protein utilization and availability using ATR-FT/IR molecular spectroscopy with chemometrics. Spectrochim. Acta A Mol. Biomol. Spectrosc. 105:304-313.

Tan-Wilson, A. L., P. M. Hartl, N. E. Delfel, and K. A. Wilson. 1985. Differential expression of Kunitz and Bowman-Birk soybean proteinase inhibitors in plant and callus tissue. Plant Physiol. 78:310-314.

Theodoridou, K. and P. Yu. 2013. Application potential of ATR-FT/IR molecular spectroscopy in animal nutrition: Reveal protein molecular structures of canola meal and presscake, as affected by heat processing methods, in relationship with their protein digestive behavior and utilization for dairy cattle. J. Agric. Food Chem. 61:5449-5458.

Xin, H., X. Zhang, and P. Yu. 2013a. Using synchrotron radiation-based infrared microspectroscopy to reveal microchemical structure characterization: frost damaged wheat vs. normal wheat. Int. J. Mol. Sci. 14:16706-16718.

Xin, H., K. C. Falk, and P. Yu. 2013b. Studies on *Brassica carinata* seed. 1. Protein molecular structure in relation to protein nutritive values and metabolic characteristics. J. Agric. Food Chem. 61:10118-10126.

Yan, X., N. A. Khan, F. Zhang, L. Yang, and P. Yu. 2014. Microwave irradiation induced changes in protein molecular structures of barley grains: relationship to changes in protein chemical profile, protein subfractions, and digestion in dairy cows. J. Agric. Food Chem. 62:6546-6555.

Yu, P., J. J. Mckinnon, C. R. Christensen, and D. A. Christensen. 2004. Using synchrotron-based FTIR micro-spectroscopy to reveal chemical features of feather protein secondary structure: comparison with other feed protein sources. J. Agric. Food Chem. 52:7353-7361.

Yu, P. and W. G. Nuez-Ortín. 2010. Relationship of protein molecular structure to metabolisable proteins in different types of dried distillers grains with solubles: A novel approach. Br. J. Nutr. 104:1429-1437.

Zhang, X. and P. Yu. 2012. Molecular basis of protein structure in combined feeds (hulless barley with bioethanol coproduct of wheat dried distillers grains with solubles) in relation to protein rumen degradation kinetics and intestinal availability in dairy cattle. J. Dairy Sci. 95:3363-3379.

18

Effects of Different Cutting Height on Nutritional Quality of Whole Crop Barley Silage and Feed Value on Hanwoo Heifers

Dong Hyeon Kim[1], Sardar M. Amanullah[1,2], Hyuk Jun Lee[1], Young Ho Joo[1],
Ouk Kyu Han[3], Adegbola T. Adesogan[4], and Sam Churl Kim[1,*]

[1] Division of Applied Life Science (BK21Plus, Institute of Agriculture and Life Science),
Gyeongsang National University, Jinju 52828, Korea

ABSTRACT: The present study evaluated the effects of different cutting height on nutritive value, fermentation quality, *in vitro* and *in vivo* digestibility of whole crop barley silage. Whole crop barley forage (Yuyeon hybrid) was harvested at height of 5, 10, and 15 cm from the ground level. Each cutting height was rolled to make round bale and ensiled for 100 days. After 100 days of ensiling, pH of silage was lower ($p < 0.05$) in 5 cm, but no difference between 10 and 15 cm of cutting height. The content of lactate and lactate to acetate ratio were increased ($p < 0.05$) in 5 cm of cutting height, whereas the acetate content was higher ($p < 0.05$) in 10 and 15 cm than that of 5 cm cutting height. Aerobic stability was greater ($p < 0.05$) in silages of 10 and 15 cm of cutting height. Three total mixed rations (TMR) were formulated with silages from the three different cutting heights (TMR5, TMR10, and TMR15) incorporated as forage at 70:30 ratio with concentrate (dry matter [DM] basis). *In vitro* dry matter digestibility was higher ($p < 0.05$) in the TMR5 and TMR10 than that in TMR15, whereas *in vitro* neutral detergent fiber digestibility was higher ($p < 0.05$) in the TMR10 and TMR15 than that in TMR5. Concentration of NH_3-N was highest ($p < 0.05$) in the TMR10 followed by TMR15 and TMR5. Total volatile fatty acid was decreased ($p < 0.05$) with increased cutting height. The digestibility of DM and neutral detergent fiber were highest ($p < 0.05$) in TMR15, than those in TMR5 and TMR10, whereas acid detergent fiber digestibility was higher ($p < 0.05$) in TMR5 than that in TMR10. The results showed that increasing cutting height, at least up to 10 to 15 cm, of whole crop barley forage at harvest (Yuyeon) may be beneficial for making silage for TMR formulation and increasing digestibility of DM and NDF. (**Key Words:** Aerobic Stability, Barley Silage, Cutting Height, Digestibility, Hanwoo Heifer)

INTRODUCTION

The forages for feeding cattle in South Korea comprises rice straw (36%), cultivated forage crop (44%), and imported forage (20%) (MIFAFF, 2011). Except for rice straw that is low quality forage, barley as winter forage occupies the largest share of the domestic forage production. Therefore, improvement of nutritional quality of barley forage may help to increase animal productivity and thus farm profitability. Increasing cutting height at harvest is one of the good options

for improving nutritional quality of forage or silage. Higher yield of dry matter (DM) and amount of grain from forage (Neylon and Kung, 2003) are considered as quality indicators for whole crop forage or silage. Recently, high digestibility of DM and fiber is also considered indispensable for good quality of forage (Lynch et al., 2015). Above characteristics may be brought into whole crop forage or silage by increasing the cutting height to make whole crop silage (Weller, 1992; Neylon and Kung, 2003; Caetano et al., 2011; Lynch et al., 2015). The potential benefit of increasing cutting height is to reduce the proportion of the fibrous bottom part which has poor digestibility (Caetano et al., 2011). Several studies reported that harvesting the forage at higher cutting height can improve nutritive value of silage and animal performance (Sinclair et al., 2003; Kennington et al., 2005; Wu and Roth, 2005). However, most of studies have been done with corn silage (Neylon and Kung, 2003; Kennington et al., 2005; Wu and Roth, 2005; Caetano et al.,

* Corresponding Author: Sam Churl Kim.
E-mail: kimsc@gnu.ac.kr
[2] Bangladesh Livestock Research Institute, Savar, Dhaka-1341, Bangladesh.
[3] National Institute of Crop Science, RDA, Suwon 16429, Korea.
[4] Department of Animal Sciences, IFAS, University of Florida, Gainesville, FL 32608, USA.

2011; Lynch et al., 2015) and very rarely with barley. Therefore, the present study was conducted to evaluate effects of different cutting height on chemical composition, fermentation characteristics of whole crop barley (WCB) silage and, *in vitro* and *in vivo* nutrient digestibility of total mixed rations (TMR) made up of silages of different cutting height. Our hypothesis was that the increase in cutting height at forage harvest could improve the silage quality and nutrient digestibility of WCB silage.

MATERIALS AND METHODS

Silage experiment

Forage production and silage making: The Yuyeon hybrid of barley was used in this experiment, which was developed in Korea by traditional crossbreeding having ruminant palatable awns and better silage quality (Park et al., 2008). The WCB was cultivated at the Animal Research Unit, Gyeongsang National University, Jinju, Korea with a seed rate of 220 kg/ha. It was planted in October 2011 and harvested in May 2012 at soft dough stage of maturity. The forage was mowed at a height of 5, 10, and 15 cm, from the ground level. Approximately 500 kg forage from each cutting height was rolled to make round bale with a baler (BWR1-150, JUKAM Machinery Co., Ltd, Suncheon, Korea) followed by wrapping (27 layers of 0.03-µm poly-ethylene film) with a wrapper and ensiled in 3 replicates for 100 days. Representative samples of fresh forage were collected from each cutting height before ensiling and stored at –20°C for chemical analysis. After ensiled for 100 days, bales were opened and silages were sub-sampled for subsequent laboratory analysis (500 g), aerobic stability (1 kg) determination and storage (–20°C) for further use in later.

Preparation of silage extract: Twenty grams of fresh silage were homogenized with 200 mL of sterile double-distilled water in a blender (HM-1600PB, HANIL Electric Co., Gimpo, Korea) for 30 seconds and then filtered through 2 layers of cheesecloth. The pH of the silage extract was measured right after extraction, sampled for microbial enumeration and then stored at –20°C for analysis of fermentation products (NH_3-N, lactic acid volatile fatty acids) in silages.

Microbial enumeration: The microbial enumeration (yeast, mold, and lactic acid bacteria [LAB]) was done from silage extract immediately after silo opening. A 20 g sample of silage in 180 mL of sterile peptone water was homogenized in a blender, filtered through 2 layers of cheese cloth to get silage extract, which was considered as the first dilution and then 10-fold serial dilutions were made. One hundred µL aliquots of 4 consecutive dilutions (10^{-3} to 10^{-7}) were plated in triplicate on selective agar medium for respective microbes. Potato dextrose agar (Difco, Detroit, MI,

USA) was used for the isolation and enumeration of yeast and mold, and lactobacilli MRS agar (Difco, USA) was used for the isolation and enumeration of LAB. The agar plates of LAB were incubated at 39°C for 24 h in a CO_2 incubator (BB15, Thermo Scientific, Waltham, MA, USA) and the agar plates of yeast and mold were incubated at 35°C for 24 h in a general incubator (JS-IN-180, Johnsam Co., Boocheon, Korea).

Aerobic stability: The silage samples were transferred to open-top polyethylene container to determine aerobic stability of silages. Two thermocouple wires were placed to the center of each silages and connected to data loggers (TR-60CH, MORHAN, Hong Kong, China) along with a computer that recorded temperature at every 30 min for 12 days. The silage containers were covered with 2 layers of cheesecloth to prevent drying and dust in the air. Aerobic stability was determined by the time required to raise the silage temperature 2°C above the ambient temperature as suggested by Arriola et al. (2011).

In vitro experiment

Three TMR were formulated with WCB silages from different cutting heights (5, 10, and 15 cm) incorporated as forage at 70:30 ratio with concentrate (DM basis). Thus treatments were i) TMR5 (WCB silage from 5 cm of forage cutting height in the TMR), ii) TMR10 (WCB silage from 10 cm of forage cutting height in the TMR), and iii) TMR15 (WCB silage from 15 cm of forage cutting height in the TMR). The ingredient and chemical compositions of three TMR diets are shown in Table 5. Diets were ground in a grinder (Cutting mill, Shinmyung Electric Co., Ltd, Gimpo, Korea), to pass through a 1 mm screen and incubated with rumen fluid and Van Soest medium (Van Soest et al., 1966) to determine the *in vitro* dry matter digestibility (IVDMD) and *in vitro* neutral detergent fiber digestibility (IVNDFD) and rumen fermentation indices. Rumen fluid was collected from two non-pregnant, cannulated Hanwoo heifers fed rice straw and grain mixed at 8:2 ratio. Collection was made 2 h after morning feeding (0800 h). The collected rumen fluid was composited, filtered via 2 layers of cheesecloth and mixed with Van Soest medium in a 1:2 ratio. The incubation media was flushed continuously by carbon dioxide gas to maintain anaerobic conditions as suggested by Adesogan et al. (2005). The diet (0.5 g) and incubation medium (40 mL) were poured into the incubation bottle (Ankom Technology, Macedon, NY, USA) and then placed into the incubator at 39°C for 48 h. Five replications of each treatment were used along with three blanks. After incubation, the sample was transferred to 50 mL conical tube and the residue was separated from supernatant using centrifuge (Supra 21k, HANIL Electric Corporation, Korea) with rotor (A50S-6C No.6, HANIL Science Industrial, Korea) at 2,568×g for 15 min. Rumen pH, NH_3-N and volatile fatty acid (VFA) were analyzed from the supernatant.

In vivo experiment

Management of animals: The experiment was carried out at Junga Hanwoo Farm, Jinju, Korea. Animals were cared according to the guidelines of the National Livestock Research Institute (NLRI), Korea. Diets were formulated to meet the nutrient requirements of growing Hanwoo heifer according to the Korean Feeding Standards for Hanwoo cattle developed by NLRI, Rural Development Administration, Ministry of Agriculture and Forestry (Korean Feeding Standard, 2002).

Twelve Hanwoo heifers weighing 375±46 kg were randomly assigned into three dietary treatments. Heifers were housed in individual tie-stalls and fed TMR diets as in the *in vitro* experiment. The diets were provided at a rate of 2.2% of live weight (Table 8; 3.25 kg) at 0800 and 1700 h daily. Free access to clean drinking water was confirmed throughout the experimental period. The trial was continued for 15 days of which first 10 days were for adaptation and last 5 days were for sample collection.

Collection and sampling: The diets were formulated daily and sub-sampled for DM, crude protein (CP), ether extract (EE), crude ash, neutral detergent fiber (NDF), and acid detergent fiber (ADF) analysis. In collection period, the feed refusals and feces were collected and weighed every morning 30 min before feeding. The feces were collected into covered plastic buckets, weighed, mixed and sub-sampled (10%) separately for each animal. At the end of collection period, the samples were composited by animal, mixed and sub-sampled representatively, dried at 65°C for 48 h, ground to pass a 1-mm screen using grinder (Cutting Mill, Shinmyung Electric Co., Ltd, Gimpo, Korea) and analyzed for DM, CP, EE, NDF, and ADF. Then DM, CP, EE, NDF, and ADF digestibility was calculated. The samples were stored at –20°C until end of analysis.

Laboratory analysis

The samples were dried using an oven at 65°C for 48 h to measure the DM content and then the dried samples were ground through a 1-mm screen using a grinder for the analyses of CP, EE, crude ash, NDF, ADF analyses and *in vitro* digestibility study. Content of CP was calculated as N ×6.25, after N was quantified using N analyzer (B-324, 412, 435 and 719 S Titrino, BÜCHI, Flawil, Switzerland). The EE was analyzed using the Soxhlet method (AOAC, 1965). Crude ash concentration was determined by burning the sample in a muffle furnace at 550°C for 5 h. Contents of NDF and ADF were analyzed according to Van Soest et al. (1991) using ANKOM[200] fiber analyzer (Ankom Technology, Macedon, NY, USA). Heat-stable α-amylase and sodium sulfite were used in NDF analysis and results were expressed inclusive of ash.

The pH was measured from silage extract using a pH meter (SevenEasy pH Meter S20, Mettler Toledo, Greifensee, Switzerland) and the content of NH_3-N in silage extract was analyzed by distillation using BÜCHI apparatus (B-324, BÜCHI, Flawil, Switzerland) followed by titration with 0.1-N H_2SO_4 according to AOAC (1984). Silage extract was centrifuged at 21,500×g for 15 min at –4°C constant temperature. The supernatant was used to measure the lactate and VFA contents by an high-performance liquid chromatography system with a pump (L-2130, HITACHI, Tokyo, Japan), auto sampler (L-2200, HITACHI, Tokyo, Japan), UV detector (L-2400, HITACHI, Japan) and a column (Metacarb 87H, Varian, Middelburg, Netherlands) as described by Adesogan et al. (2004).

Statistical analysis

The data were analyzed using the general linear model procedure of SAS (2002). Data for microbial enumeration were transformed by log_{10}. The model was $Y_{ij} = \mu + T_i + e_{ij}$, where Y_{ij} = response variable, μ = overall mean, T = effect of treatment i, and e_{ij} = error effect. Tukey's test was performed to differentiate means. Significance was declared at $p \leq 0.05$.

RESULTS

Silage experiment

The DM content of fresh WCB before ensiling was increased ($p<0.05$) with the increase in cutting height, while NDF and ADF contents decreased ($p<0.05$) (Table 1). The EE content in 10 and 15 cm of cutting height increased ($p<0.05$), but CP and crude ash were remain unaffected ($p>0.05$) by increasing cutting height. After 100 days of ensiling, the DM, CP, and Crude ash were unaffected ($p>0.05$) by cutting height (Table 2). The EE concentration was increased ($p<0.05$) while NDF and ADF concentration was decreased ($p<0.05$) by increasing cutting height. The pH of silage was lower ($p<0.05$) in 5 cm, but not different between

Table 1. Chemical composition of barley silage before ensiling (%, DM)

Items	Cutting height[1] (cm)			SEM
	5	10	15	
Dry matter	33.8[b]	33.9[a]	34.6[a]	0.285
Crude protein	8.61	8.61	8.64	0.052
Ether extract	2.45[b]	3.30[a]	3.23[a]	0.149
Crude ash	9.13	9.13	9.12	0.017
Neutral detergent fiber	57.9[a]	57.4[ab]	56.0[b]	0.588
Acid detergent fiber	33.9[a]	33.4[ab]	32.8[b]	0.432

DM, dry matter; SEM, standard error mean.

[1] Barley silage had been harvested at 5, 10, and 15 cm of cutting height from the field, respectively.

[a,b] Means in the same row with different superscripts differ significantly ($p<0.05$).

Table 2. Chemical composition of barley silage ensiled for 100 days (%, DM)

Items	Cutting height[1] (cm)			SEM
	5	10	15	
Dry matter	30.6	30.9	30.6	0.996
Crude protein	9.04	9.02	9.02	0.016
Ether extract	3.34[c]	3.57[b]	3.71[a]	0.074
Crude ash	10.2	9.00	9.00	1.376
Neutral detergent fiber	52.4[a]	52.0[ab]	52.6[b]	0.363
Acid detergent fiber	31.0[a]	30.6[ab]	30.4[b]	0.301

DM, dry matter; SEM, standard error mean.
[1] Barley silage had been harvested at 5, 10, and 15 cm of cutting height from the field, respectively.
[a,b] Means in the same row with different superscripts differ significantly (p<0.05).

Table 4. Aerobic stability and microbial growth of barley silage ensiled for 100 days

Items	Cutting height[1] (cm)			SEM
	5	10	15	
Aerobic stability (h)	139.3[b]	278.3[a]	270.5[a]	17.86
Microbes (log10 cfu/g)				
Yeast	2.76	2.34	2.42	0.582
Mold	3.43	3.21	3.19	0.285
Acid bacteria	6.27	6.58	6.21	0.299

SEM, standard error mean.
[1] Barley silage had been harvested at 5, 10, and 15 cm of cutting height from the field, respectively.
[a,b] Means in the same row with different superscripts differ significantly (p<0.05).

10 and 15 cm of cutting height (Table 3). The content of lactate and lactate to acetate ratio were increased (p<0.05) in 5 cm of cutting height, whereas the acetate content was higher (p<0.05) in 10 and 15 cm than that of 5 cm cutting height. The concentration of NH_3-N in 100 days silage was not affected (p>0.05) by increasing cutting height. Aerobic stability was greater (p<0.05) in silages of 10 and 15 cm of cutting height, whereas yeast, mold and LAB counts were not affected (p>0.05) by cutting height (Table 4).

In vitro and *in vivo* experiment

The ingredient compositions of concentrate mixture used in different TMR are presented in Table 5 and the ingredients and chemical compositions of formulated TMRs are illustrated in Table 6. There were no differences (p>0.05) in contents of DM, CP, EE, and crude ash in different TMRs used as treatments. On the other hand, NDF and ADF were found to be decreased (p<0.05) in the TMRs containing silages from higher cutting heights (Table 6). The concentrations of NDF vs ADF in TMR5, TMR10, and TMR15 were observed as 51.9% vs 27.8%, 50.9% vs 27.4%, and 50.1% vs 26.9%, respectively.

The *in vitro* digestibility and rumen fermentation indices

are described in Table 7. The IVDMD was higher (p<0.05) in the TMR5 (47.9%) and TMR10 (48.8%) than that in TMR15 (46.3%), whereas IVNDFD was higher (p<0.05) in the TMR10 (42.9%) and TMR15 (43.5%) than that in TMR5 (38.7%). The pH was lower (p<0.05) in TMR10 (6.98) than in the others, but there were no differences between TMR5 (7.06) and TMR15 (7.00). Concentration of NH_3-N (mg/100 mL) was highest (p<0.05) in the TMR10 (34.8) followed by TMR15 (33.6) and TMR5 (27.5). Total VFA was decreased (p<0.05) with increased cutting height (62.7, 59.7, and 53.3 mM/L in TMR5, TMR10, and TMR15, respectively). Molar proportions (%) of acetate, butyrate and acetate to propionate ratio were highest (p<0.05) in TMR5 (58.6, 9.14 and 2.59, respectively). The valerate was increased (p<0.05) in TMR10 and propionate concentration was increased (p<0.05) in TMR 10 and TMR5.

Table 5. Composition of concentrate mixed into the TMR using the *in vitro* and *in vivo* experiment (%, DM)

Ingredient	%
Corn meal	15.0
Barley meal	9.00
Soybean meal	12.5
Rice bran	14.6
Wheat bran	19.0
Corn gluten feed	9.50
Soy bean hull	8.30
Corn hull	0.50
Corn cob	8.00
Corn gluten meal	1.00
Salt dehydrated	0.40
Molasses	1.50
Vitamin and mineral premix[1]	0.70

TMR, total mixed rations; DM, dry matter.
[1] One kilogram of the diet contained the following: vitamin A, 450,000 IU; vitamin D_3 350,000 IU; vitamin E, 20,000 IU; vitamin K_3, 500 mg; vitamin B_1, 300 mg; vitamin B_{12}, 15 mg; pantothenic acid, 50 mg; niacin, 20 mg; biotin, 20 mg; folic acid, 10 mg; $FeSO_4$, 4,000 mg; $CoSO_4$, 100 mg; $CuSO_4$, 5,000 mg; $MnSO_4$, 2,500 mg; $ZnSO_4$, 2,000 mg; I, 500 mg; Se(Na), 100 mg.

Table 3. Fermentation indices of barley silage ensiled for 100 days

Items	Cutting height[1] (cm)			SEM
	5	10	15	
pH	3.86[b]	4.00[a]	4.02[a]	0.045
NH_3-N (% of DM)	0.08	0.09	0.09	0.005
NH_3-N (% of total N)	5.84	5.81	5.85	0.190
Volatile fatty acid (% of DM)				
Lactate	6.31[a]	5.53[b]	5.51[b]	0.117
Acetate	1.28[b]	1.43[a]	1.44[a]	0.017
Lactate:acetate ratio	4.93[a]	3.87[b]	3.83[b]	0.057

DM, dry matter; SEM, standard error mean.
[1] Barley silage had been harvested at 5, 10, and 15 cm of cutting height from the field, respectively.
[a,b] Means in the same row with different superscripts differ significantly (p<0.05).

Table 6. Ingredients and chemical composition of TMR using *in vitro* and *in vivo* experiment[1] (%, DM)

	TMR5	TMR10	TMR15	SEM
Ingredient				
Barley silage (5 cm of cutting height)	70			
Barley silage (10 cm of cutting height)		70		
Barley silage (15 cm of cutting height)			70	
Concentrate	30	30	30	
Chemical composition				
Dry matter	38.7	38.2	38.2	0.679
Crude protein	13.3	13.4	13.3	0.089
Ether extract	3.25	3.30	3.23	0.153
Crude ash	10.8	10.7	10.4	0.189
Neutral detergent fiber	51.9[a]	50.9[ab]	50.1[b]	0.283
Acid detergent fiber	27.8[a]	27.4[ab]	26.9[b]	0.182

TMR, total mixed rations; DM, dry matter; SEM, standard error mean.

[1] TMR5, TMR10, and TMR15 means TMR based on barley silage had been harvested at 5, 10, and 15 cm of cutting height from the field, respectively.

[a,b] Means in the same row with different superscripts differ significantly (p<0.05).

Feed intakes and apparent total tract digestibility of nutrients in heifers are described in Table 8. Feed intakes and the digestibility of CP were not affected (p>0.05) by TMRs. The digestibility of DM and NDF were higher (p<0.05) in TMR15 (61.5% and 58.6%), compared to those in TMR5 (56.5% and 49.3%) and TMR10 (57.8% and 51.0%), whereas ADF digestibility was higher (p<0.05) in TMR5 (45.5%) than that in TMR10 (40.7%). The EE digestibility was increased (p<0.05) with the increase in cutting height.

Table 7. Effect of total mixed ration with barley silage on *in vitro* digestibility and rumen fermentation indices[1]

	TMR5	TMR10	TMR15	SEM
IVDMD (% of DM)	47.9[a]	48.8[a]	46.3[b]	1.267
IVNDFD (% of DM)	38.7[b]	42.9[a]	43.5[a]	0.407
pH	7.06[a]	6.98[b]	7.00[a]	0.034
NH$_3$-N (mg N/100 mL)	27.5[c]	34.8[a]	33.6[b]	0.330
Total VFA (mM/L)	62.7[a]	59.7[b]	53.3[c]	0.607
Acetate (% of molar)	58.6[a]	52.7[b]	52.5[b]	1.133
Propionate (% of molar)	22.6[b]	31.8[a]	31.6[a]	0.435
Iso-butyrate (% of molar)	2.12	2.08	1.97	0.114
Butyrate (% of molar)	9.14[a]	7.49[b]	6.84[b]	0.531
Iso-valerate (% of molar)	3.88[a]	3.31[b]	3.32[b]	0.159
Valerate (% of molar)	2.82[b]	3.10[a]	2.54[c]	0.065
Acetate:propionate ratio	2.59[a]	1.66[b]	1.66[b]	0.043

SEM, standard error mean; IVDMD, *in vitro* dry matter digestibility; IVNDFD, *in vitro* neutral detergent fiber digestibility.

[1] TMR5, TMR10, and TMR15 means TMR based on barley silage had been harvested at 5, 10, and 15 cm of cutting height from the field, respectively.

[a,b,c] Means in the same row with different superscripts differ significantly (p<0.05).

DISCUSSION

Silage experiment

The increased DM content of fresh forages with the increase in cutting height in this study is related to the higher DM content of the upper part of WCB compared to the lower part. The grain and spike in the upper part in WCB contributed higher DM content (Ji et al., 2007). Similarly, whole crop wheat (446 vs 477 g/kg) and corn (33.9% vs 41.5%) were also reported to have higher DM content with higher cutting height compared to the lower cutting height (Neylon and Kung, 2003; Sinclair et al., 2003). However, the difference in DM content in fresh forage due to cutting height difference before ensiling did not persist after ensiling (Table 2). The CP, NDF, and ADF concentration of barley forage was different from the values reported by Yun et al. (2009) with same barley hybrid, which might be due to different soil characteristics and harvesting stage of barley forage (Ji et al., 2007; Song et al., 2011). The decreased NDF and ADF content of both forage and silage with increasing cutting height was in agreement with most other previous studies. Walsh et al. (2008) reported a decrease in NDF (465 vs 437 g/kg) and ADF (230 vs 194 g/kg) content when increasing cutting height of barley forage. Sinclair et al. (2003) reported that increasing cutting height of whole crop wheat reduced the NDF content (433 vs 384 g/kg). In case of whole plant corn, the NDF content was numerically decreased by cutting height, whereas ADF content (25.3% vs 23.4%) was reduced significantly (Neylon and Kung, 2003; Lynch et al., 2015).

Lactate has the major role in lowering pH of silage (Shaver, 2003; Zahiruddini et al., 2004). The pH was lowest in 5 cm of cutting height in this study because of the higher lactate content in that silage. Oude Eferink et al. (2001) reported that it is possible to convert lactate to acetate under anaerobic conditions. Through unknown mechanisms, lactate was converted to acetate in the higher cutting height silages. As a result, concentration of acetate increased in 10 and 15 cm of cutting height than 5 cm of cutting height.

Table 8. Digestibility of total mixed ration with barley silage on Hanwoo heifers[1] (DM basis)

	TMR5	TMR10	TMR15	SEM
Feed intakes (kg/d)	3.25	3.25	3.25	
Digestibility (%)				
Dry matter	56.5[b]	57.8[b]	61.5[a]	0.981
Crude protein	56.0	55.8	55.4	1.040
Ether extract	73.0[c]	78.0[b]	80.2[a]	0.890
Neutral detergent fiber	49.3[b]	51.0[b]	58.6[a]	1.109
Acid detergent fiber	45.5[a]	40.7[b]	41.9[ab]	1.021

SEM, standard error mean.

[1] TMR5, TMR10, and TMR15 means TMR based on barley silage had been harvested at 5, 10, and 15 cm of cutting height from the field, respectively.

[a,b] Means in the same row with different superscripts differ significantly (p<0.05).

Aerobic stability can be improving via the growth inhibition of yeast and mold in silage (Weissbach, 1996). Acetate is one of the antifungal factors for inhibition of yeast and mold (Courtin and Spoelstra, 1990). The yeast and mold counts were not affected in this study at the day of silo opening. However, the increased concentration of acetate in 10 and 15 cm silage might have inhibited yeast and mold growth and thereby increased aerobic stability in those silages. This result was in agreement with other barley studies (Kung and Ranjit, 2001; Taylor et al., 2002).

In vitro and *in vivo* experiment

The decreased concentration of NDF and ADF in TMR (Table 6) with the incorporation of barley silages of increased cutting height is directly related to the chemical composition of respective silages. Silage was incorporated into the TMR at a rate of 70% (DM basis) while the concentrate mixture was same for all TMRs. Therefore, the fiber concentration of TMRs was dominated by the composition of silages. Weller et al. (1995) reported a decrease in ADF content of silage with the increase of forage cutting height.

It is commonly found that increasing cutting height leads to increase digestibility of whole crop silage due to the proportionate reduction of the fibrous lower part of the plant (Tolera and Sundstøl, 1999) and an increase in the grain portion (Weller et al., 1995). Variety specific increase in IVDMD of maize silage with increased cutting height was reported by Bernard et al. (2004). However, in the present study, the IVDMD was observed lowest in TMR15, which contained the silage of highest (15 cm) cutting height. However, contrasting results in total tract apparent digestibility of DM in Hanwoo heifers was found in *in vivo* study (Table 8) compared to *in vitro*. Thus the DM digestibility from *in vivo* study was in agreement with previous findings (Bernard et al., 2004). The reason for increased NDF digestibility in TMR15 both in *in vitro* and *in vivo* experiment may be related to the increased portion of grains in that treatment. Usually grains contain more digestible nutrients than that in roughage. The reason for increased total tract apparent digestibility of ADF in TMR5 is not clear. However, as effects of cutting height on digestibility may also affected by varietal difference (Bernard et al., 2004), it may happen that fibers from the lower portion of Yuyeon barley forage are also easily digestible. Future study is needed to find digestibility of different nutrients from different portions of Yuyeon barley for a better explanation. In agreement with the present study, increasing cutting height of wheat or barley did not affect feed intake (DM intake) as reported previously (Sinclair et al., 2003; Walsh et al., 2008).

The higher production of total VFA, acetate and butyrate in TMR5 may be related to the higher IVDMD in this treatment. With the increase in cutting height, a shift to increase propionate production was (in TMR10 and TMR15) observed, that might be partially due to the increased IVNDFD in those treatments. France and Dijkstra (2005) reported when high sugar content exists in the rumen, there is a shift of fermentation pattern from acetate to propionate, and in most of cases, acetate to propionate ratio was decreased by increased propionate. Decreased acetate to propionate ratio in TMR10 and TMR15 after *in vitro* incubation is therefore in agreement with the above statement.

CONCLUSION

Present study revealed that increasing cutting height increased DM concentration in harvested forage, which, however, was not observed in the silage after 100 day of ensiling. Fiber concentration in terms of ADF and NDF was decreased both in harvested forage and silage. A dramatic increase in aerobic stability of silage was achieved by increasing cutting height, which has great importance when the silage is incorporated into TMR. Significant improvement was observed in DM and NDF digestibility in the TMR incorporated with silage of higher cutting height (15 cm). Considering all above findings, it can be concluded that increasing cutting height, at least up to 10-15 cm, of WCB forage at harvest (Yuyeon) may be beneficial for making silage for TMR formulation and increasing digestibility of DM and NDF.

ACKNOWLEDGMENTS

This work was carried out with the support of "Cooperative Research Program for Agriculture Science & Technology Development (Project No. PJ011012032016)" Rural Development Administration, Republic of Korea.

REFERENCES

Archer, J. A., E. C. Richardson, R. M. Herd, and P. F. Arthur. 1999. Potential for selection to improve efficiency of feed use in beef cattle: a review. Aust. J. Agric. Res. 50:147-162.

Arriola, K. G., S. C. Kim, C. R. Staples, and A. T. Adesogan. 2011. Effect of applying bacterial inoculants containing different types of bacteria to corn silage on the performance of dairy cattle. J. Dairy Sci. 94:3973-3979.

Adesogan, A. T., N. K. Krueger, M. B. Salawu, D. B. Dean, and C. R. Staples. 2004. The influence of treatment with dual-purpose bacterial inoculants or soluble carbohydrates on the fermentation and aerobic stability of bermudagrass. J. Dairy Sci. 87:3407-3416.

Adesogan, A. T., N. K. Krueger, and S. C. Kim. 2005. A novel, wireless, automated system for measuring fermentation gas production kinetics of feeds and its application to feed characterization. Anim. Feed Sci. Technol. 123-124:211-223.

AOAC (Association of Official Analytical Chemist). 1965. Official Method of Analysis, 10th edn. AOAC, Washington, DC, USA.

AOAC (Association of Official Analytical Chemist). 1984. Official Method of Analysis, 14th edn. AOAC, Washington, DC, USA.

Bernard, J. K., J. W. West, D. S. Trammell, and G. H. Cross. 2004. Influence of corn variety and cutting height on nutritive value of silage fed to lactating dairy cows. J. Dairy Sci. 87:2172-2176.

Caetano, H., M. D. S. de Oliveira, J. E. de Freitas Jr., A. C. de Rêgo, F. P. Rennó, and M. V. de Carvalho. 2011. Evaluation of corn cultivars harvested at two cutting heights for ensilage. R. Bras. Zootec. 40:12-19.

Courtin, M. G. and S. F. Spoelstra. 1990. A simulation model of the microbiological and chemical changes accompanying the initial stage of aerobic deterioration of silage. Grass Forage Sci. 45:153-165.

Eun, J. S. and K. A. Beauchemin. 2006. Supplementation with combinations of exogenous enzymes: Effects on in vitro fermentation of alfalfa hay and corn silage. Final report to Dyadic International Inc. Agriculture and Agrifood Research Centre Report, Lethbridge, Canada.

France, J. and J. Dijkstra. 2005. Volatile fatty acid production. In: Quantitative Aspects of Ruminant Digestion and Metabolism. (Eds. J. Dijkstra, J. M. Forbes, and J. France). CABI Publishing, Wallingford, UK. pp. 157-176.

Ji, H. C., J. I. Ju, and H. B. Lee. 2007. Feed value and yield of whole crop barley varieties depend on organic content. J. Korean Grassl. Sci. 27:263-268.

Kennington, L. R., C. W. Hunt, J. I. Szasz, A. V. Grove, and W. Kezar. 2005. Effect of cutting height and genetics on composition, intake, and digestibility of corn silage by beef heifers. J. Anim. Sci. 83:1445-1454.

KFS (Korean Feeding Standard). 2002. Korean Feeding Standard for Korean Cattle (Hanwoo). National Livestock Research Institute, Rural Development Administration, Ministry for Food, Agriculture, Forestry, and Fisheries, Gwacheon, Korea.

Krueger, N. A. and A. T. Adesogan. 2008. Effects of different mixtures of fibrolytic enzymes on digestion and fermentation of bahiagrass hay. Anim. Feed. Sci. Technol. 145:84-94.

Krueger, N. A., A. T. Adesogan, C. R. Staples, W. K. Krueger, D. B. Dean, and R. C. Littell. 2008. The potential to increase digestibility of tropical grasses with a fungal, ferulic acid esterase enzyme preparation. Anim. Feed Sci. Technol. 145:95-108.

Kung, L. Jr. and N. K. Ranjit. 2001. The effect of lactobacillus buchneri and other additives on the fermentation and aerobic stability of barley silage. J. Dairy Sci. 84:1149-1155.

Lynch, J. P., J. Baah, and K. A. Beauchemin. 2015. Conservation, fiber digestibility, and nutritive value of corn harvested at 2 cutting heights and ensiled with fibrolytic enzymes, either alone or with a ferulic acid esterase-producing inoculant. J. Dairy Sci. 98:1214-1224.

MIFAFF. 2011. Counterplan on Increasing of Forage Production. Ministry for Food, Agriculture, Forestry and Fisheries, Gwacheon, Korea. pp. 2-5.

Neylon, J. M. and L. Kung, Jr. 2003. Effects of cutting height and maturity on the nutritive value of corn silage for lactating cows. J. Dairy Sci. 86:2163-2169.

Nsereko, V. L., B. K. Smiley, W. M. Rutherford, A. Spielbauer, K. J. Forrester, G. H. Hettinger, E. K. Harman, and B. R. Harman. 2008. Influence of inoculating forage with lactic acid bacterial strains that produce ferulate esterase on ensilage and ruminal degradation of fiber. Anim. Feed Sci. Technol. 145:122-135.

Oba, M. and M. S. Allen. 1999. Evaluation of the importance of the digestibility of neutral detergent fiber from forage: effects on dry matter intake and milk yield of dairy cows. J. Dairy Sci. 82:589-596.

Oude Elferink, S. J. W. H., J. Kooneman, J. C. Gottschal, S. F. Spoelstra, F. Faber, and F. Driehuis. 2001. Anaerobic conversion of lactic acid to acetic acid and 1, 2-propanediol by Lactobacillus buchneri. Appl. Environ. Microb. 67:125-132.

Park, T. I., O. K. Han, J. H. Seo, J. S. Choi, K. H. Park, and J. G. Kim. 2008. New barley cultivars with improved morphological characteristics for whole crop forage in Korea. J. Korean. Grassl. Forage Sci. 28:193-202.

SAS (Statistical Analysis System) Institute Inc. 2002. SAS/STAT User's Guide: Version 9. SAS Institute Inc., Cary, NC, USA.

Shaver, R. D. 2003. Practical application of new forage quality tests. In: Proceedings of the 6th Western Dairy Management Conference, Reno, NV, USA. pp. 22-25.

Sinclair, L. A., R. G. Wikinson, and D. M. R. Ferguson. 2003. Effects of crop maturity and cutting height on the nutritive value of fermented whole crop wheat and milk production in dairy cows. Livest. Prod. Sci. 81:257-269.

Song, T. H., O. K. Han, S. K. Yun, T. I. Park, and H. J. Kim. 2011. Effect of harvest time on yield and feed value of whole crop barleys with different awn types. J. Korean Grassl. Forage Sci. 31:361-370.

Taylor, C. C., N. J. Ranjit, J. A. Mills, J. M. Neylon, and L. Kung Jr. 2002. The effect of treating whole-plant barley with lactobacillus buchneri 40788 on silage fermentation, aerobic stability, and nutritive value for dairy cows. J. Dairy Sci. 85:1793-1800.

Tolera, A. and F. Sundstøl. 1999. Morphological fractions of maize stover harvested at different stages of grain maturity and nutritive value of different fractions of the stover. Anim. Feed Sci. Technol. 81:1-16.

Van Soest, P. J., J. B. Robertson, and B. A. Lewis. 1991. Methods for dietary fiber, neutral detergent fiber and non-starch polysaccharides in relation to animal nutrition. J. Dairy Sci. 74:3583-3597.

Van Soest, P. J., R. H. Wine, and L. A. Moore. 1966. Estimation of the true digestibility of forages by the in vitro digestion of cell walls. In: Proceedings of the 10th International Grassland Congress. Finish Grassl. Assoc., Helsinki, Finland. pp. 438-441.

Walsh, K., P. O'Kiely, A. P. Moloney, and T. M. Boland. 2008. Intake, digestibility, rumen fermentation and performance of beef cattle fed diets based on whole-crop wheat or barley harvested at two cutting heights relative to maize silage or ad libitum concentrates. Anim. Feed Sci. Technol. 144:257-278.

Weissbach, F. 1996. New developments in crop conservation. In: Proceedings of the 11th International Silage Conference, IGER (Institute of Grassland and Environmental Research), Aberystwyth, UK. pp. 11-25.

Weller, R. F. 1992. The national whole crop cereals survey. In:

Whole-crop Cereals (Eds. B. A. Stark and J. M. Wilkinson). Chalcombe Publications, Canterbury, UK. pp. 137-156.

Weller, R. F., A. Cooper, and M. S. Dhanoa. 1995. The selections of winter wheat varieties for whole-crop cereal conservation. Grass Forage Sci. 50:172-177.

Wu, Z. and G. Roth. 2005. Considerations in managing cutting height of corn silage: Extension publication DAS 03-72. Pennsylvania State University, College Park, MD, USA.

Yun, S. K., T. I. Park, J. H. Seo, K. H. Kim, T. H. Song, K. H. Park, and O. K. Han. 2009. Effect of harvest time and cultivars on forage yield and quality of whole crop barley. J. Korean Grassl. Forage Sci. 29:121-128.

Zahiruddini, H., J. Baah, W. Absalom, and T. A. McAllister. 2004. Effect of an inoculant and hydrolytic enzymes on fermentation and nutritive value of whole crop barley silage. Anim. Feed Sci. Technol. 117:317-330.

19

The Use of Fermented Soybean Meals during Early Phase Affects Subsequent Growth and Physiological Response in Broiler Chicks

S. K. Kim, T. H. Kim, S. K. Lee, K. H. Chang[1], S. J. Cho[1], K. W. Lee, and B. K. An*

Department of Animal Science and Technology, Konkuk University, Seoul 143-701, Korea

ABSTRACT: The objectives of this experiment was to evaluate the subsequent growth and organ weights, blood profiles and cecal microbiota of broiler chicks fed pre-starter diets containing fermented soybean meal products during early phase. A total of nine hundred 1-d-old chicks were randomly assigned into six groups with six replicates of 25 chicks each. The chicks were fed control pre-starter diet with dehulled soybean meal (SBM) or one of five experimental diets containing fermented SBM products (*Bacillus* fermented SBM [BF-SBM], yeast by product and *Bacillus* fermented SBM [YBF-SBM]; *Lactobacillus* fermented SBM 1 [LF-SBM 1]; *Lactobacillus* fermented SBM 2 [LF-SBM 2]) or soy protein concentrate (SPC) for 7 d after hatching, followed by 4 wk feeding of commercial diets without fermented SBMs or SPC. The fermented SBMs and SPC were substituted at the expense of dehulled SBM at 3% level on fresh weight basis. The body weight (BW) during the starter period was not affected by dietary treatments, but BW at 14 d onwards was significantly higher (p<0.05) in chicks that had been fed BF-SBM and YBF-SBM during the early phase compared with the control group. The feed intake during grower and finisher phases was not affected (p>0.05) by dietary treatments. During total rearing period, the daily weight gains in six groups were 52.0 (control), 57.7 (BF-SBM), 58.5 (YBF-SBM), 52.0 (LF-SBM 1), 56.7 (LF-SBM 2), and 53.3 g/d (SPC), respectively. The daily weight gain in chicks fed diet containing BF-SBM, YBF-SBM, and LF-SBM 2 were significantly higher values (p<0.001) than that of the control group. Chicks fed BF-SBM, YBF-SBM, and LF-SBM 2 had significantly lower (p<0.01) feed conversion ratio compared with the control group. There were no significant differences in the relative weight of various organs and blood profiles among groups. Cecal microbiota was altered by dietary treatments. At 35 d, chicks fed on the pre-starter diets containing BF-SBM and YBF-SBM had significantly increased (p<0.001) lactic acid bacteria, but lowered *Coli*-form bacteria in cecal contents compared with those fed the control diet. The number of *Bacillus* spp. was higher (p<0.001) in all groups except for LF-SBM 1 compared with control diet-fed chicks. At 7 d, jejunal villi were significantly lengthened (p<0.001) in chicks fed the fermented SBMs vs control diet. Collectively, the results indicate that feeding of fermented SBMs during early phase are beneficial to the subsequent growth performance in broiler chicks. BF-SBM and YBF-SBM showed superior overall growth performance as compared with unfermented SBM and SPC. (**Key Words:** Fermented Soybean Meals, Growth Performance, Cecal Microbial Population, Villi Length, Broiler Chicks)

INTRODUCTION

Considerable changes in the growth performance of modern broilers have been achieved via intensive genetic selection (Zuidhof et al., 2014). Due to a continuous shortening of the growth period, emphasis has been placed on the early development after hatching. It is well known that the market weights of modern broilers has a linear relationship with their weight at the early phase of rearing (Nir et al., 1993). Therefore, the early feeding regime should be optimized to ensure the best start, which will result in better growth performance. But, the newly hatched chicks have an immature digestive system and are less efficient in utilizing nutrients (Batal and Parsons, 2002) and thus they could possibly benefit from diets containing easily digested ingredients.

Soybean meal (SBM) is the most commonly used vegetable protein source in feed industry. However, a huge variation of nutritional quality and a variety of anti-nutritional factors have limited the application of SBM in

* Corresponding Author: B. K. An.
E-mail: abk7227@hanmail.net
[1] CJ Cheiljedang Ltd., Seoul 100-400, Korea.

broiler feed (Lee et al., 2009). The carbohydrate fraction of SBM, which is primarily composed of non-starch polysaccharide and galactooligosaccharide is poorly utilized by poultry due to a lack of endogenous enzyme (Choct at al., 2010). Therefore, the value of nitrogen-corrected metabolizable energy (MEn) of SBM is about 30% less for poultry as compared with swine due to its oligosaccharide fraction (NRC, 1994; 1998). Fermentation with microorganisms has been used to help degrade these anti-nutritional factors of SBM. The fermentation process is thought to reduce residual anti-nutritional factors such as trypsin inhibitor and to increase in small-size peptides (Hong et al., 2004). In addition, it is known that feeding of fermented SBM significantly increased the activities of digestive enzymes (trypsin, lipase, and protease) in intestinal contents of broiler chickens (Feng et al., 2007b).

In practice, supplementation of fermented SBMs is only applicable in pre-starter diet due to high cost burden. But there were no studies investigating the effect of fermented SBMs added into the pre-starter diet of broiler chicks on their subsequent growth. The objective of this experiment was to evaluate the subsequent growth and physiological response of broiler chicks fed pre-starter diets including fermented SBM products.

MATERIALS AND METHODS

Animals, diets and management

On the day of hatch, the male broiler chicks (Ross 308) were obtained from a local hatchery and individually weighed. A total of nine hundred chicks were randomly allotted to six groups with six replicates of 25 chicks each. Commercially available fermented SBMs were provided by the manufacturers. *Bacillus* fermented SBM (BF-SBM), yeast by product and *Bacillus* fermented SBM (YBF-SBM), *Lactobacillus* fermented SBM 1 (LF-SBM 1), *Lactobacillus* fermented SBM 2 (LF-SBM 2) and soy protein concentrate (SPC, Archer Daniels Midland Company, Chicago, IL, USA) were used for this study. BF-SBM (http://www.cjingredient. com/product/soytide.asp) is produced by the fermentation process with *Bacillus subtillis* and YBF-SBM is a two-stage-fermented SBM which brings probiotic effects. These are known to contain few anti-

nutritional factors, such as oligosaccharides, trypsin inhibitor and lectins. LF-SBM 1 (http://www.dabombprotein.com.tw) and LF-SBM-2 (http:www.dachan.com) are fermented products that are produced by anaerobic fermentation of SBMs using *Lactobacillus* and known to contain at low levels of oligosaccharides. The chicks were fed control pre-starter diet with dehulled SBM or one of five experimental diets containing fermented SBM products or SPC for 7 d after hatching and switched to the grower and finisher diets without fermented SBMs and SPC for 4 wks. The fermented SBM products and SPC used in this study were provided by commercial feed manufactures and were substituted at the expense of dehulled SBM at 3% level on fresh weight basis.

The fermented SBM products and SPC were first analyzed for moisture, crude protein (CP), ether extract, crude fiber by the methods of AOAC (1995) and amino acid contents by amino acid analyzer (Sykam S433, Sykam GmbH, Eresing, Germany), following hydrolysis in 6 N HCl for 22 h at 100°C (Spackman et al., 1958). Methionine and cysteine were determined from samples that had been oxidized in performic acid prior to acid hydrolysis according to the method of Moore (1963). Proximal compositions of the fermented SBM products and SPC are shown in Table 1. All pre-starter test diets were formulated to be equal in the contents of nitrogen-corrected true metabolizable energy (TMEn), CP and available amino acids and processed as a small crumble form (Table 2).

The chicks were initially reared at 33°C; the room temperature was gradually decreased by 4°C weekly until a final temperature of 23°C was reached. Lighting was kept at 23/1 light/dark cycle throughout the experimental period. The chicks were allowed to have free access to diet and water. About 100 to 150 g of experimental diets per chick were provided every day at 08:00. All animal care procedures were approved by Institutional Animal Care and Use Committee in Konkuk University.

Sampling and measurements

The chicks and feed were weekly weighed per pen to calculate live body weight (BW), weight gain, feed intake and feed conversion ratio (FCR) on a pen basis. At 7 d of

Table 1. Proximal compositions of the dehulled soybean meal, fermented soybean meal products and soy protein concentrate

	Dehulled SBM	BF-SBM	YBF-SBM	LF-SBM 1	LF-SBM 2	SPC
Moisture (%)	12.3	6.21	6.25	4.74	7.04	6.31
Crude protein (%)	47.4	56.37	55.23	50.40	51.29	66.67
Ether extract (%)	2.5	0.99	1.56	1.43	1.64	0.11
Crude fiber (%)	3.7	3.49	3.96	4.21	3.75	2.40
Crude ash (%)	6.0	7.20	7.10	6.72	6.11	6.59

BF-SBM, *Bacillus* fermented SBM (CJ Corporation, Seoul, Korea); YBF-SBM, yeast by product and *Bacillus* fermented SBM (CJ Corporation, Seoul, Korea); LF-SBM, *Lactobacillus* fermented SBM 1 (DaBomb Protein Corporation, Tainan, Taiwan); LF-SBM 2, *Lactobacillus* fermented SBM 2 (Total Nutrition Technologies Co., Ltd, Tainan, Taiwan); SPC, soy protein concentrate (Archer Daniels Midland Company, Chicago, IL, USA).

Table 2. Formula and chemical compositions of pre-starter diets[1]

	Control	BF-SBM	YBF-SBM	LF-SBM 1	LF-SBM 2	SPC
Corn	46.12	46.05	46.08	45.88	45.97	46.24
Wheat	10.00	10.00	10.00	10.00	10.00	10.00
Corn gluten meal	1.72	1.72	1.72	1.73	1.72	1.71
Wheat bran	3.19	4.37	4.24	4.02	4.00	5.11
Soybean oil	4.00	4.00	4.00	4.00	4.00	4.00
Soybean meal (imported)	20.76	19.63	19.74	20.14	20.09	18.70
Soybean meal (dehulled)	10.00	7.00	7.00	7.00	7.00	7.00
Fermented soybean meal	0.00	3.00	3.00	3.00	3.00	0.00
Soy protein concentrate	0.00	0.00	0.00	0.00	0.00	3.00
Lysine-HCl, 78%	0.16	0.17	0.16	0.17	0.17	0.17
Dicalcium phosphate	1.93	1.92	1.92	1.92	1.92	1.91
DL-methionine, 98%	0.25	0.25	0.25	0.25	0.25	0.25
Limestone	1.20	1.22	1.22	1.21	1.21	1.23
Choline-Cl, 50%	0.08	0.09	0.09	0.09	0.09	0.10
Salt	0.30	0.30	0.30	0.30	0.30	0.30
Vitamin mixture	0.12	0.12	0.12	0.12	0.12	0.12
Mineral mixture	0.15	0.15	0.15	0.15	0.15	0.15
$NaHCO_3$	0.02	0.02	0.02	0.02	0.02	0.02
L-threonine, 98%	0.02	0.02	0.02	0.02	0.02	0.02
Calculated nutrient content						
Crude protein (%)	20.50	20.50	20.50	20.50	20.50	20.50
Ca (%)	6.50	6.50	6.50	6.50	6.50	6.50
Available P (%)	0.45	0.45	0.45	0.45	0.45	0.45
Lysine (%)	1.20	1.20	1.20	1.20	1.20	1.20
Methionine (%)	0.57	0.57	0.57	0.57	0.57	0.57
Total TSAA (%)	0.92	0.92	0.92	0.92	0.92	0.92
TMEn (kcal/kg)	3,080	3,080	3,080	3,080	3,080	3,080

BF-SBM, *Bacillus* fermented SBM (CJ Corporation, Seoul, Korea); YBF-SBM, yeast by product and *Bacillus* fermented SBM (CJ Corporation, Seoul, Korea); LF-SBM 1, *Lactobacillus* fermented SBM 1 (DaBomb Protein Corporation, Tainan, Taiwan); LF-SBM 2, *Lactobacillus* fermented SBM 2 (Total Nutrition Technologies Co., Ltd, Tainan, Taiwan); SPC, soy protein concentrate (Archer Daniels Midland Company, Chicago, IL, USA); TSAA, total sulfur amino acid; TMEn, nitrogen-corrected true metabolizable energy.

[1] Vit. and Min. mixture provided the following nutrients per kg of diet: vitamin A, 40,000 IU; vitamin D_3, 8,000 IU; vitamin E, 10 IU; vitamin K_3, 4.0 mg; vitamin B_1, 4.0 mg; vitamin B_2, 12.0 mg; vitamin B_6, 6.0 mg; vitamin B_{12}, 0.02 mg; niacin, 60.0 mg; pantothenic acid, 20 mg; folic acid, 2.0 mg; biotin, 0.02 mg; Fe, 30.0 mg; Zn, 25.0 mg; Mn, 20.0 mg; Cu, 5.0 mg; Se, 0.1 mg.

age, 8 chicks per treatment were selected, weighed individually and humanely euthanized by cervical dislocation. Immediately after euthanasia, small intestine was excised and 1-cm long segments were sampled from upper one-third region of the jejunum and distal two-third region of the ileum for histological measurement (Caruso et al., 2012). The segments from jejunum and ileum were fixed in 10% neutralized buffered formalin (BBC Biochemistry, Mount Vernon, WA, USA) and embedded in paraffin. The 4μm sections were stained with standard hematoxylin and eosin solution, and the cross sections were examined using a light microscope (Leica Microsystems DM 1000, Leica Microsystems, Wetzlar, Germany). Images were captured with a camera attached to the microscope (Leica Microsystems DFC 295, Leica Microsystems, Germany).

At the end of the experimental period, 8 chicks with similar BW per treatment were selected and weighed individually. The blood was drawn from wing vein using sterilized syringes for determination of the blood profiles. The concentrations of serum total protein, albumin, and blood urea nitrogen (BUN) were measured according to the colorimetric methods using biochemical analyzer (Hitachi modular system, Hitachi Ltd., Tokyo, Japan). At necropsy, liver, spleen, bursa of Fabricius, abdominal fat, right breast muscle were immediately removed and weighed. The relative weights of these organs were expressed as a percentage of live body weight.

The cecal content from each chick was aseptically sampled for microbial test at 7 d of age and the end of experiment. The cecal digesta homogenates in phosphate-buffered saline were serially diluted from 10^{-1} to 10^{-7}. Dilutions were subsequently plated on duplicate selective agar media for enumeration of target bacterial strains. Total microbes, *coliforms*, *Lactobacillus* spp. and *Bacillus* spp. were enumerated using plate count agar, MacConkey agar,

MRS agar and Mannitol egg yolk polymyxin (MYP) agar, respectively. Each plate was incubated at 37°C, for 24 to 72 h anaerobically or aerobically, and colonies were then counted. Results obtained were presented as base-10 logarithm colony-forming units per gram of cecal digesta.

Statistical analysis

Data were analyzed by the general linear model procedure of the SAS (SAS, 2002) with pen means as the experimental units for evaluating growth performances, and individual birds as the experimental unit for the other criteria. The significant differences of obtained means were determined using Duncan's multiple range test at the level of p<0.05 (Duncan, 1955).

RESULTS AND DISCUSSION

The changes in weekly BW of broiler chicks fed diets with fermented SBM products are shown in Figure 1. The BW at 7 d of age was not different between dietary treatments. The BW in chicks fed the pre-starter diets containing BF-SBM and YBF-SBM were significantly higher (p<0.05) compared with that in the control group at 14 d onwards. Chicks fed the pre-starter diets containing BF-SBM, YBF-SBM, and LF-SBM 2 gained more weight compared with the control-diet fed chicks during growing

phase and total rearing period.

Growth performance (feed intake, daily weight gain, and FCR) of broiler chicks fed diets with fermented SBM products during early phase is shown in Table 3. The feed intake during grower and finisher phases was not affected by dietary treatments. The daily weight gain in chicks fed diet containing YBF-SBM had the highest value (p<0.01) during grower phase and followed by BF-SBM group. The changes in daily body weight gain during finisher phase and total rearing period were similar to that of grower phase. The daily weight gains in chicks fed diets containing YBF-SBM, BF-SBM, and LF-SBM 2 were significantly higher (p<0.001) than those of control and group with LF-SBM1 during total rearing period. At 21 d of age, there was no significant difference in FCR among groups. When data were pooled for total rearing period, the FCR in groups with BF-SBM, YBF-SBM, and LF-SBM 2 were significantly lower (p<0.01) than those of control and group with SPC.

The earlier study by Chah et al. (1975) reported that diets with full fat soybean fermented by *Aspergillus* improved the growth performance and feed efficiency in broiler chicks. The replacing conventional SBM with fermented SBM in diets enhanced the BW gain in chicks (Hirabayashi et al., 1998; Feng et al., 2007a). It has been also suggested that fermented SBM may replace fish meal in diets fed to broiler chicks without reducing growth

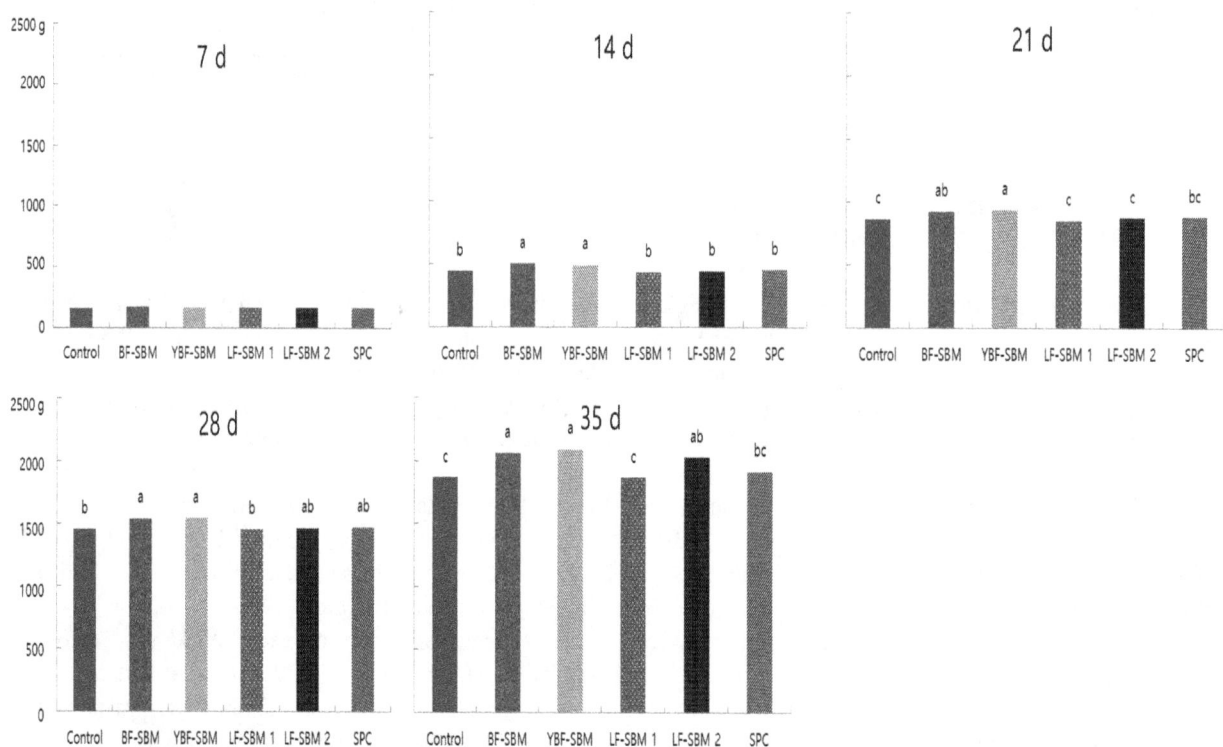

Figure 1. The change in weekly body weight of broiler chicks fed diets with fermented soybean meal (SBM) products during 7 d after hatching. BF-SBM, *Bacillus* fermented SBM; YBF-SBM, yeast by product and *Bacillus* fermented SBM; LF-SBM 1, *Lactobacillus* fermented SBM 1; LF-SBM 2, *Lactobacillus* fermented SBM 2; SPC, soy protein concentrate. [a-c] Means within a row without a common superscript letter differ (p<0.05).

Table 3. Growth performance of broiler chicks fed diets with fermented SBM products during 7 d after hatching

	Control	BF-SBM	YBF-SBM	LF-SBM 1	LF-SBM 2	SPC	SEM	p-value
Feed intake (g/bird)								
1-21 d	60.5	60.04	60.9	58.9	59.3	59.7	0.88	0.631
22-35 d	132.3	140.9	140.5	130.1	138.3	137.2	2.79	0.058
1-35 d	88.7	91.5	91.6	86.5	89.8	89.9	1.26	0.080
Body weight gain (g/bird)								
1-21 d	39.1^c	42.0^{ab}	42.6^a	38.5^c	39.5^c	39.9^{bc}	0.78	0.006
22-35 d	71.5^b	80.5^a	80.9^a	70.6^b	80.9^a	72.8^b	2.39	0.006
1-35 d	52.0^c	57.7^a	58.5^a	52.0^c	56.7^{ab}	53.5^{bc}	1.12	<0.001
FCR (feed/gain)								
1-21 d	1.54	1.43	1.43	1.53	1.50	1.50	0.04	0.171
22-35 d	1.85^{ab}	1.75^{bc}	1.74^{bc}	1.85^{ab}	1.71^c	1.89^a	0.05	0.050
1-35 d	1.70^a	1.58^{bc}	1.57^c	1.66^{ab}	1.58^{bc}	1.69^a	0.03	0.008

BF-SBM, *Bacillus* fermented SBM (CJ Corporation, Seoul, Korea); YBF-SBM, yeast by product and *Bacillus* fermented SBM (CJ Corporation, Seoul, Korea); LF-SBM 1, *Lactobacillus* fermented SBM 1 (DaBomb Protein Corporation, Tainan, Taiwan); LF-SBM 2, *Lactobacillus* fermented SBM 2 (Total Nutrition Technologies Co., Ltd, Tainan, Taiwan); SPC, soy protein concentrate (Archer Daniels Midland Company, Chicago, IL, USA); SEM, standard error of the means.
[a-c] Means within a row without a common superscript letter differ (p<0.05).

performance, but not conventional SBM (Li et al., 2014). Thus, instead of conventional SBM or fish meal, fermented SBMs are usually used as protein sources in diets for starter chicks. In this study, the use of fermented SBMs during early phase affected subsequent growth performance in broiler chicks, except for LF-SBM 1. The interpretation of difference in growth performance among fermented SBMs has remained difficult because all test diets were formulated to be equal in the contents of TMEn, CP, and available amino acids. However, the improvement of subsequent growth may be attributed to beneficial changes seen in microbial profiles due to feeding of some fermented SBMs. The fermentation process is thought to reduce residual anti-nutritional factors such as trypsin inhibitor and increase in small-size peptides. Feeding of fermented SBM significantly increased the activities of digestive enzymes in intestinal contents of broiler chickens (Feng et al., 2007b). It is also likely that the improvement of subsequent growth in chicks that had been fed some fermented SBMs during early phase may be associated with the diminution of anti-nutritional factors or changes in peptide size by proper fermentation.

At 35 d, no significant differences in relative weights of liver, spleen, bursa of Fabricius, breast meat and abdominal fat were observed between dietary treatments (Table 4). At 35 d, the concentrations of serum total protein, albumin and BUN were not affected by any of dietary treatments (Table 5). Feng et al. (2007a) reported that the serum urea nitrogen in chicks fed diet with higher levels of fermented SBM (295 g/kg feed during 0 to 3 wks and 270 g/kg feed during 4 to 6 wks) was significantly decreased as compared with that of control with conventional SBM. The decreased level of urea nitrogen represents that the chicks fed diets with fermented SBM utilized dietary nitrogen efficiently. It is likely that the observed discrepancy may be attributed to differences in feeding levels and duration of treatment as we added fermented SBM in starter diet at the level of 3%.

The cecal microbial populations of broiler chicks fed diets with fermented SBM products during early phase are shown in Table 6. At 7 d of age, the number of total microbes in control group was significantly lower (p<0.001) than those of other groups. Whereas, the number of total microbes in groups with YBF-SBM, LF-SBM 2, and SPC were significantly higher (p<0.001) than those of other groups at termination of experiment. At 7 d of age, the number of *lactic acid bacteria* in groups with YBF-SBM

Table 4. Carcass characteristics of broiler chicks fed diets with fermented SBM products during 7 d after hatching

	Control	BF-SBM	YBF-SBM	LF-SBM 1	LF-SBM 2	SPC	SEM	p-value
Liver (g/100 g BW)	2.2	2.4	2.4	2.6	2.5	2.4	0.14	0.445
Spleen (g/100 g BW)	0.11	0.09	0.09	0.08	0.09	0.09	0.01	0.812
Bursa of Fabricius (g/100 g BW)	0.14	0.14	0.14	0.16	0.14	0.16	0.01	0.804
Right breast muscle (g/100 g BW)	9.0	9.3	9.8	9.1	8.9	9.1	0.27	0.217
Abdominal fat (g/100 g BW)	1.3	1.2	1.3	1.3	1.3	1.1	0.09	0.657

BF-SBM, *Bacillus* fermented SBM (CJ Corporation, Seoul, Korea); YBF-SBM, yeast by product and *Bacillus* fermented SBM (CJ Corporation, Seoul, Korea); LF-SBM 1, *Lactobacillus* fermented SBM 1 (DaBomb Protein Corporation, Tainan, Taiwan); LF-SBM 2, *Lactobacillus* fermented SBM 2 (Total Nutrition Technologies Co., Ltd, Tainan, Taiwan); SPC, soy protein concentrate (Archer Daniels Midland Company, Chicago, IL, USA); SEM, standard error of the means.

Table 5. The blood profiles of broiler chicks fed diets with fermented SBM products during 7 d after hatching

	Control	BF-SBM	YBF-SBM	LF-SBM 1	LF-SBM 2	SPC	SEM	p-value
Total protein (g/dL)	3.4	3.5	3.3	3.5	3.2	3.1	0.15	0.249
Blood urea nitrogen (mg/dL)	3.1	3.0	3.2	3.2	3.1	3.3	0.18	0.888
Albumin (g/dL)	1.4	1.3	1.3	1.3	1.4	1.3	0.06	0.370

BF-SBM, *Bacillus* fermented SBM (CJ Corporation, Seoul, Korea); YBF-SBM, yeast by product and *Bacillus* fermented SBM (CJ Corporation, Seoul, Korea); LF-SBM 1, *Lactobacillus* fermented SBM 1 (DaBomb Protein Corporation, Tainan, Taiwan); LF-SBM 2, *Lactobacillus* fermented SBM 2 (Total Nutrition Technologies Co., Ltd, Tainan, Taiwan); SPC, soy protein concentrate (Archer Daniels Midland Company, Chicago, IL, USA); SEM, standard error of the means.

Table 6. The cecal microbial profiles of broiler chicks fed diets with fermented SBM products during 7 d after hatching

	Control	BF-SBM	YBF-SBM	LF-SBM 1	LF-SBM 2	SPC	SEM	p-value
Total microbes (log 10cfu/g)								
7 days	5.3^b	5.9^a	5.9^a	5.9^a	6.1^a	6.1^a	0.11	<0.001
35 days	5.2^b	5.2^b	5.5^a	5.1^b	5.4^a	5.4^a	0.06	<0.001
Lactic acid bacteria (log 10cfu/g)								
7 days	5.5^b	5.5^b	6.4^a	4.8^c	4.6^c	4.9^{bc}	0.19	<0.001
35 days	4.4^b	5.1^a	5.0^a	4.3^b	4.3^b	4.3^b	0.06	<0.001
Coliforms (log 10cfu/g)								
7 days	5.4	5.5	5.4	5.4	5.5	5.3	0.14	0.411
35 days	5.0^a	4.4^b	4.2^b	5.2^a	5.1^a	5.2^a	0.15	<0.001
Bacillus spp. (log 10cfu/g)								
35 days	5.9^b	7.3^a	7.2^a	6.4^b	7.1^a	6.9^a	0.16	<0.001

BF-SBM, *Bacillus* fermented SBM (CJ Corporation, Seoul, Korea); YBF-SBM, yeast by product and *Bacillus* fermented SBM (CJ Corporation, Seoul, Korea); LF-SBM 1, *Lactobacillus* fermented SBM 1 (DaBomb Protein Corporation, Tainan, Taiwan); LF-SBM 2, *Lactobacillus* fermented SBM 2 (Total Nutrition Technologies Co., Ltd, Tainan, Taiwan); SPC, soy protein concentrate (Archer Daniels Midland Company, Chicago, IL, USA); SEM, standard error of the means.
[a-c] Means within a row without a common superscript letter differ (p<0.05).

had the highest value (p<0.001). The number of *lactic acid bacteria* in groups with BF-SBM and YBF-SBM were significantly higher (p<0.001) than other groups at the end of experiment. The number of *coliform* bacteria in groups with BF-SBM and YBF-SBM were significantly lower (p<0.001) than other groups at 35 d, but not at 7 d. At 35 d, the number of *Bacillus* spp. in groups with fermented SBM and SPC were significantly higher (p<0.001) than that of control, except for LF-SBM 1. In this study, fermented SBMs such as BF-SBM, YBF-SBM, and LF-SBM 2, positively modulated the populations of cecal microflora, resulting in a potential probiotic effect. Jeong and Kim (2015) also found that dietary supplementation of fermented SBMs led to increased fecal *Lactobacillus* counts in weaned piglets. These findings suggest that fermented SBMs are the carrier of probiotics and can be used valuable feed

ingredients in growing animals.

Villi length of 7-d-old chicks fed diets with fermented SBM products is shown in Table 7. The jejunal villi length in chicks fed diets with fermented SBMs and SPC was significantly longer (p<0.001) compared with the control group. Ileal villi length in groups with YBF-SBM and LF-SBM1 tended to be higher than that of the control group, but the statistical difference was not noted. On the other hand, SPC-fed chicks exhibited the lowest ileal villi length (p<0.05). Our study is in line with the previous studies. For example, Mathivanna et al. (2006) reported that the increased length and width of ileal villi were observed in broiler chicks fed diet with 0.5% and 1.0% fermented SBM. The positive changes of duodenal and jejunal villus height were also observed in chickens fed diet containing higher levels of fermented SBM (Feng et al., 2007b). Similarly,

Table 7. Effect of fermented soybean meal (SBM) on intestinal villi length in 7-d-old broiler chicks during 7 d after hatching

	Control	BF-SBM	YBF-SBM	LF-SBM 1	LF-SBM 2	SPC	SEM	p-value
Jejunum	412.6^b	536.1^a	522.3^a	572.9^a	498.8^a	533.3^a	24.9	<0.001
Ileum	424.4^{ab}	395.9^{ab}	435.7^a	434.5^a	407.0^{ab}	369.2^b	19.3	0.044

BF-SBM, *Bacillus* fermented SBM (CJ Corporation, Seoul, Korea); YBF-SBM, yeast by product and *Bacillus* fermented SBM (CJ Corporation, Seoul, Korea); LF-SBM 1, *Lactobacillus* fermented SBM 1 (DaBomb Protein Corporation, Tainan, Taiwan); LF-SBM 2, *Lactobacillus* fermented SBM 2 (Total Nutrition Technologies Co., Ltd, Tainan, Taiwan); SPC, soy protein concentrate (Archer Daniels Midland Company, Chicago, IL, USA); SEM, standard error of the means.
[a,b] Means within a row without a common superscript letter differ (p<0.05).

fermented diet-fed weanling piglets exhibited higher villi height and villus height to crypt depth ratio (Scholten et al., 2002). The improvement of intestinal morphology in fermented SBMs-fed young animals may be likely due to the higher lactic acid level and the diminution of anti-nutritional factors presented in fermented feed (Hong et al., 2004; Mathivanna et al., 2006). It is suggested that an increased villi length is paralleled by an increased digestive and absorptive function of small intestine, which can lead to better nutrient absorption (Caspary, 1992) and overall growth performance (Feng et al., 2007b). Thus, some fermented SBMs used in this study had beneficial effects on villi morphology in starter chicks and these might be partially responsible for the subsequent growth.

CONCLUSION

The results indicate that feeding of some fermented SBMs during early phase are beneficial to the subsequent growth performance in broiler chicks through good impacts on intestinal morphology and microbial profiles. Especially, BF-SBM and YBF-SBM showed superior overall growth performance as compared with unfermented SBM and SPC.

ACKNOWLEDGMENTS

This paper was supported by the KU Research Professor Program of Konkuk University and National Institute of Animal Science, RDA (No. PJ01189801).

REFERENCES

AOAC (Association of Official Analytical Chemists) International. 1995. Official Methods of Analysis, 16th edn. Arlington, VA, USA.

Batal, A. B. and C. M. Parsons. 2002. Effect of fasting versus feeding oasis after hatching on nutrient utilization in chicks. Poult. Sci. 81:853-859.

Caspary, W. F. 1992. Physiology and pathophysiology of intestinal absorption. Am. J. Clin. Nutr. 55:299S-308S.

Caruso, M., A. Demonte, and V. A. Neves. 2012. Histomorphometric study of role of lactoferrin in atrophy of the intestinal mucosa of rats. Health 4:1362-1370.

Chah, C. C., C. W. Carlson, G. Semeniuk, I. S. Palmer, and C. W. Hesseltine. 1975. Growth-promoting effects of fermented soybeans for broilers. Poult. Sci. 54:600-609.

Choct, M., Y. Dersjant-Li, J. McLeish, and M. Peisker. 2010. Soy oligosaccharides and soluble non-starch polysaccharides: A review of digestion, nutritive and anti-nutritive effects in pigs and poultry. Asian Australas J. Anim. Sci. 23:1386-1398.

Ducan, D. B. 1955. Multiple range and multiple F test. Biometrics 11:1-42.

Feng, J., X. Liu, Z. R. Xu, Y. Y. Liu, and Y. P. Lu. 2007a. Effects of *Aspergillus oryzae* 3.042 fermented soybean meal on growth performance and plasma biochemical parameters in broilers. Anim. Feed Sci. Technol. 134:235-242.

Feng, J., X. Liu, Z. R. Xu, Y. Z. Wang, and J. X. Liu. 2007b. Effects of fermented soybean meal on digestive enzyme activities and intestinal morphology in broilers. Poult. Sci. 86:1149-1154.

Hirabayashi, M., T. Matsui, H. Yano, and T. Nakajima. 1998. Fermentation of soybean meal with *Aspergillus usamii* reduces phosphorus excretion in chicks. Poult. Sci. 77:552-556.

Hong, K. J., C. H., Lee, and S. W. Kim. 2004. *Aspergillus oryzae* GB-107 fermentation improves nutritional quality of food soybeans and feed soybean meals. J. Med. Food 7:430-435.

Jeong, J. S. and I. H. Kim. 2015. Comparative efficacy of up to 50% partial fish meal replacement with fermented soybean meal or enzymatically prepared soybean meal on growth performance, nutrient digestibility and fecal microflora in weaned pigs. Anim. Sci. J. 86:624-633.

Lee, B. K., J. Y. Kim, J. S. Kim, S. J. You, B. K. An, E. J. Kim, and C. W. Kang. 2009. Nutritional value of soybean meal from various geographic origin and effect of their dietary supplementation on performance of broilers. J. Anim. Sci. Technol. 51:217-224.

Li, C. Y., J. J. Lu, C. P. Wu, and T. F. Lien. 2014. Effects of probiotics and bremelain fermented soybean meal replacing fish meal on growth performance, nutrient retention and carcass traits of broilers. Livest. Sci. 163:94-101.

Mathivanan, R., P. Selvaraj, and K. Nanjappan. 2006. Feeding of fermented soybean meal on broiler performance. Int. J. Poult. Sci. 5:868-872.

Moore, S. 1963. On the determination of cystine as cysteic acid. J. Biol. Chem. 238:235-237.

Nir, I., Z. Nitsan, and M. Mahagna. 1993. Comparative growth and development of digestive organs and of some enzymes in broiler and egg type chicks after hatching. Br. Poult. Sci. 34:523-532.

NRC (National Research Council). 1994. Nutrient Requirements of Poultry. 9th edn. National Academy Press, Washington, DC, USA.

NRC (National Research Council). 1998. Nutrient Requirements of Swine. 10th rev. edn. National Academy Press, Washington, DC, USA.

SAS (Statistical Analysis System) Institute. 2002. SAS User's Guide. Statistics, version. 8edn. Statistical Analysis System Institute. Inc., Cary, NC, USA.

Scholten, R. H. J., C. M. C. Van der Peet-Schwering, L. A. den Hartog, M. Balk, J. W. Schrama, and M. W. A. Verstegen. 2002. Fermented wheat in liquid diets: Effects on gastrointestinal characteristics in weanling piglets. J. Anim. Sci. 80:1179-1186.

Spackman, D. H., W. H. Stein, and S. Moore. 1958. Automatic recording apparatus for use in the chromatography of amino acid. Anal. Chem. 30:1190-1206.

Zuidhof, M. J., B. L. Schneider, V. L. Carney, D. R. Korver, and F. E. Robinson. 2014. Growth, efficiency, and yield of commercial boilers from 1957, 1873, and 2005. Poult. Sci. 93:2970-2982.

20

Dietary Phytoncide Supplementation Improved Growth Performance and Meat Quality of Finishing Pigs

Han Lin Li, Pin Yao Zhao, Yan Lei, Md Manik Hossain, Jungsun Kang[1], and In Ho Kim*

Department of Animal Resource and Science, Dankook University, Cheonan 330-714, Korea

ABSTRACT: We conducted this 10-wk experiment to evaluate the effects of dietary phytoncide, Korean pine extract as phytogenic feed additive (PFA), on growth performance, blood characteristics, and meat quality in finishing pigs. A total of 160 pigs ([Landrace×Yorkshire]×Duroc, body weight (BW) = 58.2±1.0 kg) were randomly allocated into 1 of 4 treatments according to their BW and sex, 10 replicate pens per treatment with 4 pigs per pen were used (2 barrows and 2 gilts). Dietary treatments were: CON, control diet; PT2, CON+0.02% PFA; PT4, CON+0.04% PFA; PT6, CON+0.06% PFA. Overall, average daily gain (ADG) was higher in PT4 ($p<0.05$) than in PT6, average daily feed intake (ADFI) was lower in PT6 than in CON ($p<0.05$). Besides ADFI decreased linearly ($p<0.05$) with the increased level of phytoncide and gain:feed ratio in PT4 treatment was higher ($p<0.05$) than CON treatment. During 5 to 10 weeks and overall, quadratic ($p<0.05$) effect was observed in ADG among the treatments. At the end of this experiment, pigs fed with PT4 diet had a greater ($p<0.05$) red blood cell concentration compared to the pigs fed CON diet. Water holding capacity increased linearly ($p<0.05$) with the increased level of phytoncide supplementation. Moreover, firmness, redness, yellowness, and drip loss at day 3 decreased linearly ($p<0.05$) with the increase in the level of phytoncide supplementation. In conclusion, inclusion of phytoncide could enhance growth performance without any adverse effects on meat quality in finishing pigs. (**Key Words:** Blood Profile, Finishing Pig, Growth Performance, Meat Quality, Phytoncide)

INTRODUCTION

Long-term storage of animal feed in high humidity and high temperature condition is a major problem of livestock farm production and management, because nutrients in feed can be decomposed by harmful microorganisms. For example, *Aspergillus* and other fungi may produce toxins such as aflatoxins, while pathogenic bacteria, such as *Salmonella enteritidis* and *Escherichia coli*, present in feed or feedstuffs for animal consumption. All of these can be dangerous for animal health (Cole and Cox, 1981). The diet of pig consists mainly of cereal grains, which often easily contaminated with aflatoxin-producing molds. It is reported that pigs to be one of the most sensitive animals to the effects of aflatoxin (Shi et al., 2005). Thus, aflatoxin-contaminated feed is a serious threat to the swine industry.

Phytoncide are natural volatile compounds emitted by trees and plants as a protective mechanism against harmful insects, animals, and microorganisms. Trees and plants synthesize phytoncide as secondary metabolites of photosynthesis. Phytoncide are enriched in terpenoids, phenylpropanoids, and alkaloids; major ingredients include monoterpenoids, including α-pinene, careen, and myrcene (Durham et al., 1994; Nose, 2000; Oikawa, 2005). As a mean of suppressing activity of microorganisms, such as bacteria and fungi, relatively large amounts of phytoncide were released by Korean pines and other conifers into the environment (Pečiulytė et al., 2010). The suppressive activity relies on essential oil, terpenoids, alkaloid, and phenylpropanoids of phytoncide (Durham et al., 1994). Due to them, phytogenic feed additive (PFA) have a potential capacity to intercalate into the bacterial membrane, disintegrate the membrane, and cause ion leakage (Burt, 2004; Hashemi and Davoodi, 2010). Moreover, this can interfere with regulatory networks of signal transduction and gene expression in aflatoxin biosynthesis, or by

---

* Corresponding Author: In Ho Kim.

E-mail: inhokim@dankook.ac.kr

[1] Genebiotech. Co. Ltd., Seoul 06774, Korea.

blocking the enzymatic activity of a biosynthetic enzyme (Holmes et al., 2008).

These previous findings suggested that phytoncide as a potential feed additive. In support of this, one study described the antioxidative potential of phytoncide and their effects in reducing oxidative stress responses in stroke-prone spontaneously hypertensive rats (Kawakami et al., 2004). Moreover, stress level was known to have negative correlation with meat quality (Terlouw, 2005). According to these previous studies, we could hypothesis that dietary phytoncide supplementation may help to improve growth performance and meat quality in finishing pigs. However, the effect of dietary phytoncide supplementation in finishing pigs has not yet been investigated. Thus, this study was conducted to determine the effect of dietary phytoncide supplementation on growing performance, blood characteristics, and meat quality in finishing pigs.

MATERIALS AND METHODS

Preparation of phytoncide and experimental animals

The phytoncide (PHYLUS Company, Seoul, Korea) used in this study was extracted from Korean pine. It contained 20% active substance (flavonoid, phenolic compounds, alkaloid, tannin, terpene, saponin) and 80% carrier (dextrin). About 40% of active substance were essential oils. The experimental protocol was approved by the animal care and use committee of Dankook University (Ethics Approval Number: DK-1152).

A total of 160 finishing pigs ([Landrace×Yorkshire]× Duroc, body weight [BW] = 58.2±1.0 kg) were randomly allocated into 1 of 4 treatments according to their BW and sex. Ten replicate pens per treatment with 4 pigs per pen were used (2 barrows and 2 gilts). Dietary treatments included: CON, control diet; PT2, CON+0.02% PFA; PT4, CON+0.04% PFA; PT6, CON+0.06% PFA. All pigs were housed in pens (1.8×1.8 m) with a self-feeder and nipple drinker that allowed pig *ad libitum* access to feed and water throughout the whole experimental period. Temperature was maintained at 24°C to 26°C. Dietary phytoncide was supplemented in diets by replacing the same amount of corn in our current study. As showed in Table 1, all diets used in present study were formulated to meet or little exceed the nutrient recommendations of NRC (2012). Diets were formulated to the same concentrations of lysine, crude protein (CP), metabolizable energy, calcium, and phosphorus (Table 1).

Sampling and measurements

BW and feed intake were measured at the start of experiment and at the end of 5 and 10 wk to calculate average daily gain (ADG), average daily feed intake (ADFI) and gain:feed. Samples of diets were analyzed for

Table 1. Formula and chemical composition of experimental diet (as-fed basis)

Items	
Ingredients (%)	
Corn	67.81
Soybean meal (48%)	17.88
Rice bran	5.00
Molasses, sugar beets	5.00
Tallow	2.00
Dicalcium phosphate	0.75
Calcium carbonate	0.94
L-Lys·HCl (78%)	0.30
Salt	0.15
Vit-perimix*	0.07
Min-premix*	0.10
Calculated nutrient composition	
Metabolizable energy (kcal/kg)	3,308
Crude protein (%)	15.63
Lysine (%)	0.97
Calcium (%)	0.60
Phosphorus (%)	0.56
Analyzed nutrient composition	
Metabolizable energy (kcal/kg)	3,301
Crude protein (%)	15.50
Lysine (%)	0.968
Calcium (%)	0.59
Phosphorus (%)	0.52
Crude fat (%)	5.33
Crude fiber (%)	4.14

* Provided per kg of complete diet: 4,000 IU of vitamin A; 800 IU of vitamin D_3; 17 IU of vitamin E; 2 mg of vitamin K; 4 mg of vitamin B_2; 1 mg of vitamin B_6; 16 µg of vitamin B_{12}; 11 mg of pantothenic acid; 20 mg of niacin; 0.02 mg of biotin; 220 mg of Cu ($CuSO_4$ $5H_2O$); 175 mg of Fe ($FeSO_4$ H_2O); 191 mg of Zn ($ZnSO_4$ H_2O); 89 mg of Mn (MnO); 0.3 mg of I (CaI_2); 0.5 mg of Co ($CoSO_4$ $7H_2O$); 0.3 mg of Se (Na_2SeO_3 $5H_2O$).

CP (procedure 984.13; AOAC, 2006), calcium (procedure 968.08; AOAC, 2006), phosphorus (procedure 946.06; AOAC, 2006), and crude fiber (procedure 978.10; AOAC, 2006). Crude fat were determined according to the method of Thiex et al. (2003) and Kjeldahl N was determined according to the method of Thiex et al. (2002). The concentrations of neutral detergent fiber and acid detergent fiber were determined according to the method of Van Soest et al. (1991). The concentration of crude fiber was analyzed using heat-stable α-amylase and sodium sulfite without correction for insoluble ash as adapted for an Ankom Fiber Analyzer (Ankom Technology, Macedon, NY, USA). Samples of diets were analyzed for metabolizable energy via an adiabatic oxygen bomb calorimeter (Parr Instruments Co., Moline, IL, USA).

Blood samples were collected via jugular vein into either 5-mL vacuum or 5-mL K3EDTA vacuum tube (Becton Dickinson Vacutainer Systems, Franklin Lakes, NJ,

USA) from 2 pigs each pen on the initial and final day of the experimental period. Concentrations of red blood cell (RBC) and white blood cell (WBC) in whole blood samples were measured using the automatic blood analyzer (ADVIA 120, Bayer, Tarrytown, NY, USA). Serum samples were centrifuged (3,000×g) for 15 min at 4°C, and then concentrations of cortisol was measured using the automatic biochemistry analyzer (HITACHI 747, Tokyo, Japan).

At the end of the experiment, 10 pigs per treatment (1 pig per pen with a BW of pen average) were transported to the abattoir for slaughter. The carcasses were placed in a conventional chiller at 4°C. After 24 h chill period, carcasses were fabricated into primal cuts. Meat samples, which included lean and fat, were taken via perpendicular cut loins into 2-cm-thick chop beginning from the 10th and 11th ribs region. Back fat thickness was measuring midline fat thickness by using a real-time ultrasound instrument (Piglot 105, SFK Technology, Herlev, Denmark) according to method of Wang et al. (2009). The pH of longissimus muscle (LM) was measured in 24 H post-mortem with an insertion of glass electrode (Radiometer, Lyon, France) connected to a pH-meter (NWKbinar pH, K-21, Landsberg, Germany). The electrode was calibrated at 20°C in buffers at pH value of 4.00 and 7.00. Surface LM color (Minolta L*,a*,b*) was measured in triplicate on a freshly-cut surface with a Minolta Chromameter (Minolta CR 301, Tokyo, Japan). The water holding capacity (WHC) was measured according to the methods of Kauffman et al. (1986). In brief, 0.2 g sample was pressed at 3,000 psi for 3 min on 125-mm-diameter filter paper. The areas of the pressed sample and expressed moisture were delineated and then determined with a digitizing area-line sensor (MT-10S; M.T. Precision Co. Ltd., 123 Tokyo, Japan). A ratio of water: meat areas was calculated, giving a measure of WHC (the smaller ratio indicate the higher the WHC). The proportion of LM acceptable for Pork Composition and Quality Assessment Procedures (NPPC, 1991) was determined via the selection of LM with acceptable color, firmness, and marbling (all measures 3 or greater, based on a scale of 1 to 5; NPPC, 1991). Drip loss was measured using approximately 2 g of meat sample according to the plastic bag method, which was described by Honikel (1998).

Statistical analyses

For current experiment, effects of treatment (0, 0.02%, 0.04%, 0.06% phytoncide) were analyzed by analysis of variance (ANOVA) using the Mixed Models procedure of SAS (SAS Institute, 2003). The pen was used as the experimental unit for growth performance analysis, while data of blood profile and meat quality were based on individual pigs. The initial BW was used as a covariate for ADFI and ADG, and initial values were used as a covariate

for blood profile. The complete model of blood profiles included the main effects of treatment, litter and time, and the treatment×time interaction. Differences among treatment means were determined using the Tukey's multiple range test. Probabilities less than 0.05 were considered to be significant. After ANOVA analysis, linear and quadratic polynomial contrasts were performed to determine the effects of different level of phytoncide (0.02%, 0.04%, and 0.06%) in the diet.

RESULTS

Growth performance

During 0 to 5 weeks, pigs fed the PT4 diet had a greater (p<0.05) ADG than pigs fed the PT6 diet (Table 2). From 5 to 10 weeks, pigs fed the PT4 diet had a greater G:F (p<0.05) than pigs fed the CON diet. During the entire experimental period, ADG was increased with supplementation with 0.04% phytoncide (p<0.05) compared with the inclusion of 0.06% phytoncide. Furthermore, ADFI in the PT6 group was lower (p<0.05) than that in the CON group, and G:F in PT4 treatment was higher (p<0.05) than with CON.

Moreover, throughout the experimental period, the ADFI decreased linearly (p<0.05) with the increased level of phytoncide, and a significant quadratic effect (p<0.05) was observed in G:F among treatments. During 5 to 10 weeks and overall, a significant quadratic effect (p<0.05) was observed in ADG among treatments.

Blood profiles

Blood profiles are presented in Table 3. At the end of the experiment, pigs fed the PT4 diet had a higher (p<0.05) RBC concentration than the pigs fed the CON diet. Also, the cortisol concentration was decreased (p<0.05) with PT2 and PT4 treatment versus the CON treatment. Time effect was observed in cortisol and WBC, while no treatment×time interaction effect was observed. Moreover, a significant quadratic effect was observed in cortisol, RBC, and WBC among treatments.

Meat quality

As shown in Table 4, supplementation with 0.02% and 0.06% phytoncide increased (p<0.05) the sensory color compared with CON treatment, and inclusion of 0.04% and 0.06% phytoncide decreased (p<0.05) sensory firmness versus CON treatment. WHC in PT6 was higher (p<0.05) than that in the PT2 and CON treatments. No difference (p<0.05) was observed in loin muscle area, pH, meat color, or drip loss (p>0.05). Moreover, firmness, redness, yellowness, and drip loss at day 3 decreased linearly (p< 0.05) with the increase in the level of phytoncide

Table 2. Effect of phytoncide on growth performance in finishing pigs

Item	CON[1]	PT2[1]	PT4[1]	PT6[1]	SE[2]	p-value		
						Treatment	Linear	Quadratic
BW (kg)								
Initial[3]	58.2	58.3	58.2	58.2	0.1	0.80	0.01	0.84
Final[3]	116.9[b]	119.2[ab]	123.0[a]	114.6[b]	2.9	0.01	0.70	0.04
0 to 5 week								
ADG (g)	886[ab]	838[ab]	917[a]	786[b]	34.4	<0.01	0.63	0.34
ADFI (g)	2,280	2,112	2,135	2,077	65.5	0.43	0.04	0.67
Gain:feed	0.389	0.397	0.430	0.378	0.022	0.76	0.72	0.03
5 to 10 week								
ADG (g)	792	902	934	862	49.3	0.11	0.82	0.01
ADFI (g)	2,829	2,785	2,658	2,520	103.4	0.69	<0.001	0.55
Gain:feed	0.280[b]	0.324[ab]	0.351[a]	0.342[ab]	0.014	<0.01	0.24	<0.001
0 to 10 week								
ADG (g)	839[ab]	870[ab]	925[a]	806[b]	28.8	0.03	0.47	<0.001
ADFI (g)	2,555[a]	2,448[ab]	2,397[ab]	2,299[b]	66.3	0.05	<0.001	0.81
Gain:feed	0.328[b]	0.355[ab]	0.386[a]	0.351[ab]	0.015	0.01	0.39	<0.001

SE, standard error; BW, body weight; ADG, average daily gain; ADFI, average daily feed intake.

[1] CON, basal diet; P2, basal diet+0.02% phytoncide; P4, basal diet+0.04% phytoncide; P6, basal diet+0.06% phytoncide.

[2] Each mean represents 40 observations per treatment.

[3] Initial, the 1st day of experiment; final, the last day (d 70) of experiment.

[a,b] Means in the same row with different superscripts differ (p<0.05), initial BW was used as a covariate.

supplementation, and WHC increased linearly (p<0.05) with the increased level of phytoncide supplementation.

DISCUSSION

As an alternative feeding strategy to growth promoters, PFA have received considerable attentions recently. Various trials have demonstrated positive effects of PFA in swine and poultry (Wang et al., 2008; Windisch et al., 2008; Van Krimpen et al., 2010). Wang et al. (2008) reported that 0.04% dietary phytogenic feed additive, which was made of essential oils, flavonoids, pungent substances, and mucliages, did not affect the ADFI of lactating sows. Whereas, Zhang et al. (2012) observed that weaning pigs

fed with 0.2% phytoncide showed a greater G:F ratio. Moreover, Yan et al. (2010) reported that 0.01% essential oil used as PFA increased ADG and G:F at different nutrient densities in pigs. Consistent with these previous studies, our study demonstrated that 0.04% phytoncide increased ADG and G:F ratio in finishing pigs. Essential oils, as a main content of phytogenic additive, may have beneficial effects such as antiviral and antimicrobial activity, and are capable of stimulating enzyme activity and immune function in livestock (Williams and Losa, 2001; Yan et al., 2010). Also, positive of essential oils on digestive system and stimulating effect on the output of digestive enzymes from the pancreas, gut mucosa, and increase bile flow have been reported (Jamroz and Kamel, 2002; Jamroz et al., 2005;

Table 3. Effect of phytoncide on blood profile in finishing pigs

Item	CON[1]	PT2[1]	PT4[1]	PT6[1]	SE[2]	p-value				
						Treatment	Time	Treatment×time	Linear	Quadratic
Cortisol (μg/dL)										
Initial[3]	1.20	1.20	1.13	1.00	0.24	<0.01	<0.01	0.57	0.08	0.57
Final[3]	2.20[a]	1.00[c]	1.20[bc]	2.00[ab]	0.27				0.82	0.02
Red blood cell (10[6]/μL)										
Initial[3]	6.59	6.43	6.64	6.82	0.34	<0.01	0.36	0.44	0.63	0.46
Final[3]	4.67[b]	6.63[ab]	7.15[a]	6.54[ab]	0.31				0.71	<0.001
White blood cell (10[3]/μL)										
Initial[3]	22.69	21.68	22.08	21.77	2.47	0.09	0.03	0.64	0.27	0.14
Final[3]	19.17	19.47	22.10	17.89	1.42				0.52	0.01

[1] CON, basal diet; PT2, basal diet+0.02% phytoncide; PT4, basal diet+0.04% phytoncide; PT6, basal diet+0.06% phytoncide.

[2] Standard error. Each mean represents 20 observations per treatment.

[3] Initial, the 1st day of experiment; final, the last day (d 70) of experiment.

[a,b] Means in the same row with different superscripts differ (p<0.05).

Table 4. Effect of phytoncide on meat quality in finishing pigs

Item	CON[1]	PT2[1]	PT4[1]	PT6[1]	SE[2]	p-value		
						Treatment	Linear	Quadratic
Sensory evaluation								
Color	1.88[b]	2.17[a]	1.96[ab]	2.21[a]	0.077	0.05	0.14	0.57
Marbling	1.79	1.63	1.75	1.92	0.134	0.12	0.40	0.03
Firmness	2.25[a]	1.96[ab]	1.92[b]	1.92[b]	0.10	0.02	0.05	0.11
Lean meat area (cm^2)	44.78	46.84	42.52	44.21	1.97	0.23	0.23	0.42
pH	5.24	5.25	5.32	5.23	0.04	0.71	0.07	0.10
Meat color								
Lightness (L*)	55.98	56.25	57.33	56.45	1.13	0.57	0.14	0.29
Redness (a*)	21.08	20.05	18.38	17.08	1.13	0.12	0.01	0.57
Yellowness (b*)	10.60	9.45	8.83	8.53	0.72	0.26	0.04	0.37
Water holding capacity (WHC, %)	56.94[c]	64.73[b]	64.93[ab]	68.18[a]	1.03	<0.01	0.02	0.76
Drip loss (%)								
3 d	10.66	10.34	9.21	8.65	1.37	0.35	0.03	0.39
6 d	11.82	13.53	11.72	10.26	1.65	0.40	0.17	0.09

[1] CON, basal diet; PT2, basal diet+0.02% phytoncide; PT4, basal diet+0.04% phytoncide; PT6, basal diet+0.06% phytoncide.
[2] Standard error. Each mean represents 10 observations per treatment.
[a,b,c] Means in the same row with different superscripts differ (p<0.05).

Jang et al., 2007), which may explain the improvement of growth performance of pigs in our present experiment. Also, specific capabilities of phytogenic compounds is to enhance the activities of digestive enzymes (bile, pancreatic lipase and amylase etc.) and nutrient absorption (Windisch et al., 2008), which may be another explanation for our results.

The quality of plant materials, selection of particular plants, and forms for administration (i.e., extracts, oils, and dried herbs) also influence their effects on pig performance. As mentioned above, the active substances in phytoncide have a potential capacity to intercalate into the bacterial membrane, disintegrate the membrane, and cause ion leakage, it means that PFA could inhibit the biosynthesis or expression of aflatoxin, or block the enzymatic activity of a biosynthetic enzyme, which including superoxide dismutase, xanthine oxidase, glutathione peroxidase, and two microsomal reductases (Burt, 2004; Holmes et al., 2008; Hashemi and Davoodi, 2010). Those may be the reason why it could promote growth performance of pigs by protecting feed quality. However, inconsistent results have been obtained regarding the effects of PFA on ADFI in animals. Yan et al. (2011a) noted that 1 g/kg herbal extract (*Houttuynia cordata* or *Taraxacum officinale*) used as PFA increased ADFI during a 10-week trial. Wang et al. (2007) and Wenk et al. (2003) reported that herbs used as PFA could improve the flavor and palatability of feed, subsequently increasing the total feed intake and growth performance. Yan et al. (2010) demonstrated that 0.01% essential oil, including thyme, rosemary, oreganum extracts, and kaolin covered by starch had a tendency to lower ADFI. One reasonable explanation for the lower ADFI might be the tranquilizing effect of PFA, which reduced the

spontaneous physical activity of the pigs. Physical activity influences heat production, which can cause additional expenditure of energy (Wang et al., 2007). Moreover, pungent substances of some plants (e.g., capsaicin, allitricin, and piperine), which are the active components of PFA, may reduce the palatability of feed and restrict their use for animal feeding purpose (Zhong et al., 2011). In line with our study, previous studies also showed dose-related depressions of palatability of pigs fed with essential oils from the herbs thyme and oregano, as well as from caraway and fennel (Jugl-Chizzola et al., 2006; Schone et al., 2006). Therefore, according to these previous studies the elevate effect of essential may depressed as pungent substances concentration increased.

Czech et al. (2009) reported that pigs fed a mixture with a supplement of 8 g/kg herbal mixture during the whole fattening period were characterized by a significantly higher value of red blood cell system indices, including serotonin, hemoglobin, and RBC. Yan et al. (2011b) confirmed this result, describing increased RBC concentrations in pigs fed 250 or 500 mg/kg of herb extract mixture. Pungent substances (e.g. garlic, black pepper, and chilli), the active components of PFA, could activate blood circulation and metabolic processes (Rodas, 2006). Consistent with that, Iranloye (2002) showed that feeding rats with 200 mg/kg garlic juice daily increased the RBC and WBC. Moreover, Wang et al. (2008) observed that RBC and WBC were increased significantly with PFA treatment. Phytoncide had also been shown to reduce stress responses by lowering stress-related hormones (Kawakami et al., 2004).

Collectively, these findings indicate that phytoncide as PFA can increase blood circulation, metabolic processes, and the immune system of pigs.

Phytoncide can reduce stress responses in stroke-prone spontaneously hypertensive rats (Kawakami et al., 2004). Nam and Uhm (2008) documented that phytoncide decreased serum cortisol levels significantly in college students. Psychological stress affects immune function and predicts infectious disease susceptibility in humans and animals (Marsland et al., 2002). Plant extracts could stimulate T cell-mediated immune responses (Islam et al., 2004). Additionally, flavonoids, one bioactive substance within phytoncide, exhibit inhibitory effects against multiple viruses and improve immunity (Critchfield et al., 1996). In line with previous studies, the concentration of stress-related hormone, cortisol, was reduced significantly with phytoncide supplementation in our present study, indicating that stress was reduced and immunity was enhanced with phytoncide supplementation (Koopmans et al., 2005). In contrast, the effects of PFA on blood profiles have been inconsistent, which might be attributable to interactions between PFA and different feed formulae.

Unacceptable water-holding capacity costs the meat industry millions of dollars annually. Product weight losses due to purge can average as much as 1% to 3% in fresh retail cuts (Offer and Knight, 1988a) and can be as high at 10% in pale soft exudative products (Melody et al., 2004). In addition to the loss of salable weight, purge loss also entails the loss of a significant amount of protein (Offer and Knight, 1988b). On average, purge can contain approximately 112 mg of protein per mL of fluid, mostly water-soluble, sarcoplasmic proteins (Savage et al., 1990). In the current study, phytoncide treatments showed beneficial effects on WHC. It is clear that early postmortem events including the rate and extent of pH decline, proteolysis, and even protein oxidation, are key to influencing the ability of meat to retain moisture (Huff-Lonergan and Lonergan, 2005). Consequently, the antioxidant activities of phytoncide may be a factor causing the higher WHC. Moreover, cortisol was known to increase associated with stress, and stress would decrease meat quality (Terlouw, 2005), in line with our study, cortisol decreased and meat quality increased with 0.04% of phytoncide inclusion.

In conclusion, our results indicate that Korean pine extract supplementation can promote growth performance and elevate meat quality. So that dietary Korean pine extract, phytoncide, could be considered as a potential natural growth promoter in finishing pigs.

ACKNOWLEDGMENTS

This work was supported by a grant from the Next-Generation BioGreen 21 Program (No. PJ01115902), Rural Development Administration, Republic of Korea.

REFERENCES

AOAC. 2006. Official Methods of Analysis. 18th edn. Association of Official Analytical Chemists, Arlington, VA, USA.

Burt, S. 2004. Essential oils: Their antibacterial properties and potential applications in foods—A review. Int. J. Food Microbiol. 94:223-253.

Cole, R. J. and R. H. Cox. 1981. Handbook of Toxic Fungal Metabolites. Academic Press, New York, USA.

Critchfield, J. W., S. T. Butera, and T. M. Folks. 1996. Inhibition of HIV activation in latently infected cells by flavonoid compounds. AIDS Res. Hum. Retroviruses 12:39-46.

Czech, A., E. Kowalczuk, and E. R. Grela. 2009. The effect of a herbal extract used in pig fattening on the animals' performance and blood components. Ann. Univ. Mariae. Curie. Sklodowska Sect. EE Zootech. 27:25-33.

Durham, D. G., X. Liu, and R. M. E. Richards. 1994. A triterpene from Rubus pinfaensis. Phytochemistry 36:1469-1472.

Hashemi, S. R. and H. Davoodi. 2010. Phytogenics as new class of feed additive in poultry industry. J. Anim. Vet. Adv. 9:2295-2304.

Holmes, R. A., R. S. Boston, and G. A. Payne. 2008. Diverse inhibitors of aflatoxin biosynthesis. Appl. Microbiol. Biotechnol. 78:559-572.

Honikel, K. O. 1998. Reference methods for the assessment of physical characteristics of meat. Meat Sci. 49:447-457.

Huff-Lonergan, E. and S. M. Lonergan. 2005. Mechanisms of water-holding capacity of meat: The role of postmortem biochemical and structural changes. Meat Sci. 71:194-204.

Iranloye, B. O. 2002. Effect of chronic garlic feeding on some haematological parameters. ABNF J. 5:1-2.

Islam, S. N., P. Begum, T. Ahsan, S. Huque, and M. Ahsan. 2004. Immunosuppressive and cytotoxic properties of Nigella sativa. Phytother. Res. 18:395-398.

Jamroz, D. and C. Kamel. 2002. Plant extracts enhance broiler performance. In non ruminant nutrition: Antimicrobial agents and plant extracts on immunity, health and performance J. Anim. Sci. 80 (Suppl. 1):41.

Jamroz, D., A. Wiliczkiewicz, T. Wertelecki, J. Orda, and J. Skorupinska. 2005. Use of active substances of plant origin in chicken diets based on maize and locally grown cereals. Br. Poult. Sci. 46:485-493.

Jang, I. S., Y. H. Ko, S. Y. Kang, and C. Y. Lee. 2007. Effect of commercial essential oils on growth performance, digestive enzyme activity, and intestinal microflora population in broiler chickens. Anim. Feed Sci. Technol. 134:304-315.

Jugl-Chizzola, M., E. Ungerhofer, C. Gabler, W. Hagmuller, R. Chizzola, K. Zitterl-Eglseer, and C. Franz. 2006. Testing of the palatability of Thymus vulgaris L. and Origanum vulgare L. as flavouring feed additive for weaner pigs on the basis of a choice experiment. Berl. Munch. Tierarztl. Wochenschr. 119:238-243.

Kauffman, R. G., G. Eikelenboom, P. G. Van der Wal, B. Engel, and M. Zaar. 1986. A comparison of methods to estimate water holding capacity in post-rigor porcine muscle. Meat Sci. 18:307-322.

Kawakami, K., M. Kawamoto, M. Nomura, H. Otani, T. Nabika, and T. Gonda. 2004. Effects of phytoncide on blood pressure under restraint stress in SHRSP. Clin. Exp. Pharmacol. Physiol. 31:S27-S28.

Koopmans, S. J., M. Ruis, R. Dekker, H. van Diepen, M. Korte, and Z. Mroz. 2005. Surplus dietary tryptophan reduces plasma cortisol and noradrenaline concentrations and enhances recovery after social stress in pigs. Physiol. Behav. 85:469-478.

Marsland, A. L., E. A. Bachen, S. Cohen, B. Rabin, and S. B. Manuck. 2002. Stress, immune reactivity and susceptibility to infectious disease. Physiol. Behav. 77:711-716.

Melody, J. L., S. M. Lonergan, L. J. Rowe, T. W. Huiatt, M. S. Mayes, and E. Huff-Lonergan. 2004. Early postmortem biochemical factors influence tenderness and water-holding capacity of three porcine muscles. J. Anim. Sci. 82:1195-1205.

Nam, E. S. and D. C. Uhm. 2008. Effects of phytoncides inhalation on serum cortisol level and life stress of college students. Korean. J. Adult. Nurs. 20:697-706.

Nose, K. 2000. Role of reactive oxygen species in the regulation of physiological functions. Biol. Pharm. Bull. 23:897-903.

NPPC (National Pork Producers Council). 1991. Procedures to Evaluate Market Hogs. 3rd edn. National Pork Producers Council, Des Moines, IA, USA.

NRC. 2012. Nutrient Requirement of Swine. 11th edn. The National Academy Press, Washington, DC, USA.

Offer, G. and P. Knight. 1988a. The structural basis of water-holding capacity in meat. Part 1: general principles and water uptake in meat processing. In Developments in Meat Science (Ed. R. Lawrie). 4th edn. Elsevier, Oxford, UK.

Offer, G. and P. Knight. 1988b. The structural basis of water-holding capacity in meat. Part 2: Drip Losses. In Developments in Meat Science (Ed. R. Lawrie). 4th edn. Elsevier, Oxford, UK.

Oikawa, S. 2005. Sequence-specific DNA damage by reactive oxygen species: Implications for carcinogenesis and aging. Environ. Health Prev. Med. 10:65-71.

Pečiulytė, D., I. Nedveckytė, V. Dirginčiūtė-Volodkienė, and V. Būda. 2010. Pine defoliator Bupalus piniaria L. (Lepidoptera: Geometridae) and its entomopathogenic fungi. 1. Fungi isolation and testing on larvae. Ekologija 56:34-40.

Rodas, D. B. 2006. The use of botanical feed additives in nursery and sow diets in the US. In: Proceedings of Delacon Performing Nature Symposium. Vienna, Austria. pp. 1-4.

SAS. 2003. SAS version 9.1. SAS Institude Inc Cary NC, USA.

Savage, A. W. J., P. D. Warriss, and P. D. Jolley. 1990. The amount and composition of the proteins in drip from stored pig meat. Meat Sci. 27:289-303.

Schone, F., A. Vetter, H. Hartung, H. Bergmann, A. Biertumpfel, G. Richter, S. Muller, and G. Breitschuh. 2006. Effects of essential oils from fennel (Foeniculi aetheroleum) and caraway (Carvi aetheroleum) in pigs. J. Anim. Physiol. Anim. Nutr. 90:500-510.

Shi, Y. H., Z. R. Xu, J. L. Feng, M. S. Xia, and C. H. Hu. 2005. Effects of modified montmorillonite nanocomposite on

growing/finishing pigs during aflatoxicosis. Asian Austalas. J. Anim. Sci. 18:1305-1309.

Terlouw, C. 2005. Stress reactions at slaughter and meat quality in pigs: genetic background and prior experience: A brief review of recent findings. Livest. Prod. Sci. 94:125-135.

Thiex, N. J., H. Manson, S. Anderson, and J. Persson. 2002. Determination of crude protein in animal feed, forage, grain, and oilseeds by using block digestion with a copper catalyst and steam distillation into boric acid: collaborative study. J. AOAC Int. 85:309-317.

Thiex, N. J., S. Anderson, and B. Gildemeister. 2003. Crude fat, diethyl ether extraction, in feed, cereal grain, and forage (Randall/Soxtec/submersion method): Collaborative study. J. AOAC Int. 86:888-898.

Van Krimpen, M. M., G. P. Binnendijk, F. H. M. Borgsteede, and C. P. H. Gaasenbeek. 2010. Anthelmintic effects of phytogenic feed additives in Ascaris suum inoculated pigs. Vet. Parasitol. 168:269-277.

Van Soest, P. J., J. B. Robertson, and B. A. Lewis. 1991. Methods for dietary fiber, neutral detergent fiber, and nonstarch polysaccharides in relation to animal nutrition. J. Dairy Sci. 74:3583-3597.

Wang, J. P., H. J. Kim, Y. J. Chen, J. S. Yoo, J. H. Cho, D. K. Kang, Y. Hyun, and I. H. Kim. 2009. Effects of delta-aminolevulinic acid and vitamin C supplementation on feed intake, backfat, and iron status in sows. J. Anim. Sci. 87:3589-3595.

Wang, Q., H. J. Kim, J. H. Cho, Y. J. Chen, J. S. Yoo, B. J. Min, Y. Wang, and I. H. Kim. 2008. Effects of phytogenic substances on growth performance, digestibility of nutrients, faecal noxious gas content, blood and milk characteristics and reproduction in sows and litter performance. J. Anim. Feed Sci. 17:50-60.

Wang, Y., Y. J. Chen, J. H. Cho, J. S. Yoo, Q. Wang, Y. Huang, H. J. Kim, and I. H. Kim. 2007. The effects of dietary herbs and coral mineral complex on growth performance, nutrient digestibility, blood characteristics and meat quality in finishing pigs. J. Anim. Feed Sci. 16:397-407.

Wenk, C. 2003. Herbs and botanicals as feed additives in monogastric animals. Asian Austalas. J. Anim. Sci. 16:282-289.

Windisch, W., K. Schedle, C. Plitzner, and A. Kroismayr. 2008. Use of phytogenic products as feed additives for swine and poultry. J. Anim. Sci. 86:E140-E148.

Williams, P. and R. Losa. 2001. The use of essential oils and their compounds in poultry nutrition. World Poult. 17:14-15.

Yan, L., J. P. Wang, H. J. Kim, Q. W. Meng, X. Ao, S. M. Hong, and I. H. Kim. 2010. Influence of essential oil supplementation and diets with different nutrient densities on growth performance, nutrient digestibility, blood characteristics, meat quality and fecal noxious gas content in grower-finisher pigs. Livest. Sci. 128:115-122.

Yan, L., Q. W. Meng, and I. H. Kim. 2011a. The effects of dietary Houttuynia cordata and Taraxacum officinale extract powder on growth performance, nutrient digestibility, blood characteristics and meat quality in finishing pigs. Livest. Sci. 141:188-193.

Yan, L., Q. W. Meng, and I. H. Kim. 2011b. The effect of an herb extract mixture on growth performance, nutrient digestibility, blood characteristics and fecal noxious gas content in growing pigs. Livest. Sci. 141:143-147.

Zhang, S., J. H. Jung, H. S. Kim, B. Y. Kim, and I. H. Kim. 2012. Influences of phytoncide supplementation on growth performance, nutrient digestibility, blood profiles, diarrhea scores and fecal microflora shedding in weaning pigs. Asian Austalas. J. Anim. Sci. 25:1309-1315.

Zhong, M., D. Wu, Y. Lin, and Z. F. Fang. 2011. Phytogenic feed additive for sows: Effects on sow feed intake, serum metabolite concentrations, igG level, lysozyme activity and milk quality. J. Agric. Sci. Technol. A1:802-810.

21

Performance, Carcass Quality and Fatty Acid Profile of Crossbred Wagyu Beef Steers Receiving Palm and/or Linseed Oil

Wisitiporn Suksombat*, Chayapol Meeprom, and Rattakorn Mirattanaphrai

School of Animal Production Technology, Institute of Agricultural Technology,

Suranaree University of Technology, Nakhon Ratchasima, 30000, Thailand

ABSTRACT: The objective of this study was to determine the effect of palm and/or linseed oil (LSO) supplementation on carcass quality, sensory evaluation and fatty acid profile of beef from crossbred Wagyu beef steers. Twenty four fattening Wagyu crossbred beef steers (50% Wagyu), averaging 640±18 kg live weight (LW) and approximately 30 mo old, were stratified and randomly assigned in completely randomized design into 3 treatment groups. All steers were fed approximately 7 kg/d of 14% crude protein concentrate with *ad libitum* rice straw and had free access to clean water and were individually housed in a free-stall unit. The treatments were i) control concentrate plus 200 g/d of palm oil; ii) control concentrate plus 100 g/d of palm oil and 100 g/d of LSO, iii) control concentrate plus 200 g/d of LSO. This present study demonstrated that supplementation of LSO rich in C18:3n-3 did not influence feed intakes, LW changes, carcass and muscle characteristics, sensory and physical properties. LSO increased C18:3n-3, C22:6n-3, and n-3 polyunsaturated fatty acids (PUFA), however, it decreased C18:1t-11, C18:2n-6, *cis*-9, *trans*-11, and *trans*-10,*cis*-12 conjugated linoleic acids, n-6 PUFA and n-6:n-3 ratio in *Longissimus dorsi* and *Semimembranosus* muscles. (**Key Words:** Linseed Oil, Growth Performance, Carcass Quality, Sensory Evaluation, Fatty Acids, Wagyu Beef Steers)

INTRODUCTION

The health benefits of α-linolenic acid (C18:3n-3) have become a leading topic for nutritional science research. The long-chain polyunsaturated fatty acids (LC-PUFA), especially linoleic acid (C18:2n-6) and C18:3n-3 have anticarcinogenic and cardioprotective roles in humans (Parodi, 1997). As a result, public health guidelines in most developed countries have recommended decreases in saturated and *trans* fatty acids and an increase in 18:3n-3, 20:5n-3, and 22:6n-3 in the human diet to reduce the incidence of chronic disease (World Health Organization, 2003). A reduction in the consumption of saturated fatty acids (SFA) and an increase in the consumption of polyunsaturated fatty acids (PUFA, fatty acids with more than one double bond) is encouraged, while monounsaturated fatty acids (fatty acids with one double

bond; MUFA) are generally regarded as beneficial for human health. The main sources of supplementary fatty acids in ruminant rations are plant oils and oilseeds. Supplementation of plant oils and oilseeds rich in omega-3 fatty acids could increase omega-3 fatty acids in beef fat.

Challenges in increasing the n-3 FA content of ruminant tissues and products differ from non-ruminants. In addition to the issues of the effects of unsaturated fatty acids (UFAs) on the stability and sensory acceptability of products, these FAs inhibit various essential anaerobic bacteria of the rumen, especially those involved in fiber digestion, biohydrogenation of UFAs, and methanogenesis (Palmquist and Jenkins, 1980). Of major concern in modifying the FA composition of animal products is ruminal biohydrogenation; under almost all feeding conditions, >80% of dietary UFAs are partially biohydrogenated to myriad *trans* unsaturated products or completely biohydrogenated to saturated products, mainly stearic acid (C18:0). Only in recent years has a general interest caused detailed investigation of biohydrogenation processes and

---

* Corresponding Author: Wisitiporn Suksombat.
E-mail: wisitpor@sut.ac.th

the associated microbiology. Most of the research aimed at improving dietary quality of beef has been focused on manipulation of animal feed with attempts to increase the content of 2 groups of fatty acids - intramuscular n-3 PUFA and conjugated linoleic acid (CLA) contents accomplished by feeding n-3 PUFA or linoleic acid (C18:2n-6) rich in ruminants' diets (Scollan et al., 2006; Palmquist, 2009). In addition, low PUFA n-6/n-3 ratio aids in the prevention of many chronic diseases. Increasing the content of PUFA and reducing SFA with the net effect of increasing PUFA/SFA and reducing n-6/n-3 ratio are priorities (Scollan et al., 2006). In general, previous studies reported the effect of different linseed form and concentration on performance and on FA composition of muscle and adipose tissue in beef cattle (Mach et al., 2006). Herdmann et al. (2010) found significant increases in the concentrations of n-3 fatty acids (C18:3n-3, C20: 5n-3 and C22:5n-3 and C22:6n-3) in meat from German Holstein Bulls fed 3% linseed oil (LSO) and 12% rapeseed cake. Therefore, supplementation of LSO enriched in C18:3n-3 would increase C18:3n-3 in muscle lipid. Thus, the objective of the present study was to examine the effect of LSO supplementation on performance and beef fatty acid profile of Wagyu crossbred beef steers.

MATERIALS AND METHODS

Animals, experimental design, and treatments

Twenty four Wagyu crossbred fattening steers (50% Wagyu), averaging 640±18 kg live weight (LW) and approximately 30 mo old, were stratified by their LW into 3 groups and each group was randomly assigned to three dietary treatments. All steers were fed approximately 7 kg/d of 14% crude protein (CP) concentrate with *ad libitum* rice straw and had free access to clean water and were individually housed in a free-stall unit. The treatments were i) control concentrate plus 200 g/d of palm oil (PO); ii) control concentrate plus 100 g/d of PO and 100 g/d of LSO; iii) control concentrate plus 200 g/d of LSO. The PO was added to the first and second group to balance energy concentration in the rations. The chemical composition of concentrate used in the experiment are presented in Table 1 while the fatty acid composition of feed and oils used in the present study are presented in Table 2.

The basal diet was formulated to meet NRC (2000) requirements. All steers received *ad libitum* rice straw, had free access to clean water, were individually housed in a free-stall unit, and individually fed according to treatments. The experiment lasted for 67 days with the first 7 days for adjustment, followed by 60 days of measurement period.

Measurements, sample collection, and chemical analysis

Feed refusals from each individual steer was weighed on 2 consecutive days weekly to calculate dry matter

Table 1. Chemical composition of the experimental diets

Items	Concentrate[1]	Palm oil	Linseed oil	Rice straw
Dry matter	93.9			92.3
------------------% of DM------------------				
Ash	7.0			10.9
Crude protein	14.6			4.0
Ether extract	4.1	100	100	0.8
Crude fiber	17.1			39.8
Neutral detergent fiber	42.6			76.3
Acid detergent fiber	26.3			52.3
Neutral detergent in soluble N	1.1			0.5
Acid detergent insoluble N	0.9			0.4
Acid detergent lignin	10.9			6.3
TDN_{1X} (%)[2]	60.2	184.2	184.2	46.1
DE_{1X} (Mcal/kg DM)[3]	2.8	7.7	7.7	2.0
ME (Mcal/kg DM)[4]	2.7	5.8	5.8	2.0
NE_M (Mcal/kg DM)[5]	1.4	4.2	4.2	0.8
NE_G (Mcal/kg DM)[6]	0.9	3.1	3.1	0.2

[1] kg/100 kg concentrate: 30 dried cassava chip, 4 ground corn, 10 rice bran, 25 palm meal, 15 coconut meal, 6 dried distillers grains with solubles, 0.5 sodium bicarbonate, 6 molasses, 1 dicalciumphosphate (16% P), 1.5 urea, 0.5 salt, and 0.5 premix. Premix: provided per kg of concentrate including vitamin A, 5,000 IU; vitamin D_3, 2,200 IU; vitamin E, 15 IU; Ca, 8.5 g; P, 6 g; K, 9.5 g; Mg, 2.4 g; Na, 2.1 g; Cl, 3.4 g; S, 3.2 g; Co, 0.16 mg; Cu, 100 mg; I, 1.3 mg; Mn, 64 mg; Zn, 64 mg; Fe, 64 mg; Se, 0.45 mg.
[2] Total digestible nutrients, TDN_{1X} (%) = tdNFC+tdCP+(tdFA×2.25)+tdNDF−7 (NRC, 2000).
[3] Digestible energy, DE_{1X} (Mcal/kg) = [(tdNFC/100)×4.2]+[(tdNDF/100)×4.2]+[(tdCP/100)×5.6]+[(FA/100)×9.4]−0.3.
[4] Metabolisable energy, ME = 0.82×DE (NRC, 2000).
[5] Net energy for maintenance, NE_M = 1.37ME−0.138ME2+0.0105ME3−1.12 (NRC, 2000).
[6] Net energy for growth, NE_G = 1.42ME−0.174ME2+0.0122ME3−1.65 (NRC, 2000).

intakes (DMI). Samples were taken and dried at 60°C for 48 hours and at the end of the experiment, feed samples were pooled to make representative samples for proximate and detergent analyses. Samples were ground through 1 mm screen and analyzed for chemical composition. Dry matter (DM) was determined by hot air oven at 60°C for 48 h while CP was analyzed by Kjeldahl method (AOAC, 1995). Ether extract was determined by using petroleum ether in a Soxtec System (AOAC, 1995). Fiber fraction, neutral detergent fiber and acid detergent fiber were determined using the method described by Van Soest et al. (1991), adapted for Fiber Analyzer. Ash content was determined by ashing in a muffle furnace at 600°C for 3 h. The chemical analysis was expressed on the basis of the final DM. Fatty acids composition of concentrates, fresh grass and rice straw were determined by gas chromatography (GC).

Meat pH (pH meter model UB-5, Denver Instrument, Gottingen, Germany) was determined in *Longissimus dorsi*

Table 2. Fatty acid composition (g/100 g fat) of concentrate, rice straw and oils used in the experiment

Fatty acids	Concentrate	Rice straw	Palm oil	Linseed oil
C8:0	0.74	ND	0.05	0.05
C10:0	1.14	ND	0.02	ND
C12:0	17.96	ND	0.19	ND
C14:0	6.38	1.28	0.96	0.06
C16:0	17.85	47.49	38.29	4.91
C18:0	2.71	8.57	4.42	3.46
C18:1 n9c	31.90	16.76	40.61	17.88
C18:2 n6c	20.33	19.88	13.66	16.73
C20:0	0.00	0.00	0.04	ND
C18:3 n3	0.35	6.03	0.26	55.87
C18:3 n6	0.66	ND	0.11	0.24
SFA	46.77	57.34	44.05	8.70
MUFA	31.90	16.76	41.07	17.96
PUFA	21.34	25.91	14.89	73.34
total n3[1]	0.35	6.03	0.43	56.20
total n6[2]	20.99	19.88	14.46	17.04
PUFA:SFA	0.46	0.45	0.34	8.43
n6/n3	60.01	3.30	33.69	0.30

ND, not detected; SFA, sum of saturated fatty acid from C8:0–C20:0; MUFA, monounsaturated fatty acid from C18:1; PUFA, sum of polyunsaturated fatty acid from C18:2–C18:3.
[1] Sum of n6 fatty acid C18:2 n-6–C18:3 n-6.
[2] n3 fatty acid from C18:3 n-3.

(LD) and *Semimembranosus* (SM) muscles at 45 min and 24 h. After dissection, the LD and SM samples were cut in to 2.5 cm thick slices, put into polyethylene bags, chilled at 4°C for 48 h and then stored in the refrigerator outside of the bag for 1 h ('blooming') before conducting color measurements using a Color Quest XE (HunterLab, Hunter Associates Laboratory, Inc., Reston, VA, USA).

Water-holding capacity was assessed via substance losses occurring during different procedures. Thawing and cooking losses were determined in the 2.5 cm thick slices of LD and SM frozen in polyethylene bags at –20°C. Thawing was performed over 24 h at 4°C. Before weighing, the sample surfaces were dried with soft paper. Afterwards, samples were sealed in heat-resistant plastic bags to be boiled in water bath (WNE 29, Memmert, Schwabach, Germany) at 80°C until an internal temperature of 70°C was reached. Samples were cooled to ambient temperature and weighed after drying the surfaces with soft paper. For the determination of the grilling loss, 2.5 cm thick slices were grilled in a convection oven (model 720, Mara, Taipei, Taiwan) at 150°C until an internal temperature of 70°C was reached. In the LD, additionally drip loss was determined. In the boiled samples, shear force were measured after cooling and drying. A steel hollow-core device with a diameter of 1.27 cm was punched parallel to the muscle fibers to obtain six pieces from each muscle sample. Measurements were carried out on a material testing

machine by Texture analyzer (TA-TX2 Texture Analyzer, Stable Micro Systems, Surrey, UK) using a Warner–Bratzler shear. A crosshead speed of 200 mm/min and a 5 kN load cell calibrated to read over a range of 0×100 N were applied.

Samples of the LD and SM were minced and analyzed in duplicate for moisture, fat, ash and protein contents according to AOAC (1995). Total lipids in feed and beef samples were extracted using chloroform-methanol (Folch et al., 1957). Fifteen grams of each sample were homogenized for 2 min with 90 mL of chloroform-methanol (2:1) (Nissei AM-8 Homogenizer, Nihon Seiki Kaisha, LTD., Tokyo, Japan). Samples were then further homogenized for 2 min with 30 mL of chloroform. Then, samples were separated in separating funnel and 30 mL of deionized water and 5 mL of 0.58% NaCl was added. Fatty acid methyl esters (FAME) were prepared by the procedure described by Ostrowska et al. (2000), briefly, placing 30 mg of the extract into a 15 mL reaction tube fitted with a teflon-lined screw cap. One and a half mL of 0.5 M NaOH in methanol was added. The tubes were flushed with nitrogen, capped, heated at 100°C for 5 min with occasional shaking and then cooled to room temperature. One ml of C17:0 internal standard (2.00 mg/mL in hexane) and 2 mL of BF_3 in methanol were added and heated at 100°C for 5 min with occasional shaking and 10 mL of deionized water were added. The solution was transferred to a 40 mL centrifuged tube and 5 mL of hexane were added for FAME extraction. The solution was centrifuged at 2,000 g, at 10°C for 20 min and then the hexane layer was dried over sodium sulfate and transferred into a vial for analyzing by GC (7890A GC System, Agilent Technologies, Inc., Wilmington, DE, USA) equipped with a 100 m×0.25 mm×0.2 μm film fused silica capillary column (SP1233, Supelco Inc, Bellefonte, PA, USA). Injector and detector temperatures were 250°C. The column temperature was kept at 70°C for 4 min, then increased at 13°C/min to 175°C and held at 175°C for 27 min, then increased at 4°C/min to 215°C and held at 215°C for 17 min, then increased at 4°C/min to 240°C and held at 240°C for 10 min.

Quantitative descriptive analysis was used for sensory evaluation (Stone et al., 1974), a test panel was selected from a number of students and faculty members of the School of Animal Production Technology, Suranaree University of Technology, who had undergone sensory evaluation training. Grilled 2.5-cm slices of LD and SM were cut into pieces of 1.3×1.3×1.9 cm and served warm. Panelists were asked to grade samples for tenderness, juiciness, flavor and overall acceptability and assessments were given individually using a structured line graph and determined on a straight line. Thus, each point on a linear scale to represent the quantity that can be measured with a ruler. Samples were served subsequently in a randomized order with respect to group and animal. The 24 samples (from 12 animals and two muscles) were tested by 8 persons.

Statistical analysis

All measured data were analysed by analysis of variance for complete randomized design using the Statistical Analysis System (SAS, 2001). Significant differences among treatment were assessed by Duncan's new multiple range test. A significant level of $p < 0.05$ was used (Steel and Torrie, 1980).

RESULTS AND DISCUSSION

Feed composition and performance

The chemical composition of the feeds are presented in Table 1 and the concentrate was formulated to meet the requirement of the steers. LSO had the highest proportion of PUFA (73.34 g/100 g fat) while PO had the highest proportion of SFA (44.05 g/100 g fat). In the concentrate, the main SFA was C12:0 (17.96 g/100 g fat) and C16:0 (17.85 g/100 g fat), whereas C18:1n-9 was the main MUFA in PO (40.61 g/100 g fat), C18:2n-6 was the main PUFA in 14% CP concentrate (20.33 g/100 g fat) and C18:3n-3 was the main PUFA in LSO (55.87 g/100 g fat) (Table 2). The supplementation of LSO was chosen to reduce SFA (8.70) and to increase n-3 FA (56.20) (Table 2).

No significant difference was found for DM and CP intakes among groups; however, the animals supplemented with LSO tended to have greater concentrate and CP intake than those fed the PO diets (p = 0.05). In particular, CP intake of the animals fed LSO was 97 and 70 g/d higher than those fed PO (Table 3). Palmquist and Jenkins (1980) concluded that fat at 5% to 10% of the diet reduced intake and digestion. Rule et al. (1989) also reported that DMI is often depressed when diets contain more than 8% fat. With diets containing lower levels of added fat, Huerta-Leidenz et al. (1991) reported no influence on daily gain, intake or feed conversion ratio when dietary whole cotton seed of 15% or 30% (3.3% and 6.6% additional fat) was supplemented. In the present trial, fat contents of experimental diets were between 5.0% and 5.1%, it is unlikely that these levels of fat affected feed intake. When individual FA intake was calculated (Table 3), cattle on LSO diets consumed more C18:3n-3 and PUFA and less C16:0, C18:1, SFA, and MUFA than those cattle on PO diet.

No remarkable changes were found for final LW and LW change among the treatments (Table 3). The amount of dietary fat did not affect LW of the steers over the course of the trial; however, LW was increased at 1.11, 1.20, and 1.20 kg/d in the animals fed PO, PO/LSO, and LSO, respectively (Table 3). Similarly, Noci et al. (2007) reported that 150 g/d sunflower oil (SFO) and 150 g/d LSO supplementation did not affected final LW and average daily gain (ADG). This is partially because total net energy (Mcal/d) consumption was balanced by treatment. Furthermore, He and Armentano (2011) supplemented a mixture of flaxseed oil and SFO at 5% of diet and reported no significant effects of supplementation on DMI, body weight, ADG, and gain per unit feed.

Carcass quality

Slaughter weight, warm carcass weight, % warm carcass, cold carcass weight, % cold carcass, beef marbling score, loin eye area (LEA) and back fat thickness were not significantly different among treatments (Table 3). Noci et al. (2007) reported that Charolais crossbred heifers fed 150 g/d SFO and 150 g/d LSO showed no differences in carcass weight and dressing percentage.

The LEA and 12th rib back fat thickness were not significantly different among treatments (Table 3). The eye muscle area can be used as a representative measure of the quantity, quality, and distribution of the muscle mass. Late-maturing muscles are used to represent the muscle tissue development rate. Thus, the *longissimus* is the most suitable muscle for analysis because, in addition to its late maturation, it is easy to measure. Zinn et al. (2000) did not observe effects on eye muscle area and fat thickness cover using Holstein steers fed diets containing protected fat or animal fat as a lipid source at up to 60.0 g/kg.

Beef quality

Initial (45 min post slaughter) and final pH (24 hour post slaughter) values were not different among the treatments (Table 4). The initial pH was considered ideal and final pH values were also found in the interval considered to be normal (5.4 to 5.8) for beef (Mach et al., 2008). The final pH corresponds to the accumulation of lactic acid resulting from the production of adenosine triphosphate from glucose encountered in the form of glycogen reserves. In general, cattle supplemented with grains possess a greater availability of glycogen at the time of slaughter and a lower final pH in the beef (Neath et al., 2007). The final pH values suggested that there was no elevated stress prior to slaughter, because acidification of the muscle occurred as expected, and that the level of substitution of oil supplement evaluated did not affect the final pH.

The chemical composition of beef composed of moisture, protein, fat and ash were not significantly different among treatments (Table 4). The amounts of fat in the muscle typically result from a balance between dietary energy and metabolic requirements of the animal (Oliveira et al., 2012). If energy intake is higher than its metabolic demands, this excess will be stored as fat. In the present study, the greater supply of lipids in the diet was not enough to increase the deposition of fat in the muscle and back fat.

Table 3. DM and CP intakes, and performances of Wagyu crossbred cattle fed palm and/or linseed oil (n = 8)

Items	200 g/d PO	100 g/d PO+100 g/d LSO	200 g/d LSO	SEM	p-value
DM intake (kg/d)					
Concentrate	5.70	6.46	6.32	0.23	0.062
Rice straw	2.87	2.79	2.82	0.12	0.883
Total	8.57	9.25	9.14	0.31	0.351
CP intake (g/d)					
Concentrate	623	726	705	29.3	0.051
Rice straw	81	74	68	4.9	0.230
Total	704	801	774	32.8	0.122
FA intake (g/d)					
C8:0	1.8	2.1	2.0	0.23	0.805
C10:0	2.7	3.0	3.0	0.11	0.764
C12:0	42.4	51.0	46.5	2.97	0.564
C14:0	17.1	18.2	16.9	0.94	0.689
C16:0	129.2[a]	101.1[b]	66.8[c]	4.91	0.001
C18:0	17.1	17.0	15.9	0.76	0.485
C18:1	159.6[a]	146.7[a]	122.2[b]	4.23	0.041
C18:2	79.4	88.7	90.6	4.38	0.392
C18:3n-3	2.7[c]	58.4[b]	114.0[a]	3.04	0.001
C18:3n-6	1.8	2.1	2.2	0.15	0.653
SFA	210.6[a]	189.4[b]	151.5[c]	6.50	0.001
MUFA	160.5[a]	165.2[a]	124.2[b]	4.65	0.039
PUFA	85.6[c]	150.5[b]	207.8[a]	5.83	0.001
Total FA intake	456.7	505.2	483.5	18.67	0.423
Initial weight (kg)	642	631	646	18.0	0.817
Final weight (kg)	709	703	718	16.7	0.801
Live weight change (kg/d)	1.11	1.20	1.20	0.06	0.481
Slaughter weight (kg)	669	663	685	23.9	0.792
Warm carcass weight (kg)	425	420	435	14.0	0.755
% warm carcass	63.5	63.4	63.4	0.53	0.984
Cold carcass weight (kg)	411	406	422	12.8	0.678
% cold carcass	61.5	61.2	61.6	0.73	0.935
Marbling score[1]	3.04	3.13	3.25	0.13	0.623
Loin eye area (cm^2)	92.46	91.89	90.22	3.88	0.764
Back fat thickness (cm)	1.14	1.20	1.18	0.18	0.927

PO, palm oil; LSO, linseed oil; SEM, standard error of the mean; DM, dry matter; CP, crude protein; FA, fatty acid; SFA, saturated fatty acid; MUFA, monounsaturated fatty acid; PUFA, polyunsaturated fatty acid.

[1] 1 = very scarce, 12 = very abundant (Japanese Meat Grading Association).

The literature suggests that the total protein content is less variable in bovine meat, with values of approximately 20% observed in the LD muscle without the fat cover, and this is independent of food, breed, the genetic group, and the physiological condition (Marques et al., 2006).

The sensory tenderness, juiciness, beef flavor, color firmness and texture were unaffected by treatments. In another study, German Holstein and Simmental bulls finished on grass or fed a concentrate of silage, barley, and cracked linseed produced beef that had a higher n-3 PUFA content than did beef from animals fed a grass-based diet, but the sensory profiles did not differ, except that meat from grass-finished beef had higher bloody and fishy notes (Nuernberg et al., 2005). On the other hand, when steers were fed diets that had similar base components, but the diets differed in the amount or composition of fatty acids through the addition of different oils, lipid and color stability were more closely associated with fatty acid composition and greater abnormal flavors and rancidity scores (Scollan et al., 2006). Scheeder et al. (2001) evaluated the beef of bulls fed different sources of fat and found that the beef of animals fed with LSO tended to be juicier and to possess a more agreeable aroma. These results may be due to the higher proportions of n-3 PUFA in these animals, triggering odor precursors that are activated by oxidation during heating. However, changes in PUFA concentrations in the present experiment would not likely have been large enough to have affected taste panel

Table 4. Beef characteristics of Wagyu crossbred cattle fed palm and/or linseed oil

Items	200 g/d PO	100 g/d PO+100 g/d LSO	200 g/d LSO	SEM	p-value
No. of cattle	8	8	8		
Longissimus dorsi					
pH 45 min.	6.32	6.25	6.34	0.14	0.659
pH 24 h	5.63	5.51	5.59	0.07	0.723
Drip loss (%)	3.87	4.26	3.77	0.23	0.458
Thawing loss (%)	4.46	4.39	4.02	0.73	0.826
Grilled cooking loss (%)	37.28	37.59	37.34	2.58	0.875
Moisture cooking loss (%)	25.88	25.26	25.95	1.23	0.784
Moisture content (%)	70.68	70.75	70.59	0.91	0.903
Crude protein (%)	21.59	21.24	21.31	0.86	0.823
Ash (%)	1.25	1.29	1.32	0.07	0.457
Fat (%)	6.48	6.72	6.78	0.70	0.769
Semimembranosus					
pH 45 min.	6.33	6.27	6.29	0.17	0.622
pH 24 h	5.62	5.51	5.53	0.07	0.705
Drip loss (%)	4.02	3.86	4.12	0.74	0.483
Thawing loss(%)	4.37	4.21	4.02	0.97	0.897
Grilled cooking loss (%)	38.62	38.49	38.03	2.33	0.885
Moisture cooking loss (%)	25.22	25.47	24.89	0.31	0.723
Moisture content (%)	70.64	70.52	70.39	0.96	0.884
Crude protein (%)	21.27	21.36	21.29	0.92	0.799
Ash (%)	1.27	1.36	1.34	0.09	0.438
Fat (%)	6.82	6.76	6.98	0.77	0.834

PO, palm oil; LSO, linseed oil; SEM, standard error of the mean.

assessments.

Beef color remained unaffected by treatments (Table 5). Values encountered in literature for L*, a*, and b* were used to measure beef color in the CIELAB space (Lightness, L*; redness, a*; yellowness, b*; CIE, 1978) being in the following ranges of variation: 33 to 41, 11.1 to 23.6, and 6.1 to 11.3, respectively. Values obtained in the present study were within the range given.

The shear forces of LD and SM muscle were unaffected (p>0.05) by the addition of LSO in the diets (Table 5). Beef tenderness is a trait considered to be of great relevance for consumers with shear force being an objective measure of tenderness. Bovine meat is considered to have an acceptable tenderness if its shear strength values are below 8 N (Swan et al., 1998). The beef in the report of Santana et al. (2014) was considered tender regardless of the lipid supplementation adopted because the average values obtained were 7.5 N. The present trial found shear force values between 3.57 and 4.33 kg/cm^2 which were considered to be tender (Table 5). These values were closely related to the values obtained from sensory perception of tenderness by trained panelists (5.29 to 5.76; Table 5). Such variations in the shear force values may be caused by differences in the thicknesses of the blades utilized in the analysis

Beef fatty acid profile

In the current study, LSO-containing diets resulted in marked alternations in beef FA composition relative to the added PO in the diet and the responses to treatment diets in both LD and SM muscles showed the same trend (Tables 6 and 7). To compare with PO diet, LSO diets significantly decreased C18:1t-11, C18:2n-6 FAs, cis-9, trans-11, and trans-10, cis-12 CLAs, and C20:4n-6 FAs, and increased C18:3n-3 and C22:6n-3 FAs whereas they had no effect on C10:0 - C16:0, C20:5n-3, and C22:5n-3 FAs. The diets containing LSO resulted in greater LD and SM C18:3n-3 FAs, represented respectively 0.86 and 0.87; and 1.59% and 1.56% of the total beef FA when the animals were fed 100 and 200 g/d LSO diets compared to 0.48% and 0.52% in the animals fed the control diets (p<0.01). The concentrations of total beef n-3 PUFA were increased in animals fed diets supplemented with LSO (p<0.01), whereas the contents of total beef n-6 PUFA were reduced when the steers were fed diets supplemented with LSO leading to a decrease (p<0.01) in the ratio of n-6 to n-3 FA in beef fat of steers fed LSO treatments.

The concentration of C18:2n-6 ranging from 4.23% to 6.91% in LD and SM muscles were somewhat slightly higher than the 4.78% reported for the subcutaneous adipose tissue of Hanwoo steers fed linseed (Kim et al., 2009), and higher than the 1.43% reported in heifers fed

Table 5. Sensory and physical evaluations of beef from Wagyu crossbred cattle fed palm and/or linseed oil

Items	200 g/d PO	100 g/d PO+100 g/d LSO	200 g/d LSO	SEM	p-value
No. of cattle	8	8	8		
Longissimus dorsi					
Tenderness	5.41	5.57	5.29	0.14	0.729
Juiciness	5.52	5.66	5.59	0.10	0.821
Beef flavor	5.44	5.28	5.51	0.16	0.904
Color	3.83	3.80	3.90	0.04	0.729
Firmness	3.13	3.11	3.09	0.06	0.695
Texture	4.11	4.19	4.15	0.08	0.817
L*	38.31	38.17	38.34	2.13	0.816
a*	23.33	23.61	23.56	1.76	0.905
b*	7.62	7.64	7.59	0.67	0.897
Shear force (kg/cm^2)	3.61	3.57	3.63	0.17	0.799
Semimembranosus					
Tenderness	5.56	5.67	5.53	0.16	0.824
Juiciness	5.69	5.76	5.67	0.13	0.698
Beef flavor	5.13	5.27	5.19	0.15	0.827
Color	3.88	3.83	3.95	0.05	0.766
Firmness	3.18	3.15	3.13	0.07	0.697
Texture	4.18	4.24	4.20	0.07	0.803
L*	42.56	42.38	41.95	2.24	0.823
a*	23.23	23.31	23.27	1.88	0.877
b*	8.86	8.92	8.67	0.77	0.865
Shear force (kg/cm^2)	4.31	4.33	4.31	0.19	0.806

PO, palm oil; LSO, linseed oil; SEM, standard error of the mean.
Tenderness, juiciness, beef flavor, color, firmness, and texture: 1 = extremely though, dry, bland, pink, firm, and fine respectively; 8 = extremely tender, juicy, intense, dark red, soft and coarse, respectively.

different vegetable oils (Noci et al., 2007). The concentration of C18:2n-6 decreased with increasing LSO, possibly because of its partial conversion to C18:0 in the rumen. The increase in the concentration of n-3 PUFA coincided with the major increase in the concentration of C18:3n-3 in the LSO diets. These increases are in agreement with the results of Kim et al. (2009) who fed cattle with linseed. This can be explained by the higher intake of C18:3n-3 of LSO cattle (58.4 and 114 g/d of the respective PO/LSO and LSO compared to 2.7 g/d of the PO cattle).

The PO was added to the first and second group to balance energy concentration in the rations, however, the FA content of PO might affect FA composition in beef. Choi et al. (2015) demonstrated that the SFA (C16:0 and C18:0) strongly promote adipogenic gene expression in intramuscular preadipocytes whereas, the MUFA (C18:1 cis-9) and the PUFA (C18:2n-6 and C18:3n-3) depress adipogenic gene expression. Supplementing the diets of beef cattle with PO, rich in C16:0 would increase adiposity without increasing C16:0 in beef. Partida et al. (2007) demonstrated an increase in C16:0 in intramuscular fat of bulls after supplementing hydrogenated PO, although the total SFA content remained similar to control bulls. An increase of oleic acid in beef improves beef palatability

(Westering and Hedrick, 1979) and increases its healthfulness (Gilmore et al., 2011). When the LSO results are compared with practical LSO-unsupplemented diets, Quinn et al. (2008) observed that C18:2n-6 and C18:3n-3 in LM muscle were increased 2 to 3 fold by adding 4% LSO to the diets of Holstein-Jersey steers whereas C20:4n-6 was reduced. Increased C18:3n-3 and decreased C20:4n-6 in both LM and SM muscles were also found when LSO was added in the present study. However, increased LSO in the diet reduced C18:2n-6 in the current study.

SFA, UFA, MUFA, and PUFA in LD and SM muscles were unaffected by dietary treatments. There was more total n-6 PUFA than total n-3 PUFA, and C18:2n-6 was the most concentrated PUFA across treatment. Feeding 100 and 200 g/d LSO increased total n-3 PUFA (p<0.01) and C18:3n-3 (p<0.01) in LD and SM muscles compared with the control treatment. Overall, feeding 200 g/d LSO led to approximate triple total n-3 PUFA compared with 200 g/d PO in LD and SM muscles. The lack of dietary effects on total PUFA in LD and SM muscles indicates that LSO addition had no effect on rates of lipolysis in the rumen. Feeding 100 and 200 g/d LSO increased C22:6n-3 (DHA) content when compared with the control treatment. The lack of diet effects on C22:5n-3 (DPA) in LD and SM relates to the limited capacity for the last steps in the n-3 PUFA

Table 6. Fatty acid composition (g/100 g fat) of *Longissimus dorsi* muscle from Wagyu crossbred cattle fed palm and/or linseed oil

Items	200 g/d PO	100 g/d PO+100 g/d LSO	200 g/d LSO	SEM	p-value
No. of cattle	8	8	8		
Longissimus dorsi					
C10:0	0.12	0.13	0.13	0.01	0.744
C12:0	3.76	3.52	2.96	0.26	0.661
C14:0	5.25	4.59	5.87	0.19	0.506
C15:0	0.90	0.81	0.73	0.07	0.368
C16:0	22.76	24.99	25.78	0.32	0.493
C16:1	4.40	3.79	3.43	0.20	0.078
C17:0	1.21	1.24	1.11	0.09	0.810
C18:0	7.28	7.48	7.97	0.31	0.349
C18:1n-9	39.23	40.35	39.86	0.38	0.274
C18:1t-11	3.35[a]	2.79[b]	2.67[b]	0.10	0.014
C18:2n-6	6.86[a]	5.72[b]	4.06[c]	0.27	0.004
CLA c9,t11	1.05[a]	0.77[b]	0.62[b]	0.06	0.007
CLA t10,c12	0.17[a]	0.13[b]	0.10[b]	0.01	0.031
C18:3n-3	0.48[c]	0.86[b]	1.59[a]	0.10	0.001
C20:4n-6	3.81[a]	2.88[b]	2.34[c]	0.12	0.002
C20:5n-3	0.08	0.11	0.27	0.03	0.523
C22:5n-3	0.28	0.37	0.55	0.03	0.132
C22:6n-3	0.23[b]	0.37[b]	0.68[a]	0.05	0.004
SFA[1]	41.28	42.76	44.55	0.48	0.983
UFA[2]	58.72	57.24	55.45	0.48	0.913
MUFA[3]	46.98	46.93	45.96	0.48	0.234
PUFA[4]	11.74	10.31	9.49	0.48	0.574
n-3 PUFA[5]	1.07[c]	1.71[b]	3.09[a]	0.19	0.004
n-6 PUFA[6]	10.67[a]	8.60[b]	6.40[c]	0.32	0.002
Total CLA[7]	1.22[a]	0.90[b]	0.72[b]	0.07	0.008
n-6:n-3	9.97[a]	5.03[b]	2.07[b]	0.79	0.001
UFA:SFA	1.42	1.34	1.24	0.03	0.981
PUFA:SFA	0.28	0.24	0.21	0.01	0.062

PO, palm oil; LSO, linseed oil; SEM, standard error of the mean; SFA, saturated fatty acid; UFA, unsaturated fatty acids; MUFA, monounsaturated fatty acid; PUFA, polyunsaturated fatty acid.
[1] Sum of saturated fatty acid from C10:0–C18:0. [2] Sum of unsaturated fatty acid from MUFA, n-3 PUFA, and n-6 PUFA.
[3] Sum of monounsaturated fatty acid from C16:1–C18:1. [4] Sum of monounsaturated fatty acid from n-3 PUFA and n-6 PUFA.
[5] Sum of n-3 polyunsaturated fatty acid from C18:3 n-3 – C22:6 n-3. [6] Sum of n-6 polyunsaturated fatty acid from C18:2 n-6 and C20:4 n-6.
[7] Sum of CLA from CLA c9,t11 and CLA t10,c12.

elongation and desaturation pathway (Raes et al., 2004). Therefore, supplementing bovines with UFAs can increase their passage to the small intestine, which allows more absorption and the possibility of changing the fatty acid profile of beef. The percentages of total CLA (*cis*-9, *trans*-11 C18:2) in LD and SM were decreased by dietary LSO treatments (p<0.01, Tables 6 and 7), confirming the previous results of Bessa et al. (2007). In addition, Noci et al. (2007) observed that the *cis*-9, *trans*-11 CLA content in intramuscular fat was higher in heifers supplemented with SFO as source C18:2n-6 then with LSO. The decreases in CLA may be due to a reduction in delta-9 desaturase activity which would have been reflected in muscle concentrations of MUFA and SFA. This simply was due to

dilution of C18:2n-6 by C18:3n-3 in the diet, as C18:2n-6 is the direct precursor of *cis*-9, *trans*-11 CLA and hence *trans*11, C18:1. The aim of the current study was to increase n-3 FAs in beef by supplementing LSO, rich in C18:3n-3 while PO was added to balance energy concentration in the rations. Choi et al. (2015) demonstrated that PO decreased C18:1*trans*- 11 and *cis*-9, *trans*- 11 CLA and increased C18:0, relative to control muscle while soy bean oil (SBO) increased C18:2n-6 and C18:3n-3, relative to control muscle. Increased C18:2n-6 supply to the diet by adding SBO did not change muscle C18:1*trans*- 11 and *cis*-9, *trans*- 11 CLA (Choi et al., 2015). However, the present study found reductions in muscle C18:1*trans*- 11 and *cis*-9,

Table 7. Fatty acid composition (g/100 g fat) of *Semimembranosus* muscle from Wagyu crossbred cattle fed palm and/or linseed oil

Items	200 g/d PO	100 g/d PO+100 g/d LSO	200 g/d LSO	SEM	p-value
No. of cattle	8	8	8		
Semimembranosus					
C10:0	0.10	0.11	0.12	0.01	0.756
C12:0	3.69	3.42	3.01	0.24	0.658
C14:0	5.08	4.61	5.76	0.16	0.511
C15:0	0.87	0.83	0.75	0.07	0.351
C16:0	22.87	25.08	25.81	0.30	0.487
C16:1	4.37	3.59	3.41	0.19	0.073
C17:0	1.19	1.25	1.14	0.08	0.804
C18:0	7.39	7.41	7.87	0.28	0.351
C18:1n-9	39.23	40.60	39.96	0.34	0.268
C18:1t-11	3.46[a]	2.83[b]	2.74[b]	0.11	0.013
C18:2n-6	6.91[a]	5.82[b]	4.23[c]	0.26	0.003
CLA c9,t11	1.07[a]	0.76[b]	0.61[b]	0.06	0.007
CLA t10,c12	0.16[a]	0.14[ab]	0.11[b]	0.01	0.029
C18:3n-3	0.52[c]	0.87[b]	1.56[a]	0.09	0.001
C20:4n-6	3.75[a]	2.74[b]	2.21[c]	0.11	0.002
C20:5n-3	0.08	0.10	0.24	0.03	0.467
C22:5n-3	0.24	0.38	0.53	0.03	0.129
C22:6n-3	0.25[b]	0.36[b]	0.66[a]	0.04	0.003
SFA[1]	41.19	42.71	44.46	0.46	0.979
UFA[2]	58.81	57.29	55.54	0.46	0.922
MUFA[3]	47.06	47.02	46.11	0.46	0.228
PUFA[4]	11.75	10.27	9.43	0.46	0.575
n-3 PUFA[5]	1.09[c]	1.71[b]	2.99[a]	0.17	0.003
n-6 PUFA[6]	10.66[a]	8.56[b]	6.44[c]	0.31	0.002
Total CLA[7]	1.23[a]	0.90[b]	0.72[b]	0.06	0.007
n-6:n-3	9.78[a]	5.01[b]	2.15[c]	0.73	0.001
UFA:SFA	1.43	1.34	1.25	0.03	0.978
PUFA:SFA	0.29	0.24	0.21	0.01	0.064

PO, palm oil; LSO, linseed oil; SEM, standard error of the mean; SFA, saturated fatty acid; UFA, unsaturated fatty acids; MUFA, monounsaturated fatty acid; PUFA, polyunsaturated fatty acid.
[1] Sum of saturated fatty acid from C10:0–C18:0. [2] Sum of unsaturated fatty acid from MUFA, n-3 PUFA, and n-6 PUFA.
[3] Sum of monounsaturated fatty acid from C16:1–C18:1. [4] Sum of monounsaturated fatty acid from n-3 PUFA and n-6 PUFA.
[5] Sum of n-3 polyunsaturated fatty acid from C18:3 n-3 – C22:6 n-3. [6] Sum of n-6 polyunsaturated fatty acid from C18:2 n-6 and C20:4 n-6.
[7] Sum of CLA from CLA c9,t11 and CLA t10,c12.

trans- 11 CLA when LSO was added. If CLA and n-3 FAs are needed to increase, a blend of C18:3n-3 and C18:2n-6 containing oils is required to supplement to the diets of finishing steers.

Treatments had no effect on total or individual SFA in LD and SM (Tables 6 and 7). The predominant SFA across all diets in LD and SM was C16:0, followed by C18:0 and C14:0. SFA relates to changes in endogenous FA synthesis that may not have been differentially affected by diet (Mapiye et al., 2013). Oliveira et al. (2012) with feeding different oils reported lower percentages (about 45%) of SFA.

CONCLUSION

Feeding a diet supplemented with 200 g/d of oil comprised of PO, PO/LSO, and LSO did not negatively affect any performance or carcass qualities of steers. The overall feed consumption of the steers was unaffected when dietary oil was provided. LSO supplement did also not influence muscle sensory and physical characteristics. LSO increased the percentage of n-3 fatty acids (mainly C18:3n3) in the LD and SM fat and lowered the n-6/n-3 ratio in beef. Thus, it can be concluded that 200 g/d LSO can be safety supplemented to diets of steers to enrich beef with potentially health beneficial FA.

ACKNOWLEDGMENTS

Authors would like to express special thanks to the Moongcharoen Farm, the Center for Scientific and Technological Equipment, Suranaree University of Technology for their great support. Financial support was provided by the Thailand National Research Council.

REFERENCES

AOAC (Association of Official Analytical Chemists). 1995. Official Method of Analysis (16th edn). Association of Official Analytical Chemists, Washington DC, USA.

Bessa, R. J. B., S. P. Alves, E. Jernimo, C. M. Alfaia, J. A. M. Prates, and J. Santos-Silva. 2007. Effect of lipid supplements on ruminal biohydrogenation intermediates and muscle fatty acid in lambs. Eur. J. Lipid Sci. Technol. 109:868-878.

Choi, S. H., G. O. Gang, J. E. Sawyer, B. J. Johnson, K. H. Kim, C. W. Choi, and S. B. Smith. 2013. Fatty acid biosynthesis and lipogenic enzyme activities in subcutaneous adipose tissue of feedlot steers fed supplementary palm oil or soybean oil. J. Anim. Sci. 91:2091-2098.

CIE (Commission Internationale de l'Eclairage). 1978. Recommendations on Uniform Color Spaces, Color Difference Equations, Psychometric Color Terms. CIE Supplement #2 to Report #15, TC-2.4, CIE publication, Paris, France.

Folch, J., M. Lees, and G. H. Sloane Stanley. 1957. A simple method for the isolation and purification of total lipids from animal tissues. J. Biol. Chem. 226:497-509.

Gilmore, L. A., R. L. Walzem, S. F. Crouse, D. R. Smith, T. H. Adams, V. Vaidyanathan, X. Cao, and S. B. Smith. 2011. Consumption of high-oleic acid ground beef increases HDL-cholesterol concentration but both high- and low- oleic acid ground beef decrease HDL particle diameter in normocholesterolemic men. J. Nutr. 141:1188-1194.

He, M. and L. E. Armentano. 2011. Effect of fatty acid profile in vegetable oils and antioxidant supplementation on dairy cattle performance and milk fat depression. J. Dairy Sci. 94:2481-2491.

Herdmann, A., J. Martin, G. Nuernberg, D. Dannenberger, and K. Nuernberg. 2010. Effect of dietary *n-3* and *n-6* PUFA on lipid composition of different tissues of German Holstein bulls and the fate of bioactive fatty acids during processing. J. Agric. Food Chem. 58:8314-8321.

Huerta-Leidenz, N. O., H. R. Cross, D. K. Lunt, L. S. Pelton, J. W. Savell, and S. B. Smith. 1991. Growth, carcass traits, and fatty acid profiles of adipose tissues from steers fed whole cottonseed. J. Anim. Sci. 69:3665-3672.

Kim, C. M., J. H. Kim, Y. K. Oh, E. K. Park, G. C. Ahn, G. Y. Lee, J. I. Lee, and K. K. Park. 2009. Effects of flaxseed diets on performance, carcass characteristics and fatty acid composition of Hanwoo steers. Asian Australas J. Anim. Sci. 22:1151-1159.

Mach, N., A. Bach, A. Velarde, and M. Devant. 2008. Association between animal, transportation, slaughterhouse practices, and meat pH in beef. Meat Sci. 78:232-238.

Mach, N., M. Devant, I. Díaz, M. Font-Furnols, M. A. Oliver, J. A. García, and A. Bach. 2006. Increasing the amount of n-3 fatty acid in meat from young Holstein bulls through nutrition. J. Anim. Sci. 84:3039-3048.

Mapiye, C., J. L. Aalhus, T. D. Turner, D. C. Rolland, J. A. Basarab, V. S. Baron, T. A. McAllister, H. C. Block, B. Uttaro, O. Lopez-Campos, S. D. Proctor, and M. E. R. Dugan. 2013. Effects of feeding flaxseed or sunflower-seed in high-forage diets on beef production, quality and fatty acid composition. Meat Sci. 95:98-109.

Marques, J. A., I. N. Prado, J. L. Moletta, I. M. Prado, J. M., Prado, L. M. A. Macedo, N. E. Souza, and M. Matsushita. 2006. Carcass and meat traits of feedlot finished heifers submitted to surgical or mechanical anoestrous. R. Bras. Zootec. 35:1514-1522.

Neath, K. E., A. N. Del Barrio, R. M. Lapitan, J. R. V. Herrera, L. C. Cruz, T. Fujihara, S. Muroya, K. Chikuni, M. Hirabayashi, and Y. Kanai. 2007. Difference in tenderness and pH decline between water buffalo meat and beef during postmortem aging. Meat Sci. 75:499-505.

Noci, F., P. Freach, F. J. Monahan, and A. P. Moloney. 2007. The fatty acid composition of muscle fat and subcutaneous adipose tissue of grazing heifers supplemented with plant oil-enriched concentrates. J. Anim. Sci. 85:1062-1073.

NRC (National Research Council). 2000. Nutrient Requirements of Beef Cattle, 7th Rev. edn. National Academy Press, Washington DC, USA.

Nuernberg, K., D. Dannenberger, G. Nuernberg, K. Ender, J. Voigt, N. D. Scollan, J. D. Wood, G. R. Nute, and R. I. Richardson. 2005. Effect of a grass-based and a concentrate feeding system on meat quality characteristics and fatty acid composition of *longissimus* muscle in different cattle breeds. Livest. Prod. Sci. 94:137-147.

Oliveira, E. A., A. A. M. Sampaio, W. Henrique, T. M. Pivaro, B. L. Rosa, A. R. M. Fernandes, and A. T. Andrade. 2012. Quality traits and lipid composition of meat from Nellore young bulls fed with different oils either protected or unprotected from rumen degradation. Meat Sci. 90:28-35.

Ostrowska, E., F. R. Dunshea, M. Muralitharan, and R. F. Cross. 2000. Comparison of silver-ion high-performance liquid chromatographic quantification of free and methylated conjugated linoleic acids. Lipids 35:1147-1153.

Palmquist, D. L. 2009. Omega-3 fatty acids in metabolism, health, and nutrition and for modified animal product foods. Prof. Anim. Sci. 25:207-249.

Palmquist D. L. and T. C. Jenkins. 1980. Fat in lactations rations: Review. J. Dairy Sci. 63:1-14.

Parodi, P. W. 1997. Milk fat conjugated linoleic acid: Can it help prevent breast cancer? In: Proceedings of Nutrition society of New Zealand, Nutrition Society of New Zealand, (Ed. G. P. Savage), Canterbury, New Zealand. pp. 137-149.

Partida, J. A., J. L. Olleta, C. Sanudo, P. Alberti, and M. M. Campo. 2007. Fatty acid composition and sensory traits of beef fed palm oil supplements. Meat Sci. 76:444-454.

Quinn, M. J., E. R. Loe, B. E. Depenbusch, J. J. Higgins, and J. S. Drouillard. 2008. The effects of flaxseed oil and derivatives on *in vitro* gas production, performance, carcass characteristics, and meat quality of finishing steers. Prof. Anim. Sci. 24:161-168.

Raes, K., S. De Smet, and D. Demeyer. 2004. Effect of dietary fatty acids on incorporation of long chain polyunsaturated fatty acids and conjugated linoleic acid in lamb, beef, and pork meat: a review. Anim. Feed Sci. Technol. 113:199-221.

Rule, D. C., W. H. Wu, J. R. Busboom, F. C. Hinds, and C. J. Kercher. 1989. Dietary canola seeds alter the fatty acid composition of bovine subcutaneous adipose tissue. Nutr. Rep. Int. 39:781-786.

SAS (Statistical Analysis System). 2001. A Handbook of Statistical Analyses Using SAS (2nd edn), SAS Institute Inc., Cary, NC, USA.

Santana, M. C. A., G. Fiorentini, P. H. M. Dian, R. C. Canesin, J. D. Messana, R. V. Oliveira, R. A. Reis, and T. T. Berchielli. 2014. Growth performance and meat quality of heifers receiving different forms of soybean oil in the rumen. Anim. Feed Sci. Technol. 194:35-43.

Scheeder, M. R. L., M. M. Casutt, M. Roulin, F. Escher, P. A. Dufey, and M. Kreuzer. 2001. Fatty acid composition, cooking loss and texture of beef patties from meat of bulls fed different fats. Meat Sci. 58:321-328.

Scollan, N., J. F. Hocquette, K. Nuernberg, D. Dannenberger, I. Richardson, and A. Moloney. 2006. Innovations in beef production systems that enhance the nutritional and health value of beef lipids and their relationship with meat quality. Meat Sci. 74:17-33.

Steel, R. G. D. and J. H. Torrie. 1980. Principles and Procedures of Statistics: A Biometeric Approach. 2nd edn. McGrowHill, New York, NY, USA.

Stone, H., J. Sidel, S. Oliver, A. Woolsey, and R. C. Singleton. 1974. Sensory evaluation by quantitative descriptive analysis. Food Technol. 28:24-34.

Swan, J. E., C. M. Esguerra, and M. M. Farouk. 1998. Some physical, chemical and sensory properties of chevon products from three New Zealand goat breeds. Small Rumin. Res. 28:273-280.

Van Soest, P. J., J. B. Robertson, and B. A. Lewis. 1991. Methods for dietary fiber, neutral detergent fiber and non-starch polysaccharides in relation to animal production. J. Dairy Sci. 74:3583-3597.

Westering, D. B. and H. B. Hedrick. 1979. Fatty acid composition of bovine lipids as influenced by diet, sex and anatomical location and relationship to sensory characteristics. J. Anim. Sci. 48:1343-1348.

WHO (World Health Organization). 2003. Diet, Nutrition and the Prevention of Chronic Diseases, Report of the joint WHO/FAO expert consultation, WHO Technical Report Series, No. 916 (TRS 916), Geneva, Switzerland.

Zinn, R. A., S. K. Gulati, A. Plascencia, and J. Salinas. 2000. Influence of ruminal biohydrogenation on the feeding value of fat in finishing diets for feedlot cattle. J. Anim. Sci. 78:1738-1746.

22

Lower ω-6/ω-3 Polyunsaturated Fatty Acid Ratios Decrease Fat Deposition by Inhibiting Fat Synthesis in Gosling

Lihuai Yu, Shunan Wang, Luoyang Ding, Xianghuan Liang[1],
Mengzhi Wang, Li Dong, and Hongrong Wang*

The College of Animal Science and Technology, Yangzhou University, Yangzhou 225009, China

ABSTRACT: The objective of the current study was to investigate the effects of dietary ω-6/ω-3 polyunsaturated fatty acid (PUFA) ratios on lipid metabolism in goslings. One hundred and sixty 21-day-old Yangzhou geese of similar weight were randomly divided into 4 groups. They were fed different PUFA-supplemented diets (the 4 diets had ω-6/ω-3 PUFA ratios of 12:1, 9:1, 6:1, or 3:1). The geese were slaughtered and samples of liver and muscle were collected at day 70. The activities and the gene expression of enzymes involved in lipid metabolism were measured. The results show that the activities of acetyl coenzyme A carboxylase (ACC), malic enzyme (ME), and fatty acid synthase (FAS) were lower ($p<0.05$), but the activities of hepatic lipase (HL) and lipoprotein lipase (LPL) were higher ($p<0.05$), in the liver and the muscle from the 3:1 and 6:1 groups compared with those in the 9:1 and 12:1 groups. Expression of the genes for FAS ($p<0.01$), ME ($p<0.01$) and ACC ($p<0.05$) were higher in the muscle of groups fed diets with higher ω-6/ω-3 PUFA ratios. Additionally, *in situ* hybridization tests showed that the expression intensities of the high density lipoprotein (*HDL-R*) gene in the 12:1 and 9:1 groups were significantly lower ($p<0.01$) than that of the 3:1 group in the muscle of goslings. In conclusion, diets containing lower ω-6/ω-3 PUFA ratios (3:1 or 6:1) could decrease fat deposition by inhibiting fat synthesis in goslings. (**Key Words:** ω-6/ω-3 Polyunsaturated Fatty Acid, Lipid Metabolism, Goose)

INTRODUCTION

Polyunsaturated fatty acids (PUFAs) not only are necessary nutrients for living organisms but also have important impacts on enzyme activity and gene expression (Sampath and Ntambi, 2005; Jump, 2008). PUFA-supplemented diets could regulate the activities of enzymes involved in lipid metabolism and inhibit the expression of genes related to fat synthesis, thus decreasing fat deposition (Clarke and Jump, 1994). Our previous study suggested that different ω-6/ω-3 PUFA ratios had different effects on fat deposition in Yangzhou goslings (Wang et al., 2010). However, the underlying mechanism is not clearly understood.

Previous studies in humans have showed that an increase in the ω-6/ω-3 PUFA ratio increase the risk for obesity. (Simopoulos, 2016). The higher ration of ω-6 PUFA could increase the membrane permeability, thus increase the cellular triglyceride (Ukropec et al., 2003). We hypothesized that the decreased fat deposition in PUFA-supplemented goslings might also be associated with changes in gene expression and enzyme activity related to fat metabolism. The activities of enzymes related to fat metabolism and the expression of genes critical for the synthesis and metabolism of lipids in gosling were examined in the present study. An *in situ* hybridization test was also conducted to measure the expression intensities of the high density lipoprotein receptor (*HDL-R*) and low density lipoprotein receptor (*LDL-R*) genes. The goal of our study was to provide a theoretical basis to help clarify the mechanisms controlling fat deposition in PUFA-supplemented gosling.

* Corresponding Author: Hongrong Wang.
E-mail: hongrongwang@sina.com
[1] Yangzhou Kangyuan Dairy CO., LTD, Yangzhou, 215009, China.

MATERIALS AND METHODS

Animals and experimental design

This experiment was conducted at the Experimental Farm of Yangzhou University. All animal handing protocols were approved by the Yangzhou University Animal Care and Use Committee. One hundred and sixty 21-day-old healthy Yangzhou goslings (80 male and 80 female) of similar weight (0.407±0.023 kg) were randomly divided into 4 groups. Each group contained 4 replicates of 10 birds each (5 males and 5 females). Birds in different groups were each fed a different diet (shown in Table 1) from day 21 to day 70. The four groups of goslings were raised in the same house with 16 small house(the ambient temperature is 25°C, humidity is 40%, stocking density is 10 birds per 3 m^2) and could have feed and water freely.

Birds were acclimated to their experimental diet over 7 days. The experimental diet made up 1/3 of the basal diet when the birds were 22 to 24 days old, 1/2 when they were 25 to 26 days old, and 2/3 when they were 27 to 28 days old. They were fed only the experimental diet after this point. Two birds (one male and one female) from each replicate of each group were randomly selected and slaughtered at the end of the experiment. All birds were fed in the same enclosure of one shelter. Additionally, no vaccinations were given throughout the experimental period.

Composition of experimental diets

A maize-peanut meal diet was designed based on previous studies on this breed (Shi et al., 2007). The basal diet used maize, soyabean meal and Lucerne hay as raw material. We produced the experimental diets by mixing different amounts of 2% peanut oil, sunflower seed oil, linseed oil, palmitic acid and oleic acid into the basal diet to adjust the saturated fatty acid:monounsaturated fatty acid: polyunsaturated fatty acid (PUFA) ratio to 1:1:1 and the ratios of ω-6/ω-3 PUFA to either 12:1, 9:1, or 6:1, 3:1. The composition and nutrient levels of the experimental diets are shown in Table 1.

Table 1. Composition and nutrient levels of the experimental diets (based on air-dried samples) (%)

Items	ω-6/ω-3 PUFA ratio			
	12:1	9:1	6:1	3:1
Ingredients[1]				
Corn	66.20	66.20	66.20	66.20
Soybean meal	17.30	17.30	17.30	17.30
Alfalfa powder	10.70	10.70	10.70	10.70
Peanut oil	1.30	1.28	1.16	1.06
Sunflower seed oil	0.16	0.15	0.16	0.08
Linseed oil	0.10	0.13	0.19	0.32
Palmitic acid	0.37	0.37	0.38	0.40
Oleic acid	0.07	0.07	0.11	0.14
CaH$_2$PO$_4$	1.20	1.20	1.20	1.20
Limestone	0.60	0.60	0.60	0.60
L-Lys·HCl	0.35	0.35	0.35	0.35
Met	0.15	0.15	0.15	0.15
NaCl	0.50	0.50	0.50	0.50
Premix 1	1.00	1.00	1.00	1.00
Total	100.00	100.00	100.00	100.00
Nutrient levels (based on DM)[2]				
CP	14.34	14.34	14.34	14.34
Metabolic energy (MJ/kg)	11.70	11.70	11.70	11.70
Crude fiber	5.27	5.27	5.27	5.27
Ca	0.78	0.78	0.78	0.78
AP	0.40	0.40	0.40	0.40
SFA	0.67	0.67	0.66	0.66
MUFA	0.67	0.67	0.68	0.68
PUFA	0.66	0.67	0.65	0.66
n-6 PUFA	0.60	0.60	0.56	0.50
n-3 PUFA	0.05	0.07	0.09	0.16
ω-6/ω-3 PUFA ratio	12.05:1	9.13:1	5.93:1	3.10:1

PUFA, polyunsaturated fatty acid; DM, dry matter; CP, crude protein; AP, available phosphorus; SFA, saturated fatty acid; MUFA, monounsaturated fatty acid; LA, lactic acid.

[1] One kg of premix contains the following: vitamin A, 10,000 IU; vitamin D$_3$, 3,000 IU; vitamin E, 30 mg; vitamin K$_3$, 2 mg; vitamin B$_1$, 5 mg; vitamin B$_2$, 7 mg; vitamin B$_6$, 5 mg; vitamin B$_{12}$, 20 μg; nicotinic acid, 38 mg; pantothenic acid, 9 mg; folic acid, 1 mg; biotin, 35 μg; choline chloride, 6 g; Cu, 5 mg; I, 0.9 mg; Fe, 100 mg; Zn, 110 mg; Mn, 100 mg; Se, 0.15 mg; Co, 0.5 mg.

[2] LA, LNA and ω-6/ω-3 PUFA ratio were all measured values, other values were calculated.

Sample collection

Thirty-two geese were slaughtered on day 70. The liver and crureus were isolated and stored in liquid nitrogen before being analyzed for gene expression and enzyme activities.

Enzyme activity tests

Two grams of tissue were mixed with specific amounts of phosphate buffer saline (pH = 7.4) and fully homogenized at 2°C to 8°C. The homogenates were centrifuged at 2,000 to 3,000 rpm for 20 min, and the supernatant was collected for enzyme activity tests. Enzyme-linked immuno sorbent assay (ELISA) kits for acetyl-CoA carboxylase (ACC, No. H232), malic enzyme (ME, No. H233), and fatty acid synthase (FAS, No. H231) were purchased from Jiancheng Company (Nanjing, China). The experiment was conducted according to the kit's instructions. The correlation coefficients (R) of a linear regression between measured enzyme standards and expected concentrations were greater than 0.99. Coomassie brilliant blue assay kits (No. A045-2) were purchased from Jiancheng Company (Nanjing, China). The supernatant of each sample was diluted with normal saline to make a 1% solution, and protein content was measured according to the kit's instructions. Enzyme activity was expressed IU/mg of tissue protein.

Test of hepatic lipase and lipoprotein lipase activity

Assay kits for hepatic lipase (HL assay kit, No. A067) and lipoprotein lipase (LPL, No. A068) were purchased from Jiancheng Company (Nanjing, China). The experiment was conducted according to the kit instructions. One activity unit was defined as the enzyme activity required to produce 1 mmol of free fatty acids per hour.

Extraction and reverse transcription of total RNA

Total RNA was extracted according to the instructions for the Trizol total RNA Extraction Reagent kit (TaKaRa Biotechnology Dalian Co., Ltd., Dalian, China). A spectrophotometer (NanoDrop 1000 spectrophotometer) was used to detect the purity and concentration of RNA. The 260/280 nm absorbance ratio was between 1.8 to 2.0 for all samples. Agarose gel electrophoresis was then used to determine RNA integrity. A 25 μL reaction system was used to reverse transcribe RNA to cDNA.

mRNA expression analysis by real-time polymerase chain reaction

We used samples that were mixed at the same concentrations as the experimental samples to optimize the polymerase chain reaction (PCR) reaction conditions using different target genes, cycle numbers and proportions of the different internal control and target genes. The PCR reaction was carried out with an ABI 7300 Real-Time PCR System (Applied Biosystems, Foster City, CA, USA) with SYBR-Green real-time PCR Master Mix (TaKaRa Biotechnology Co. Ltd, China) according to the manufacturer's specifications. The housekeeping gene *β-actin* was used as the internal control (Table 2). The reaction system volume was 25 μL and contained 2.5 μL RT product and 12.5 μL SYBR Green real-time PCR Master Mix. The final concentration of primers was 0.1 μM. The parameters of the PCR reaction were as follows: Pre-denatured at 95°C for 5 min, 40 to 45 cycles of being denatured at 95°C for 30 s, annealed at 58°C to 61°C for 30 s, and extended at 72°C for 20 s. The results were expressed

as the $2^{-\Delta\Delta C(t)}$ values,' where $\Delta\Delta C(t) = [C(t)_{ij} - C(t)\beta\text{-}actin_j] - [C(t)_{il} - C(t)\beta\text{-}actin_l]$. $C(T)_{ij}$ and $C(T)\beta\text{-}actin_j$ refer to the objective gene $(_i)$ of the sample $(_j)$ and the housekeeping gene, respectively, and $C(T)_{il}$ and $C(T)\beta\text{-}actin_l$ referred to the target gene and the housekeeping gene, respectively, of the control group. Each sample was run in triplicate, and ultrapure water was used as a negative control.

In situ hybridization test of mRNA for proteins involved in fat metabolism

In situ hybridization kits for HDL-R and LDL-R mRNA were purchased from Jiancheng Company (Nanjing, China) and used according to the kit instructions. Cells with brown granules or patches in the cytoplasm or nuclear membrane were considered positive for HDL-R or LDL-R expression. Samples were examined with an OLYMPUS microscope at 400×. From each sample, 5 complete and non-overlapping views were selected. The average number of positive cells out of 20 cells from each view were counted. The mean value of each slide was calculated and used as the value for the expression rate of the sample.

Statistical analysis

All statistical analyses were performed using SPSS 17.0. The significance of differences among treatments were evaluated by Duncan's multiple range test (Kapoor et al., 2004; Um et al., 2013). Data were analyzed using one-way analysis of variance. Statistical significance was set at $p < 0.05$. Data are presented as the mean±standard error of the mean.

RESULTS

Effects of ω-6/ω-3 PUFA ratios on the activities of enzymes involved in fat metabolism

Table 3 shows that the activities of ACC in the liver and the crureus were highest ($p < 0.05$) in the 12:1 group. The 9:1 group had lower values than the 12:1 group ($p < 0.05$) but was still higher than the other groups. The lowest ($p < 0.05$) activities were in the 6:1 and 3:1 groups. Table 3 shows that the ME activity in the livers of the 12:1 and 9:1 groups

Table 2. Sequences of polymerase chain reaction primers

Gene name	ID in GeneBank	The sequence of primers (5′- 3′)	Production (bp)
ACC	J03541	F: TCTCGCTTTATTATTGGTT	312
		R: CATTGTTGGCTATCAGGAC	
ME	AF408407	F: GCTGCAATTGGTGGTGCTT	106
		R: ACTCTGCTTTGCTGGTAGGATTG	
FAS	J04485	F: TGAAGGACCTTATCGCATTGC	195
		R: GCATGGGAAGCATTTTGTTGT	
β-actin	L08165	F: TGCGTGACATCAAGGAGAAG	300
		R: TGCCAGGGTACATTGTGGTA	

ACC, acetyl-CoA carboxylase; *ME*, malic enzyme; *FAS*, fatty acid synthase.

Table 3. The effects of ω-6/ω-3 PUFA ratios on the activities of enzymes involved in fat metabolism in goslings (U/mg)

Item	Tissue	12:1	9:1	6:1	3:1	SEM
ACC	Liver	4.95[a]	4.01[b]	3.20[c]	3.10[c]	0.18
	Leg muscle	1.83[a]	1.40[b]	0.84[c]	0.80[c]	0.09
ME	Liver	4.30[a]	3.86[ab]	3.20[c]	3.10[c]	0.20
	Leg muscle	3.40[a]	3.26[a]	2.64[c]	2.14[c]	0.17
FAS	Liver	8.84[a]	8.10[a]	5.08[c]	4.79[c]	0.98
	Leg muscle	2.88[a]	2.91[a]	1.45[c]	1.40[c]	0.23
HL	Liver	56.77[c]	64.17[bc]	71.95[ab]	75.68[a]	3.79
	Leg muscle	2.35[c]	3.01[ab]	3.50[a]	3.09[a]	0.31
LPL	Liver	2.40[c]	2.80[bc]	3.07[ab]	3.49[a]	0.21
	Leg muscle	4.75[c]	5.04[c]	5.42[bc]	6.21[a]	0.35

PUFA, polyunsaturated fatty acid; SEM, standard error of the mean; *ACC*, acetyl-CoA carboxylase; *ME*, malic enzyme; *FAS*, fatty acid synthase; HL, hepatic lipase; LPL, lipoprotein lipase.
Values in the same row with the same superscript were not significantly different (p>0.05); values in the same row with different superscripts were significantly different (p<0.05); values with different superscripts were significantly different (p<0.01).

were higher (p<0.05) than in those of the 6:1 and 3:1 groups. The activity of FAS in the liver was higher than that in the crureus. The activities of FAS in the liver and crureus were higher (p<0.05) in the 12:1 and 9:1 groups than they were in the 6:1 and 3:1 groups. However, activities of HL in the liver were highest in the 3:1 group and lowest in the 12:1 and 9:1 groups. The activity of HL in the crureus was only 1/20th of that in liver. The activity of HL in the liver and the crureus were higher in the 6:1 and 3:1 groups than in the 12:1 group (p<0.05). In addition, the activity of LPL in the liver was higher (p<0.05) in the 3:1 and 6:1 groups than (p<0.05) in the 12:1 and 9:1 groups. The activity of LPL in the crureus was higher in the 3:1 group than (p<0.05) in the 12:1, 9:1, and 6:1 groups.

Effects of ω-6/ω-3 PUFA ratios on genes related to lipid metabolism

Table 4 shows that the gene expression of FAS, ACC, and ME, enzymes related to fat synthesis, were down-regulated with a decrease in the dietary ω-6/ω-3 PUFA ratio. The gene expression of ACC in the 12:1 group was higher than in the other three groups (p<0.05), whereas the gene

expression of ACC was significantly lower in the 3:1 group than in the other three groups (p<0.05). Table 4 also shows that the expression of ME in the 12:1 and 9:1 groups was significantly higher than that in the 6:1 and 3:1 groups (p<0.01). The expression of FAS in the 12:1 group was also significantly higher (p<0.01) than that in the 6:1 and 3:1 groups.

The results of the *in situ* hybridization analysis of HDL-R mRNA (Figure 1 and Table 4) showed that the positive rates of HDL-R expression in the 12:1 and 9:1 groups were significantly lower (p<0.01) than that of the 3:1 group. The positive expression rates in the 12:1 and 9:1 groups were different from that of the 6:1 group, but the difference was not significant (p>0.05). The *in situ* hybridization images show that HDL-R labeling in the 12:1 and 9:1 groups was less intense than in the 6:1 and 3:1 groups. Figure 2 and Table 4 show that the positive rates of LDL-R expression between the different groups had a tendency to decrease with decreasing ω-6/ω-3 PUFA ratios, but these differences were not statistically significant (p>0.05).

DISCUSSION

Effects of ω-6/ω-3 PUFA ratios on enzymes involved in fatty acid synthesis

The enzyme ACC is present in many organisms. It catalyzes a reaction that restricts the rate of the first stage of fatty acid synthesis. The rate of fatty acid synthesis in animals can change based on different nutritional and hormonal conditions, and ACC plays a major role in these changes (Kim and Tae, 1994; Chow et al., 2014). The current study showed that the activity of ACC in the liver was higher than that in muscle and that the activity of ACC in the liver of the 12:1 and 9:1 groups were both significantly higher than those in the other groups. The results showed that diets with a low ω-6/ω-3 PUFA ratio may inhibit the activity of ACC in the liver and muscle.

The decarboxylation of malic acid catalyzed by ME is part of the citric acid-pyruvic acid cycle that enables acetyl-CoA to enter the cytosol from the mitochondria. The nicotinamide adenine dinucleotide phosphate (NADPH) that is produced is mainly used for fatty acid synthesis,

Table 4. The effects of ω-6/ω-3 PUFA ratio on the expression genes critical to fat metabolism in goslings

Item	Tissue	12:1	9:1	6:1	3:1	SEM
ACC	Leg muscle	5.09[a]	3.44[b]	2.84[bc]	2.39[c]	0.33
ME	Leg muscle	5.90[a]	4.86[ab]	3.45[c]	3.15[c]	0.17
FAS	Leg muscle	7.62[a]	6.92[ab]	4.97[cd]	3.98[d]	0.89
HDL-R positive rates	Leg muscle	42.71[b]	47.61[b]	55.22[ab]	58.79[a]	1.35
LDL-R positive rates	Leg muscle	55.36	54.23	51.04	51.30	4.23

PUFA, polyunsaturated fatty acid; SEM, the standard error of the mean; *ACC*, acetyl-CoA carboxylase; *ME*, malic enzyme; *FAS*, fatty acid synthase; HDL-R, high density lipoprotein receptor; LDL-R, low density lipoprotein receptor.
Values in the same row with the same superscript were not significantly different (p>0.05); values in the same row with different superscripts were significantly different (p<0.05); values with different superscripts were significantly different (p<0.01).

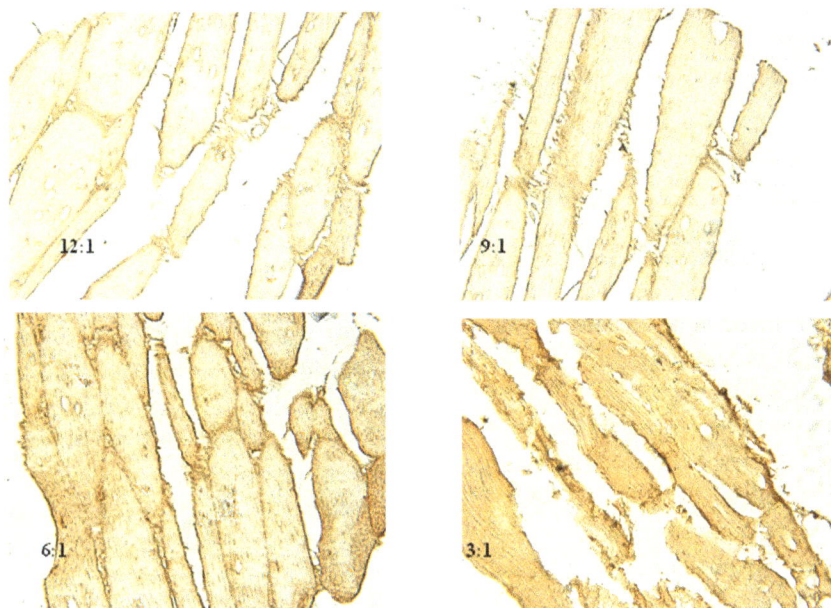

Figure 1. The gene expression of HDL-R in the leg muscle measured by *in situ* hybridization. Five to ten slides for each tissue were prepared, and images were taken using a binocular microscope (Olympus BX5; Olympus, Japan) coupled to a digital camera (Nikon H550L, Japan). The four images shown in Figure 1 were magnified 400 times. The *in situ* hybridization images show that the 12:1 and 9:1 groups had lower levels of labeling than the 6:1 and 3:1 groups. The positive rates of HDL-R expression in the 12:1 and 9:1 groups were significantly lower ($p<0.01$) than that of the 3:1 group (shown in Table 2).

especially in birds, whose source of NADPH for fatty acid synthesis relies primarily on ME activity (Mourot et al., 2000). The present study showed that the difference between the activities of ME in liver and in muscle was not significant, and the effects of different diets on liver and muscle ME activity were similar. Rates in the 12:1 and 9:1 groups were significantly higher than those in the 6:1 and 3:1 groups. Our study also showed that diets with a low ω-6/ω-3 PUFA ratio could inhibit the activity of ME.

The FAS catalyzes fatty acid synthesis. Mammalian FAS is an enzyme formed from multiple units with a molecular weight of 272 kDa (Guichard et al., 1992) and is

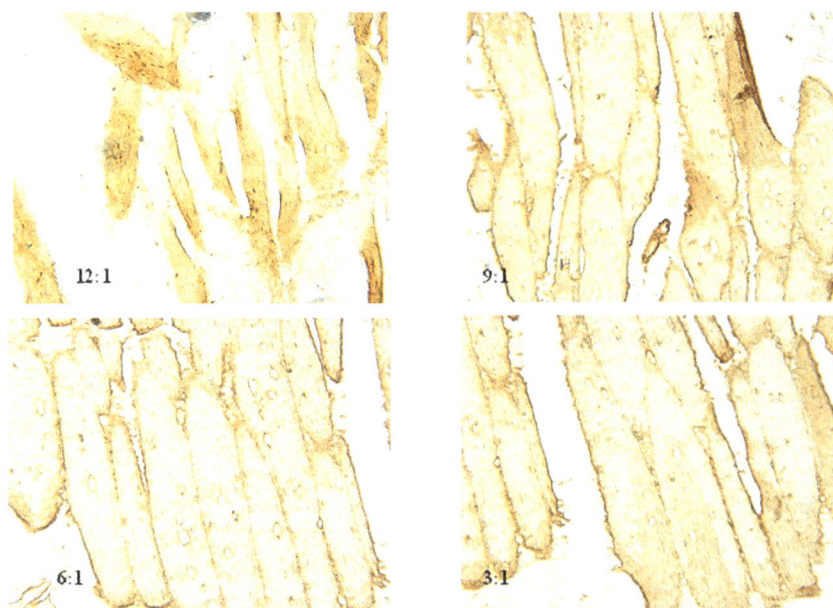

Figure 2. The gene expression of LDL-R in the leg muscle measured by *in situ* hybridization. Five to ten slides for each tissue were prepared, and images were taken using binocular microscope (Olympus BX5; Olympus, Japan) coupled to a digital camera (Nikon H550L, Japan). The four images shown in Figure 2 were all magnified 400. Figure 2 shows that the positive rates of LDL-R expression between the different groups had a tendency to decrease with decreasing ω-6/ω-3 polyunsaturated fatty acid ratios (data shown in Table 2).

involved in the process of fatty acid synthesis. Previous studies have shown that PUFAs can inhibit the activities of the enzymes involved fatty acid synthesis (Foretz et al., 1999). The results of the present study suggest that a diet containing low ω-6/ω-3 PUFA ratios (3:1 or 6:1) could inhibit this activity even more.

Effects of ω-6/ω-3 PUFA ratios on enzymes involved in lipid metabolism

The HL is mainly involved in the metabolism of small lipoprotein particles such as very low density lipoprotein (VLDL) and chylomicron emulsion (CM) and the triglyceride (TG) component of VLDL residual grains. It modulates the reaction responsible for transferring cholesterol from the peripheral tissue to the liver where VLDL is converted to low density lipoprotein (LDL). The unesterified cholesterol accumulated in high density lipoprotein (HDL) is taken up by the liver with the help of HL. This could prevent the excess accumulation of cholesterol in the peripheral tissues of the liver (Connolly et al., 1999; Perret et al., 2002). The results of the present study showed that the activity of HL in the liver was more than 20 times that in the crureus. Feeding goslings with diets of different ω-6/ω-3 PUFA ratios had significant effects. The 3:1 group had the highest HL activity suggesting that this diet promoted the absorption of VLDL and CM and of the TG component of VLDL residual grains, which may be beneficial to the health of the animal.

The LPL is synthesized in parenchymal cells in fat, myocardium, skeletal muscle and breast tissue. It also has functional similarities with HL. Its main physiological function is to catalyze the central TG of CM and VLDL and break the structure into fatty acids and monoglycerides for use in aerobic metabolism or for fat storage. In the present study, the activity of LPL in the crureus was higher than that in the liver. The ratios of ω-6/ω-3 PUFA also had significant effects on LPL. The activity of LPL was highest in the group whose dietary ω-6/ω-3 PUFA ratio was the lowest (3:1), and our results showed that this ratio improved the absorption of TG in the body. However, the results of the present study were different from those of a previous study (Zhang et al., 2009), where dietary n-6:n-3 PUFA ratios were 1.28, 5.03, 9.98, 68.62, which showed that the different n-6:n-3 PUFA ratios had no effect on the expression of LPL in the mouse liver. This might due to difference in the species and the different n-6:n-3 PUFA ratios.

Studies have shown that PUFAs can have effects on the expression of genes (Van Deursen et al., 2009) and the activities of enzymes (Clarke, 2001) involved in fatty acid metabolism by affecting sterol regulatory element binding proteins (SREBPs) and peroxisome proliferator activated receptors (PPARs). The present research also showed that the activities of enzymes involved in fatty acid synthesis and metabolism in groups with different ω-6/ω-3 PUFA ratios in their diet were different, but the regulating mechanisms were still unknown. Therefore, we tested the expression of the genes involved in these processes.

Effects of ω-6/ω-3 PUFA ratios on the expression of genes related to fat metabolism

The expression and activities of the genes involved in fatty acids metabolism, such as FAS, ACC, and ME, are closely related to the metabolism of fatty acids (Fang and Hillgartner, 1998), and their transcription has a common control element, which is directly affected by SREBP-1 (Richards et al., 2003). The expression levels of the 3 genes measured in the present study showed that the effects of different ω-6/ω-3 PUFA ratios on these genes were similar, and our results agreed with those found in the chicken (Huang et al., 2007). In this study, the expression levels of enzymes involved in lipid metabolism, such as FAS, ME, and ACC, were higher in the groups fed diets with higher ω-6/ω-3 PUFA ratios. This indicates that low ω-6/ω-3 PUFA dietary ratios could inhibit lipid metabolism.

The HDL can transport the free cholesterol accumulated in peripheral tissues and the lipoproteins in circulation to cells in the liver and can accelerate the removal of cholesterol from cells, thus playing an important role in minimizing atherosclerosis (Khera et al., 2011; Mora et al., 2011). The quantity of HDL-R and its expression reflected the turnover rate and metabolism of HDL. In the present study, the expression of HDL-R was higher in the groups fed diets with lower ω-6/ω-3 PUFA ratios than in those fed diets with higher ω-6/ω-3 PUFA ratios. This increased the activity of HDL-R and accelerated the removal of HDL-cholesterol in the circulation.

The LDL is one of the main carriers of cholesterol in the blood. It primarily transports cholesterol to peripheral tissues. The quantity of LDL-R in cells correlates to the speed of LDL removal from the circulation. The sterol regulatory element is in the region of the LDL-R gene promoter 5 (Mora et al., 2011) and the cholesterol content of cells and affect this regulator and controls the synthesis of LDL receptors (Yang et al., 1995). The results of the present study showed a higher expression of LDL-R in the groups fed diets with higher ω-6/ω-3 PUFA ratios. This might be because the lower ω-6/ω-3 PUFA ratios inhibited the expression of LDL-R, and the opposite effect may have occurred in the high ratio groups. Thus, the LDL-R activity in groups fed diets with high ω-6/ω-3 PUFA ratios increased. This may promote the expression of LDL-R, thereby clearing LDL-C and accelerating lipid metabolism. However, the differences between the groups were not significant. The reasons for the observed LDL-R results were not clear in the current work; further studies are needed to elucidate the specific mechanisms.

CONCLUSION

The activities of the enzymes involved in fatty acid metabolism, such as ACC, ME, and FAS, were lower, but the activities of enzymes in the turnover of fat, such as HL and LPL, were higher, in groups fed diets with lower ω-6/ω3 PUFA ratios (3:1 and 6:1). The expression levels of genes involved in fatty acid metabolism, such as FAS, ME, and ACC, were higher in groups fed diets with higher ω-6/ω-3 PUFA ratios. The expression of HDL-R in groups fed diets with lower ω-6/ω-3 PUFA ratios was higher than those fed diets with higher ω-6/ω-3 PUFA ratios, whereas the opposite pattern was observed in the expression of LDL-R. In conclusion, diets containing lower a ω-6/ω-3 PUFA ratios (3:1 or 6:1) could decrease fat deposition in goslings by inhibiting fat synthesis

ACKNOWLEDGMENTS

This work was supported by funding from the Priority Academic Program Development of Jiangsu Higher Education Institutions (PAPD) and the founding of the Prospective Study on Agriculture in Yangzhou (YZ2014143).

REFERENCES

Chow, J. D., R. T. Lawrence, M. E. Healy, J. E. Dominy, J. A. Liao, D. S. Breen, F. L. Byrne, B. M. Kenwood, C. Lackner, and S. Okutsu et al. 2014. Genetic inhibition of hepatic acetyl-CoA carboxylase activity increases liver fat and alters global protein acetylation. Mol. Metab. 3:419-431.

Clarke, S. D. 2001. Polyunsaturated fatty acid regulation of gene transcription: A molecular mechanism to improve the metabolic syndrome. J. Nutr. 131:1129-1132.

Clarke, S. D. and D. B. Jump. 1994. Dietary polyunsaturated fatty acid regulation of gene transcription. Annu. Rev. Nutr. 14:83-98.

Connolly, J. M., E. M. Gilhooly, and D. P. Rose. 1999. Effects of reduced dietary linoleic acid intake, alone or combined with an algal source of docosahexaenoic acid, on MDA-MB-231 breast cancer cell growth and apoptosis in nude mice. Nutr. Cancer 35:44-49.

Fang, X. and F. B. Hillgartner. 1998. Cell-specific regulation of transcription of the malic enzyme gene: characterization of cis-acting elements that modulate nuclear T3 receptor activity. Arch. Biochem. Biophys. 349:138-152.

Foretz, M., F. Foufelle, and P. Ferre. 1999. Polyunsaturated fatty acids inhibit fatty acid synthase and spot-14-protein gene expression in cultured rat hepatocytes by a peroxidative mechanism. Biochem. J. 341:371-376.

Guichard, C., I. Dugail, X. Le Liepvre, and M. Lavau. 1992. Genetic regulation of fatty acid synthetase expression in adipose tissue: Overtranscription of the gene in genetically obese rats. J. Lipid Res. 33:679-687.

Huang, J., D. Yang, and T. Wang. 2007. Effects of replacing soy-oil with soy-lecithin on growth performance, nutrient utilization and serum parameters of broilers fed corn-based diets. Asian Australas. J. Anim. Sci. 20:1880-1886.

Wang, J. F., S. Dong, S. Wang, Z. Hao, B. Xu, and H. Wang. 2010. Effects of ratios of n-6/n-3 PUFA in diet on meat quality of geese. Shanghai Anim. Husb. Vet. Commun. 6:9-11.

Jump, D. B. 2008. N-3 polyunsaturated fatty acid regulation of hepatic gene transcription. Curr. Opin. Lipidol. 19:242-247.

Khera, A. V., M. Cuchel, M. de la Llera-Moya, A. Rodrigues, M. F. Burke, K. Jafri, B. C. French, J. A. Phillips, M. L. Mucksavage, R. L. Wilensky, E. R. Mohler, G. H. Rothblat, and D. J. Rader. 2011. Cholesterol efflux capacity, high-density lipoprotein function, and atherosclerosis. New. Engl. J. Med. 364:127-135.

Kim, K. H. and H. J. Tae. 1994. Pattern and regulation of acetyl-CoA carboxylase gene expression. J. Nutr. 124:1273S-1283S.

Makar, R. S., P. E. Lipsky, and J. A. Cuthbert. 1998. Sterol-independent, sterol response element-dependent, regulation of low density lipoprotein receptor gene expression. J. Lipid Res. 39:1647-1654.

Mora, S., J. E. Buring, P. M. Ridker, and Y. Cui. 2011. Association of high-density lipoprotein cholesterol with incident cardiovascular events in women, by low-density lipoprotein cholesterol and apolipoprotein B100 levels: a cohort study. Ann. Intern. Med. 155:742-750.

Mourot, J., G. Guy, S. Lagarrigue, P. Peiniau, and D. Hermier. 2000. Role of hepatic lipogenesis in the susceptibility to fatty liver in the goose (Anser anser). Comp. Biochem. Physiol. B, Biochem. Mol. Biol. 126:81-87.

Perret, B., L. Mabile, L. Martinez, F. Tercé, R. Barbaras, and X. Collet. 2002. Hepatic lipase structure/function relationship, synthesis, and regulation. J. Lipid Res. 43:1163-1169.

Richards, M. P., S. M. Poch, C. N. Coon, R. W. Rosebrough, C. M. Ashwell, and J. P. McMurtry. 2003. Feed restriction significantly alters lipogenic gene expression in broiler breeder chickens. J. Nutr. 133:707-715.

Sampath, H. and J. M. Ntambi. 2005. Polyunsaturated fatty acid regulation of genes of lipid metabolism. Annu. Rev. Nutr. 25:317-340.

Shi, S. R., Z. Y. Wang, H. M. Yang, and Y. Y. Zhang. 2007. Nitrogen requirement for maintenance in Yangzhou goslings. Br. Poult. Sci. 48:205-209.

Simopoulos, A. P. 2016. An increase in the omega-6/omega-3 fatty acid ratio increases the risk for obesity. Nutrients 8:128.

Ukropec, J., J. E. Reseland, D. Gasperikova, E. Demcakova, L. Madsen, R. K. Berge, A. C. Rustan, I. Klimes, C. A. Drevon, and E. Sebokova. 2003. The hypotriglyceridemic effect of dietary n-3 FA is associated with increased β-oxidation and reduced leptin expression. Lipids 38:1023-1029.

Um, M. Y., K. H. Hwang, J. Ahn, and T. Y. Ha. 2013. Curcumin attenuates diet☐induced hepatic steatosis by activating AMP☐Activated Protein Kinase. Basic. Clin. Pharmacol. Toxicol. 113:152-157.

Van Deursen, D., M. Van Leeuwen, D. Akdogan, H. Adams, H. Jansen, and A. J. M. Verhoeven. 2009. Activation of hepatic lipase expression by oleic acid: possible involvement of USF1. Nutrients 1:133-147.

Yang, J., M. S. Brown, Y. Ho, and J. L. Goldstein. 1995. Three different rearrangements in a single intron truncate sterol regulatory element binding protein-2 and produce sterol-resistant phenotype in three cell lines. Role of introns in protein evolution. J. Biol. Chem. 270:12152-12161.

Zhang, L., Y. Geng, N. Xiao, M. Yin, L. Mao, G. Ren, C. Zhang, P. Liu, N. Lu, L. An, and J. Pan. 2009. High dietary n-6/n-3 PUFA ratio promotes HDL cholesterol level, but does not suppress atherogenesis in apolipoprotein E-null mice 1. J. Atheroscler. Thromb. 16:463-471.

23

Effects of Octacosanol Extracted from Rice Bran on the Laying Performance, Egg Quality and Blood Metabolites of Laying Hens

Kai Peng[a], Lei Long[a], Yuxi Wang[1], and Shunxi Wang*

College of Engineering, China Agricultural University, Beijing 100083, China

ABSTRACT: A 42-d study with 384 Hy-line brown laying hens was conducted to assess the effects of dietary octacosanol supplementation on laying performance, egg quality and blood metabolites of laying hens. Hens were randomly allocated into 4 dietary groups of 8 cages each, which were fed basal diet supplemented with 0 (Control), 9 (OCT9), 18 (OCT18), and 27 (OCT27) mg/kg diet of octacosanol isolated from rice bran, respectively. The experiment was conducted in an environmental controlled house and hens were fed twice daily for *ad libitum* intake. Laying performance was determined over the 42-d period, and egg quality as well as blood metabolites were estimated on d 21 and d 42. Diets in OCT18 and OCT27 increased ($p<0.05$) laying rate, egg weight, egg mass, egg albumen height, Haugh unit and eggshell strength on d 42, but decreased ($p<0.05$) feed conversion rate and levels of total cholesterol, triglyceride and low density lipoprotein cholesterol in the serum as compared to those of Control. Feed intake, yolk color, yolk diameter, eggshell thickness and high density lipoprotein cholesterol were similar ($p>0.05$) among treatments. Results demonstrate that supplementing 18 to 27 mg/kg diet of rice bran octacosanol can improve laying rate and egg quality and reduce blood lipid of laying hens. (**Key Words:** Octacosanol, Laying Performance, Egg Quality, Blood Metabolites, Laying Hens)

INTRODUCTION

Octacosanols are high molecular weight aliphatic alcohols that are main constituents of natural wax products such as beeswax, sugarcane wax and rice bran (Taylor et al., 2003). It has been reported that octacosanols affected lipid metabolism and enhanced stamina and energy (Kabir and Kimura, 1994; Taylor et al., 2003; Chen et al., 2007). In particular, their cholesterol-lowering effect, cytoprotective use, and ergogenic properties have been extensively investigated (Saint and McNaughton, 1986; Arruzazabala et al., 1994; Carbajal et al., 1995). Recently, an octacosanol isolated from rice bran has been reported to improve growth performance, enhance immunity and anti-oxidant status, and significantly reduce diarrhea of weanling piglets (Long

et al., 2015a) as well as promote the secretion of growth hormone and up-regulate the gene expressions of glucose transporter protein (Long et al., 2015b). Moreover, octacosanol has been shown to be very safe to animal with the median lethal dose (LD_{50}) to rats being 18,000 mg/kg body weight (BW) which is greater than that 3,000 mg/kg BW of sodium chloride (Pons et al., 1993). Octacosanols are currently used in health food, cosmetic and pharmaceutical industries. However, their potentials as natural feed additive have not been fully explored. Particularly, there is little information on effects of octacosanols on poultry productive performance. The objective of this study was to evaluate the effects of rice bran octacosanol on the laying performance, egg quality and blood metabolites of laying hens.

MATERIALS AND METHODS

Preparation of octacosanol by high vacuum distillation method

Octacosanol used in this study was obtained from a

* Corresponding Author: Shunxi Wang.
E-mail: wsx68@cau.edu.cn
[1] Agriculture and Agri-Food Canada, Lethbridge Research and Development Centre, Lethbridge, AB T1J 4B1, Canada.
[a] These two authors contribute equally to this work.

crude octacosanol extract of rice bran wax (Huzhou Shengtao Biological Co., Ltd., Zhejiang, China). The crude extracts contained approximately 13.6% (dry matter basis) octacosanol, and was processed further by high vacuum distillation (HVD) method (Long, 2014) to 53.7% of octacosanol in the product.

Animals, dietary treatments, and experimental design

Three hundreds and eighty four healthy laying hens (Hy-Line Brown, 25-week) with average initial BW of 1.03±0.10 kg and similar laying rates were used. The laying hens were randomly divided into 32 cages (12 hens per cage). The cages (0.9 m×0.6 m×0.4 m) equipped with nipple drinkers and trough feeders were located in an environmentally controlled room (24°C and 16 h of light at 20 lx/d) and raised to 30 cm above the ground. The 32 cages of hens were randomly assigned to 4 dietary treatments (8 cages per treatment) which were basal diet supplemented with 0 (Control), 9 (OCT9), 18 (OCT18), and 27 (OCT27) mg/kg diet of octacosanol. The octacosanol product (purify 53.7%) used in this study was firstly mixed with lime powder at ratio of 1:6 and then incorporated into the basal diet upon feeding at levels of 0, 100, 200, and 300 mg/kg diet to achieve corresponding levels of 0, 9, 18, and 27 mg/kg diet of octacosanol. The lime powder was used as a carrier to uniformly distribute octacosanol into diet. The treatments were arranged as completely randomised design with 8 replications (cages) per treatment. The basal diet (Table 1) was formulated to meet or exceed the nutrient recommendations of NRC (1994), and its nutrient composition was analyzed by the method of AOAC (2001). The experiment consisted of a week of adaptation followed by 6-week data collection periods. Diets and feeding management were the same for both periods. Hens were fed twice daily at 08:00 and 16:00 h for *ad libitum* intake and had free access to water throughout the entire feeding period. The animal protocol for this study was approved by the Animal Welfare Committee of China Agricultural University and was conducted in accordance with the guidelines for experimental animals. The health of hens was closely monitored by technicians and feeding staff.

Determination of laying performance and egg quality

Feed intake was measured daily by the difference between diet offered and residue collected prior to morning feeding. Eggs from individual cages were collected, counted and weighed daily. Laying rate was calculated as number of eggs per hen per day in each cage, and egg mass was calculated as the daily total egg weight dividing by number of hens in each cage. Feed conversion rate (FCR) for the 6-week experimental period was determined weekly as feed intake/egg mass. On d 21 and d 42 of the

Table 1. Composition and nutrient content of basal diet (DM basis)

Items	
Ingredient	
Corn	62.50
Soybean meal	19.33
Cottonseed meal	3.00
Rapeseed meal	3.00
Fish meal	1.50
DL-Methionine	0.12
Limestone	8.30
CaHPO$_4$	1.00
NaCl	0.25
Premix[1]	1.00
Total	100.00
Nutrient concentration	
Metabolisable energy (MJ/kg)	11.09
Crude protein (%)	16.55
Calcium (%)	3.30
Total phosphorus (%)	0.56
Available phosphorus (%)	0.37
Lysine (%)	0.75
Methionine (%)	0.39
Methionine+cysteine (%)	0.68
Threonine (%)	0.65
Tryptophan (%)	0.21
Arginine (%)	1.05

[1] Premix provides the following per kilogram of the basal diet which contained: vitamin A, 3,600 µg; vitamin D$_3$, 50 mg; vitamin K$_3$, 31 mg; vitamin E, 4.6 mg; vitamin B$_{12}$, 6 mg; d-pantothenic acid, 5 mg; folic acid, 0.1 mg; niacin, 7 mg; Cu (as CuSO$_4$•5H$_2$O), 8 mg; Zn (as ZnO), 40 mg; Fe (as FeSO$_4$•H$_2$O), 70 mg; Mn (as MnSO$_4$), 30 mg; I (as KI), 0.175 mg; Se (as Na$_2$SeO$_3$•5H$_2$O), 0.075 mg.

experiment, five eggs were randomly picked out from each cage and individually measured for egg weight, yolk diameter and albumen height with an egg analyzer (EA-01, ORKA, Ramat HaSharon, Israel), for eggshell thickness at three different points in the middle part of the egg using a dial gauge micrometer (FHK-P1, FHK, Tokyo, Japan), eggshell strength using an Eggshell Strength Tester (EFR-01, ORKA, Israel), and yolk color using method of the National Egg Products Association (NEPA) values according to Kahlenberg (1949). The Haugh unit (HU) score for each individual egg was calculated using following equation (Silversides and Scott, 2001):

$$HU = 100 \times Log_{10} (h - 1.7w^{0.37} + 7.6)$$

Where h is the albumen height (mm) and w is the weight of egg (g).

Determination of serum metabolites

Blood samples were collected from wing vein of

Table 2. Effects of dietary supplementation of octacosanol isolated from rice bran on laying performance of laying hens

Items	Treatments[1]			
	Control	OCT9	OCT18	OCT27
Feed intake (g/hen/d)	113.04±2.56	112.57±2.89	109.90±1.21	111.06±1.75
Laying rate (%)	82.31±0.72[b]	83.90±0.91[b]	88.14±0.98[a]	87.63±0.83[a]
Egg weight (g/egg)	60.45±0.55[b]	61.07±0.48[b]	61.56±0.89[a]	61.62±0.53[a]
Egg mass (g/hen/d)	49.76±0.46[b]	51.24±0.52[b]	54.26±0.55[a]	54.00±0.64[a]
Feed conversion rate (g/g)	2.27±0.13[b]	2.20±0.09[b]	2.02±0.17[a]	2.06±0.04[a]

[1] Control, basal diet; OCT9, basal diet supplemented with 9 mg/kg diet of octacosanol; OCT18, basal diet supplemented with 18 mg/kg diet of octacosanol; OCT27, basal diet supplemented with 27 mg/kg diet of octacosanol.

[a, b] Mean values within the same row not sharing a common superscript are statistically different at $p<0.05$.

randomly selected eight hens (one per cage) in each treatment using Vacutainer tubes (BD Franklin Lakes, NJ, USA) before morning feeding on d 21 and d 42 of the experiment. The blood samples were kept at room temperature for 20 min, followed by centrifugation at 8,000×g for 10 min and the resultant serum was stored at −80°C until analysis. The serum samples were analyzed for total cholesterol (TC), triglyceride (TG), low density lipoprotein cholesterol (LDLC) and high density lipoprotein cholesterol (HDLC) using corresponding commercial kits (Nanjing Jiancheng Bioengineering Institute, Nanjing, China) according to manufacturer recommended procedures.

Statistical analysis

All data were analyzed statistically by one-way analysis of variance using SPSS Statistics Base 17.0 (SPSS Inc., 2007) with treatment as main effect and cage as statistical unit and were expressed as mean±standard error of the mean. Data obtained at d 21 and d 42 were analyzed separately. Differences between treatments were tested using Duncan's multiple range test system and the significance was declared at $p<0.05$.

RESULTS AND DISCUSSION

Effect of octacosanol on laying performance and egg quality

Hens had similar feed intake over the entire experimental period regardless of the treatments (Table 2). However, hens fed OCT18 and OCT27 diet had greater ($p<0.05$) laying rate, individual egg weight and egg mass but lower ($p<0.05$) FCR than those fed OCT9 and Control diet. In addition, there was no difference ($p>0.05$) in the laying rate and egg weight between OCT18 and OCT27 groups. Eggs obtained on d 21 of the experiment had similar ($p>0.05$) albumen height, HU, yolk color, yolk diameter, and eggshell thickness, but eggs in OCT18 and OCT27 treatment had greater ($p<0.05$) eggshell strength compared to that of Control and OCT8 treatment (Table 3). At d 42, eggs from all octacosanol treatments had higher

Table 3. Effects of dietary supplementation of octacosanol isolated from rice bran on egg quality of laying hens

Items	Control	OCT9	OCT18	OCT27
Albumen height (mm)				
Day 21	6.61±0.14	6.68±0.35	6.70±0.22	6.69±0.30
Day 42	6.59±0.52[b]	6.74±0.63[a]	6.79±0.31[a]	6.76±0.46[a]
Haugh unit				
Day 21	80.92±3.17	81.09±4.25	81.02±2.16	80.96±1.47
Day 42	80.69±3.21[b]	81.49±2.19[a]	81.63±1.98[a]	81.42±2.03[a]
Yolk color				
Day 21	8.25±0.39	8.35±0.44	8.40±0.17	8.50±0.32
Day 42	8.40±0.42	8.40±0.31	8.50±0.29	8.60±0.28
Yolk diameter (mm)				
Day 21	8.89±0.21	9.09±0.08	9.20±0.08	9.01±0.09
Day 42	8.90±0.24	9.07±0.35	9.24±0.12	9.11±0.20
Eggshell strength (kg/cm²)				
Day 21	3.62±0.09[b]	3.72±0.13[ab]	3.84±0.29[a]	3.88±0.13[a]
Day 42	3.87±0.29[b]	3.98±0.62[b]	4.25±0.36[a]	4.22±0.25[a]
Eggshell thickness (mm)				
Day 21	0.41±0.31	0.42±0.09	0.43±0.14	0.41±0.23
Day 42	0.42±0.05	0.41±0.08	0.42±0.16	0.43±0.47

[a, b] Mean values within the same row not sharing a common superscript are statistically different at $p<0.05$.

(p<0.05) albumen height and HU as compared with that of Control, and eggshell strength of eggs from OCT18 and OCT27 were greater (p<0.05) than that from Control and OCT9 treatments. Similar to that found on d 21, there was no difference among treatments in yolk color, yolk diameter or eggshell thickness on d 42.

The similar feed intake of laying hens across all treatments indicated that dietary supplementation of octacosanol up to 27 mg/kg diet did not affect feed intake. On the other hand, the increased laying rate and egg weight for hens fed OCT18 and OCT27 indicated that octacosanol supplemented at the levels of 18 and 27 mg/kg diet increased egg production. This, comparing with the observation that hens fed OCT9 (9 mg/kg diet of octacosanol) had similar laying rate and egg weight to that of Control, suggests that laying performance responded to the octacosanol in a dose response manner. Information about effect of rice barn octacosanol on laying performance of laying hens is lacking, but dietary concentration of rice barn up to 10% has been shown no adverse effect on laying performance (Samli et al., 2006). Long et al. (2015a) reported that supplementing 8 mg/kg diet of the same octacosanol increased growth rate and improved feed efficiency of piglet. The same growth-promoting effect of octacosanol was also reported by Xiang et al. (2012) in rate. All these results suggested that octacosanol has the potential as natural feed additive to laying hens and piglets although the mechanisms by which octacosanols improve animal performance are unknown. Limited studies showed that octacosanol could stimulate fat metabolism, improve energy efficiency and increase protein synthesis (Kato et al., 1995; Castano et al., 2000; Singh et al., 2006). Long et al. (2015b) demonstrated that 8 mg/kg diet of octacosanol greatly promoted secretion of growth hormone of weaning piglets. Furthermore, Yang (2012) found that octacosanol increased the gene expressions of glucose transporter-4, glutamine synthetase and adenosine monophosphate and activated protein kinase thereby improved energy metabolism and increased protein synthesis in rats. All of these could have positive effect on animal productive performance and therefore further research in this area is needed.

HU is a measurement of the internal quality of an egg, which is positively correlated to its weight and albumin height (Loh et al., 2014). Higher albumin height and HU values indicate better quality and longer shelf life of eggs (Dai et al., 2012). To authors' knowledge, there is no information about the effects of octacosanol on egg quality of laying hens. This study demonstrated that octacosanol supplemented at the dietary concentrations of 9 to 27 mg/kg diet significantly increased albumen height and HU of eggs at d 42, but not at 21 d of the experiment. This suggests that octacosanol supplementation improved the egg internal quality via increasing albumin height. The fact that this egg quality-promoting effect octacosanol was not observed for eggs collected at 21 d of the experiment may indicate that there is lag time between initiation of octacosanol supplementation and animal productive response to octacosanol. Kabir and Kimura (1993) reported that absorption of octacosanol is very low and mainly excreted through feces. Moreover, the same authors suggested that octacosanol may be partly oxidized and degraded to fatty acids through beta-oxidation. This suggests that it may need time for octacosanol to be accumulated to sufficient concentration in the body to exert its effects on HU. This study showed that supplementing 18 to 27 mg octacosanol/kg diet did not affect eggshell thickness but increased eggshell strength, which suggests that external quality of egg was increased. Eggshell quality is closely related to calcium metabolism in laying hens (Gordon and Roland, 1998). It is not known whether octacosanol has the direct effect on calcium absorption/metabolism or through other mechanism. However, it has been reported that dietary octacosanol supplementation at the level of 8 mg/kg diet significantly increased growth hormone concentration in the blood of piglets (Long et al., 2015b). Growth hormone has been reported to have positive effect on calcium absorption and retention in animal (Braithwaite, 1975).

Effect of octacosanol on blood fat metabolism related metabolites

Concentration of TC in the serum was lower (p<0.05) for hens fed OCT18 and OCT27 diets than for hens fed Control and OCT9 diets and TC concentration of OCT9 was also lower (p<0.05) than that of Control on d 21 and d 42 of the experiment (Table 4). Similarly, OCT18 and OCT27 hens had lower (p<0.05) serum LDLC than Control

Table 4. Effects of dietary supplementation of octacosanol isolated from rice bran on serum metabolites of laying hens

Items (mmol/L)	Control	OCT9	OCT18	OCT27
TC				
Day 21	5.39±1.52[a]	4.28±1.50[b]	3.24±0.95[c]	3.18±0.76[c]
Day 42	4.82±1.78[a]	3.60±1.59[b]	2.89±1.37[c]	2.76±0.98[c]
TG				
Day 21	4.25±0.92	3.58±0.70	3.56±0.83	3.47±0.65
Day 42	4.75±0.96[a]	3.41±0.60[b]	3.26±0.58[b]	3.21±0.73[b]
HDLC				
Day 21	0.97±0.20	1.02±0.15	1.04±0.28	0.99±0.31
Day 42	1.03±0.18	1.06±0.14	1.10±0.05	1.05±0.21
LDLC				
Day 21	1.52±0.37[a]	1.37±0.25[a]	0.74±0.16[b]	0.68±0.20[b]
Day 42	2.14±0.43[a]	2.09±0.23[a]	1.07±0.30[b]	1.12±0.25[b]

TC, total cholesterol; TG, triglyceride; HDLC, high density lipoprotein cholesterol; LDLC, low density lipoprotein cholesterol.
[a,b,c] Mean values within the same row not sharing a common superscript are statistically different at p<0.05.

and OCT9 hens, but TC concentration of OCT9 was similar (p>0.05) to that of Control on both dates. In contrast, all hens irrespective of the treatments had similar (p>0.05) HDLC concentration in the serum on both dates. All octacosanol diets fed hens had lower (p<0.05) TG than that control hens on d 42, but this difference was not observed on d 21 of the experiment.

Previous researches have shown that dietary supplementation of octacosanol reduced blood cholesterol and TG in various test models (Hernandez et al., 1992; Kato et al., 1995; Castano et al., 2000; 2002; Taylor et al., 2003). In this study, the similar blood cholesterol and TG lowering effect of dietary octacosanol was also observed in laying hens. The exact mechanism by which supplementary octacosanol reduced cholesterol and TG in laying hens is not known but may partly attributable to its ability to regulate some enzymatic activities in lipid metabolism (Kato et al., 1995; Mas et al., 2004). However, whether cholesterol content in egg yolk is affected by the dietary octacosanol is not known and is of interest to be further studied in terms of its high relevance to human health.

CONCLUSION

Supplementation of octacosanol up to 27 mg/kg diet increased laying rate, egg weight and improved egg quality, but reduced serum TC and total fat concentrations. The responses of these variables are dose-dependent and optimum supplementation level is about 18 mg/kg diet under the conditions of this study. The information from this preliminary study suggest that rice bran octacosanol could be a potential natural feed additive to laying hens but further study is needed to elucidate the mechanisms by which octacosanol improve laying performance of laying hens.

ACKNOWLEDGMENTS

This study was conducted with financial support from Special Fund for Agro-scientific Research in the Public Interest of China (No. 201203015). The authors greatly appreciate for advice and technical assistance of Feng Yuan. We also acknowledge Jing Sun for assistance with sample collection and care of the laying hens.

REFERENCES

AOAC (Association of Official Analytical Chemists). 2001. Official Methods of Analysis, 17th edn. Association of Official Analytical Chemists, Washington, DC, USA.

Arruzazabala, M. L., D. Carbajal, R. Mas, V. Molina, S. Valdes, and A. Laguna. 1994. Cholesterol-lowering effects of policosanol in rabbits. Biol. Res. 27:205-208.

Braithwaite, G. D. 1975. The effect of growth hormone on calcium metabolism in the sheep. Br. J. Nutr. 33:309-314.

Carbajal, D., V. Molina, S. Valdes, L. Arruzazbala, and R. Mas. 1995. Anti-ulcer activity of higher primary alcohols of beeswax. J. Pharm. Pharmacol. 47:731-733.

Castano, G., R. Mas, M. L. Arruzazabala, M. Noa, J. Illnait, J. C. Fernandez, V. Molina, and A. Menendez. 1999. Effects of policosanol and pravastatin on lipid profile, platelet aggregation and endothelemia in older hypercholesterolemic patients. Int. J. Clin. Pharmacol. Res. 19:105-116.

Castano, G., R. Menendez, R. Mas, A. Amor, J. L. Fernandez, R. L. Gonzalez, M. Lezcay, and E. Alvarez. 2002. Effect of policosanol and lovastatin on lipid profile and lipid peroxidation in patients with dyslipidemia associated with type 2 diabetes mellitus. Int. J. Clin. Pharmacol. Res. 22:89-99.

Chen, F., G. H. Zhao, T. Y. Cai, Q. J. Feng, and M. R. Pei. 2007. Effect of octacosanol on the behavior and neuroendocrine index of rat forced by cold water swimming. Acta Nutrimenta Sinica 29:408-410.

Dai, L., L. Y. Gu, Zhu, Q. M., S. Zhu, A. T. Zhang, X. T. Zou, and C. H. Hu. 2012. Dietary valine level affects performance, egg quality and serum biochemical indices in laying hens. Chin. J. Anim. Nutr. 24:654-660.

Gordon, R. W. and D. A. Roland Sr. 1998. Influence of supplemental phytase on calcium and phosphorus utilization in laying hens. Poult. Sci. 77:290-294.

Hernandez, F., J. Illnait, R. Mas, G. Castano, L. Fernandez, M. Gonzalez, N. Cordovi, and J. C. Fernandez. 1992. Effect of policosanol on serum lipids and lipoproteins in healthy volunteers. Curr. Ther. Res. 51:568-575.

Kabir, Y. and S. Kimura. 1993. Biodistribution and metabolism of orally administered octacosanol in rats. Ann. Nutr. Metab. 37:33-38.

Kabir, Y. and S. Kimura. 1994. Distribution of radioactive octacosanol in response to exercise in rats. Mol. Nutr. Food Res. 38:373-377.

Kahlenberg, O. J. 1949. A quick and reliable gauge of yolk color. Food Ind. 21:467-470.

Kato, S., K. I. Karino, S. Hasegawa, J. Nagasawa, A. Nagasaki, M. Eguchi, T. Ichinose, K. Tago, H. Okumori, K. Hamatani, M. Takahashi, J. Ogasawara, and S. Masushige. 1995. Octacosanol affects lipid metabolism in rats fed on a high fat diet. Br. J. Nutr. 73:433-441.

Loh, T. C., D. W. Choe, H. L. Foo, A. Q. Sazili, and M. H. Bejo. 2014. Effects of feeding different postbiotic metabolite combinations produced by *Lactobacillus plantarumstrains* on egg quality and production performance, faecal parameters and plasma cholesterol in laying hens. BMC Vet. Res. 10:149.

Long, L. 2014. The Development and Application Effect Study of Octacosanol as a New Feed Additive. Ph. D. Thesis, China Agricultural University, Beijing, China.

Long, L., M. Z. Gao, K. Peng, J. Sun, and S. X. Wang. 2015a. Effects of octacosanol extracted from rice bran on production performance and blood parameters in weanling piglets. J. Chin. Cereal Oil. Assoc. 30:94-100.

Long, L., S. G. Wu, J. Sun, J. Wang, H. J. Zhang, and G. H. Qi. 2015b. Effects of octacosanol extracted from rice bran on blood hormone levels and gene expressions of glucose transporter protein-4 and adenosine monophosphate protein kinase in weaning piglets. Anim. Nutr. 1:293-298.

Mas, R., G. Castano, J. Femandez, R. Gamez, J. Illnait, L. Fernandez, E. Lopez, M. Measa, E. Alvarez, and S. Mendoza. 2004. Long term effects of policosanol on obese patients with type II hypercholesterolemia. Asia Pac. J. Clin. Nutr. 13:S102.

NRC (National Research Council). 1994. Nutrient Requirements of Poultry, 9th edn. National Academy Press, Washington, DC, USA.

Pons, P., A. Jimenez, M. Rodriguez, J. Illnait, R. Mas, L. Fernandez, and J. C. Fernandez. 1993. Effects of policosanol in elderly hypercholesterolemic patients. Curr. Ther. Res. 53:265-269.

Saint, J. M. and L. McNaughton. 1986. Octacosanol ingestion and its effects on metabolic responses to submaximal cycle ergometry, reaction time and chest and grip strength. Int. Clin. Nutr. Rev. 6:81-87.

Samli, H. E., N. Senkoylu, H. Akyurek, and A. Agma. 2006. Using rice bran in laying hen diets. J. Cent. Eur. Agric. 7:135-140.

Silversides, F. G. and T. A. Scott. 2001. Effect of storage and layer age on quality of eggs from two lines of hens. Poult. Sci. 80:1240-1245.

Singh, D. K., L. Li, and T. D. Porter. 2006. Policosanol inhibits cholesterol synthesis in hepatoma cells by activation of AMP-kinase. J. Pharmacol. Exp. Ther. 318:1020-1026.

SPSS (Statistical Packages for Social Sciences). 2007. Statistical Packages for Social Sciences. (version 17.0) SPSS Inc., Chicago, IL, USA.

Taylor, J. C., L. Rapport, and G. B. Lockwood. 2003. Octacosanol in human health. Nutrition 19:192-195.

Xiang, Y., H. Yang, L. L. Li, and X. Wu. 2012. Rat body GS gene expression regulating by octacosanol. J. Pingyuan Univ. 29:47-51.

Yang, H. 2012. The Preparation of Octacosanol and the Mechanism of the Impact on Energy Metabolism. M. Sc. Thesis, Changsha University, Changsha, China.

24

Effects of Dietary Octacosanol on Growth Performance, Carcass Characteristics and Meat Quality of Broiler Chicks

L. Long[1,2], S. G. Wu[1], F. Yuan[2], J. Wang[1], H. J. Zhang[1], and G. H. Qi[1,]*

[1] Key Laboratory of Feed Biotechnology of Ministry of Agriculture, Feed Research Institute, Chinese Academy of Agricultural Sciences, Beijing 100081, China

ABSTRACT: Octacosanol, which has prominent physiological activities and functions, has been recognized as a potential growth promoter in animals. A total of 392 1-d-old male Arbor Acres broiler chicks with similar body weight were randomly distributed into four dietary groups of seven replicates with 14 birds each supplemented with 0, 12, 24, or 36 mg octacosanol (extracted from rice bran, purity >92%)/kg feed. The feeding trial lasted for six weeks and was divided into the starter (day 1 to 21) and the grower (day 22 to 42) phases. The results showed that the feed conversion ratio (FCR) was significantly improved in broilers fed a diet containing 24 mg/kg octacosanol compared with those fed the control diet in the overall phase (day 1 to 42, $p = 0.042$). The average daily gain and FCR both showed linear effects in response to dietary supplementation of octacosanol during the overall phase ($p = 0.031$ and 0.018, respectively). Broilers fed with 24 or 36 mg/kg octacosanol diet showed a higher eviscerated yield, which increased by 5.88% and 4.26% respectively, than those fed the control diet ($p = 0.030$). The breast muscle yield of broilers fed with 24 mg/kg octacosanol diet increased significantly by 12.15% compared with those fed the control diet ($p = 0.047$). Eviscerated and breast muscle yield increased linearly with the increase in dietary octacosanol supplementation ($p = 0.013$ and 0.021, respectively). Broilers fed with 24 or 36 mg/kg octacosanol diet had a greater ($p = 0.021$) pH_{45min} value in the breast muscle, which was maintained linearly in response to dietary octacosanol supplementation ($p = 0.003$). There was a significant decrease ($p = 0.007$) in drip loss value between the octacosanol-added and the control groups. The drip loss showed linear ($p = 0.004$) and quadratic ($p = 0.041$) responses with dietary supplementation of octacosanol. These studies indicate that octacosanol is a potentially effective and safe feed additive which may improve feed efficiency and meat quality, and increase eviscerated and breast muscle yield, in broiler chicks. Dietary supplementation of octacosanol at 24 mg/kg diet is regarded as the recommended dosage in the broilers' diet. (**Key Words:** Octacosanol, Growth Performance, Carcass Characteristics, Meat Quality, Broiler Chick)

INTRODUCTION

Octacosanol ($HO\text{-}CH_2\text{-}[CH_2]_{26}CH_3$), a long-chain aliphatic alcohol, is the main component of a natural wax product that exists in wheat germ oil, rice bran oil, fruits or leaves (Taylor et al., 2003; de Oliveira et al., 2012). Octacosanol has been reported to exhibit a variety of important biological activities in humans and rodents, including antifatigue properties (Kim et al., 2004), antioxidant activities (Ohta et al., 2008), cholesterol-lowering effects (Hernandez et al., 1992), cytoprotective function (Carbajal et al., 1996) and ergogenic properties (Oliaro-Bosso et al., 2009). Studies have shown that the addition of dietary octacosanol has a beneficial effect on metabolic responses to submaximal cycle ergometry, grip and chest strength, and reaction time in human subjects (Saint-John and McNaughton, 1986), and has enhanced the running performance and related biochemical parameters in exercise-trained rats (Kim et al., 2004). Other studies on the effects of octacosanol on lipid metabolism have revealed that the addition of octacosanol leads to a significant decrease in the weight of perirenal adipose tissue of high-fat diet rats (Taylor et al., 2003) and reduces low-density lipoprotein (LDL), triacylglycerol and cholesterol content in

---

* Corresponding Author: G. H. Qi.
E-mail: qiguanghai@caas.cn
[2] Tianjin Naer Biotechnology Co., Ltd., Tianjin 300457, China.

the plasma of mice, healthy volunteers or patients (Aneiros et al., 1995; Menendez et al., 2000; Xu et al., 2007). Decreased cholesterol synthesis or enhanced LDL catabolism may be the mechanism of action here (Gouni-Berthold and Berthold, 2002). In terms of antioxidant function, octacosanol protects against CCl_4-induced liver injury in rats by attenuating disrupted hepatic reactive oxygen species metabolism (Ohta et al., 2008), and inhibits in *vitro* copper ion-induced rat lipoprotein peroxidation (Menendez et al., 1999). In addition, cytoprotection and anti-inflammatory properties have also been confirmed (Carbajal et al., 1996; de Oliveira et al., 2012). Mixtures of long-chain aliphatic alcohols (e.g. policosanol, of which octacosanol is the main component) have been shown to exert similar effects (Arruzazabala et al., 1994; Singh et al., 2006). These studies on the physiological function of octacosanol or policosanol were mainly conducted on rodents or human volunteers.

In view of its prominent physiological functions and high level of safety, with $LD_{50}>18,000$ mg/kg (Pons et al., 1993), octacosanol has been widely applied to health food, functional beverage, medicine and other fields. Due to its versatile functions, octacosanol may be developed as a potential feed additive for domestic animals. In our previous study, we found that octacosanol increased the growth performance of weaning piglets (Long et al., 2015), which indicated that the addition of octacosanol was effective in animal production. The objective of this study was, therefore, to test the efficacy of dietary octacosanol supplementation on growth performance, carcass characteristics and meat quality in broiler chicks, offering the poultry industry a new direction.

MATERIALS AND METHODS

Birds and management

The animal protocol for the present study was approved by the Animal Care and Use Committee of Feed Research Institute of the Chinese Academy of Agricultural Sciences. A total of 392 1-d-old male Arbor Acres broiler chicks with similar body weight (BW) was obtained from the Beijing Huadu Broiler Company (Beijing, China). All chicks were housed in a standardized chicken house with double-floor cages containing a tube feeder, a nipple drinker line and built-up soft-wood shavings, and exposed to a 23 h of light for the first two weeks and 20 h of light thereafter. The initial temperature in the chicken house was 34°C for the first week and was reduced by 2°C each consecutive week until the temperature reached 24°C. Relative humidity was set at 50% throughout the trial. Diets, in pellet form, and fresh water were supplied *ad libitum*, and ventilation was controlled. All management of broilers was in accordance with the recommendations of the Arbor Acres Broiler

Management Handbook (Aviagen Inc., 2014).

Experimental design and the diets

The broilers were randomly divided into four groups with seven replicates of 14 birds each. Broilers were fed with a basal diet (Table 1) supplemented with 0, 12, 24, or 36 mg octacosanol/kg diet. Octacosanol (purity>92%), extracted from rice bran, was provided by the Huzhou Shuanglin Shengtao Vegetable Oil Factory (Zhejiang, China). Basal diets, divided into two phases according to the ages of the chicks (starter phase, 1 to 21 d; grower phase, 22 to 42 d), were formulated according to the NRC (1994) and the Chinese Feeding Standard of Chicken (Ministry of Agriculture of China, 2004).

Growth performance

The data relating to the BW and feed intake of the chicks during the starter phase (d 1 to 21) and the grower phase (d 22 to 42) were recorded. The average daily gain (ADG), average daily feed intake (ADFI) and feed conversion ratio (FCR, the ratio of feed to gain) were calculated for each period and cumulatively. These

Table 1. Dietary composition and nutrient levels of the basal diets (as-fed basis)

	Starter phase (d 1 to 21)	Grower phase (d 22 to 42)
Ingredient (%)		
Corn	60.80	63.90
Soybean meal	32.90	28.95
Soybean oil	2.00	3.00
Dicalcium phosphate	1.62	1.50
Limestone	1.55	1.60
Salt	0.36	0.36
DL-methionine	0.23	0.16
L-lysine·HCl	0.25	0.24
Vitamin-mineral premix[1]	0.29	0.29
Total	100.00	100.00
Nutrient level[2]		
AME (MJ/kg)	12.12	12.50
CP (%)	20.30 (20.85)	18.56 (18.02)
Ca (%)	0.96 (1.02)	0.94 (0.86)
Total P (%)	0.64 (0.68)	0.60 (0.62)
Available P (%)	0.41	0.39
Lysine (%)	1.10	1.00
Methionine (%)	0.52	0.41
Methionine+cysteine (%)	0.83	0.72

AME, apparent metabolizable energy; CP, crude protein.
[1] Supplied per kg diet: vitamin A, 12,500 IU; vitamin D_3, 2,500 IU; vitamin E, 15 IU; vitamin K_3, 2.5 mg; vitamin B_1, 3.2 mg; vitamin B_2, 6.5 mg; vitamin B_{12}, 0.025 mg; nicotinic acid, 30 mg; calcium pantothenate, 12 mg; biotin, 0.03 mg; folic acid, 1.0 mg; iron, 80 mg; copper, 8 mg; manganese, 100 mg; zinc, 75 mg; iodine, 0.35 mg; and selenium, 0.15 mg.
[2] The value in parentheses indicates the analyzed value. Others are calculated values.

parameters were corrected according to mortality.

Carcass characteristics

On the last day of the experiment, two chicks were selected from each replicate and killed after a 12 h fasting for slaughter mensuration. The sacrificed chicks were exsanguinated and deplumed immediately. Next, the eviscerated weight and the internal organ weight of the heart, liver and spleen were measured. Breast muscles (including pectoralis major and minor), leg muscles (including thigh and drumstick muscles) and abdominal fat (including leaf fat surrounding the cloaca and abdominal fat surrounding the gizzard) were removed and weighed. Eviscerated yield, breast muscle yield, leg muscle yield, abdominal fat percentage and relative internal organ weight were all calculated as percentages of BW. The average value of the two broilers was regarded as the replicate value.

Meat quality

The qualities of the breast muscle were evaluated. The meat colour at three points of a meat piece was measured with a spectrocolorimeter (Minolta-CR200, Tokyo, Japan) at 45 min after slaughter. The meat quality was evaluated according to the Commission Internationale de L'Eclairage (CIE) L*a*b* coordinates (where L* indicates relative lightness, a* indicates relative redness, and b* indicates relative yellowness). At the same time, the pH values at a depth of 2.5 cm below the surface for each sample were measured at 45 min, as initial pH, and at 24 h, as ultimate pH, after slaughter using a pH meter (Star A, Thermo Fisher Scientific, Waltham, MA, USA). Next, drip loss was

measured using approximately 2 g of breast muscle sample according to the method described by Honikel (1998).

Statistical analysis

All data were performed using SPSS 17.0 for Windows software and expressed as mean values with pooled standard errors of the means using one-way analysis of variance. The difference between group means was separated by the least significant difference (Duncan's) multiple range test. The dose-response effect of supplemental octacosanol was computed using orthogonal polynomial contrast for linear and quadratic effects. Differences were considered statistically significant at p<0.05 unless otherwise stated.

RESULTS

Growth performance

The results of growth performance of broilers are presented in Table 2. No significant differences in growth performance were observed between the octacosanol-added and control groups during the starter phase (d 1 to 21, p>0.05). In the grower phase (d 22 to 42), dietary supplementation of octacosanol improved BW and FCR compared with the control group, and BW increased linearly (p = 0.031) with dietary supplementation of octacosanol, while FCR decreased linearly (p = 0.016). During the overall phase (d 1 to 42), chicks fed with 24 mg/kg octacosanol diet showed comparably lower FCR, which decreased by 5.17%, compared with those fed the control diet (p = 0.042). Average daily gain and FCR both

Table 2. Effect of dietary octacosanol supplementation on growth performance of broilers[1]

| Items | Dietary octacosanol addition (mg/kg) | | | | Pooled SEM | p-value | | |
	0	12	24	36		ANOVA[2]	Linear[3]	Quadratic[3]
Starter phase (d 1 to 21)								
BW d 21 (g)	729	725	740	731	3.22	0.450	0.458	0.719
ADG (g)	32.62	32.44	33.14	32.74	0.15	0.445	0.453	0.716
ADFI (g)	48.83	48.07	48.39	49.28	0.52	0.874	0.742	0.458
FCR (feed:gain, g:g)	1.496	1.483	1.460	1.506	0.02	0.815	0.964	0.414
Grower phase (d 22 to 42)								
BW d 42 (g)	2,272	2,310	2,351	2,339	12.52	0.114	0.031	0.302
ADG (g)	73.50	75.47	76.71	76.55	0.61	0.222	0.060	0.374
ADFI (g)	144.95	145.67	142.29	144.04	0.69	0.349	0.325	0.709
FCR (feed:gain, g:g)	1.973	1.935	1.857	1.877	0.02	0.061	0.016	0.370
Whole phase (d 1 to 42)								
ADG (g)	53.07	53.96	54.94	54.64	0.30	0.111	0.031	0.298
ADFI (g)	96.90	96.86	95.34	96.41	0.39	0.486	0.409	0.488
FCR (feed:gain, g:g)	1.829[b]	1.796[ab]	1.737[a]	1.764[ab]	0.01	0.042	0.018	0.188

SEM, standard error of the mean; ANOVA, analysis of variance; BW, body weight; ADG, average daily gain; ADFI, average daily feed intake; FCR, feed conversion ratio.

[1] n = 7 per treatment, with 14 broilers per replicate. [2] One-way ANOVA of all treatment groups.

[3] Linear and quadratic effects of increasing inclusion levels of octacosanol.

[a,b] Values within a row with different superscript letters are significantly different (p<0.05).

Table 3. Effect of dietary octacosanol supplementation on carcass characteristics of broilers at d 42[1]

Items[2]	Dietary octacosanol addition (mg/kg feed)				Pooled SEM	p-value		
	0	12	24	36		ANOVA[3]	Linear[4]	Quadratic[4]
Eviscerated yield (%)	70.70[b]	72.55[ab]	74.86[a]	73.71[a]	0.54	0.030	0.013	0.126
Breast muscle yield (%)	18.03[b]	19.11[ab]	20.22[a]	19.62[ab]	0.29	0.047	0.021	0.126
Leg muscle yield (%)	15.98	16.54	16.86	16.34	0.24	0.662	0.543	0.293
Abdominal fat (%)	1.77	1.73	1.70	1.70	0.01	0.201	0.048	0.437
Heart weight (g/kg BW)	4.92	4.66	4.51	4.99	0.08	0.098	0.929	0.081
Liver weight (g/kg BW)	16.19	16.56	16.37	16.82	0.19	0.702	0.340	0.911
Spleen weight (g/kg BW)	1.05	1.06	1.04	1.09	0.01	0.562	0.445	0.557

SEM, standard error of the mean; ANOVA, analysis of variance; BW, body weight.
[1] n=7 per treatment, with 2 birds per replicate. [2] All the indicators were calculated as percentage of BW.
[3] One-way ANOVA of all treatment groups. [4] Linear and quadratic effects of increasing inclusion levels of octacosanol.
[a,b] Values within a row with different superscript letters are significantly different (p<0.05).

showed linear effects in response to dietary addition of octacosanol during the overall phase (p = 0.031 and 0.018, respectively). No significant difference was observed in ADFI among all groups (p>0.05).

Carcass characteristics

The effect of dietary supplementation of octacosanol on the carcass characteristics of broilers is presented in Table 3. Broilers fed with dietary supplemental 24 or 36 mg/kg octacosanol showed higher eviscerated yield, which increased by 5.88% and 4.26% respectively, than those fed the control diet (p = 0.030). Compared with the control group, breast muscle yield of the broilers was significantly increased by 12.15% when the broilers were fed with 24 mg/kg octacosanol diet (p = 0.047). There was a linear increase in eviscerated (p = 0.013) and breast muscle yield (p = 0.021) with the dietary addition of octacosanol. No differences were observed in leg muscle yield and abdominal fat percentage among different groups (p>0.05). However, the abdominal fat content decreased linearly in response to dietary octacosanol supplementation (linear effect, p = 0.048). No significant differences were found in the heart, liver or spleen weight among all groups (p>0.05).

Meat quality

Broilers fed with the 24 or 36 mg/kg octacosanol diets had a higher pH_{45min} value (p = 0.021) in the breast muscle, which remained higher linearly—in response to the dietary addition of octacosanol (p = 0.003)—than those fed the control diet (Table 4). No differences were observed in pH_{24h} value among all treatments (p>0.05) or in L*, a*, or b* value among all treatments (p>0.05). However, the a* value increased linearly in response to dietary octacosanol supplementation (linear effect, p = 0.051). In addition, a significant difference (p = 0.007) in drip loss was observed between the octacosanol-added groups and the control group. Compared with the control diet, the chicken drip loss decreased by 2.97%, 5.03%, and 3.89%, which showed linear (p = 0.004) and quadratic (p = 0.041) effects with dietary supplementation of octacosanol.

DISCUSSION

Octacosanol or policosanol is considered to be a safe agent (Arruzazabala et al., 1994). As it is an additive, the effect on the growth performance of animals is the most valuable index. This study revealed that the dietary addition of octacosanol improved the growth performance of broiler

Table 4. Effect of dietary octacosanol supplementation on meat quality of broilers at d 42[1]

Items	Dietary octacosanol addition (mg/kg feed)				Pooled SEM	p-value		
	0	12	24	36		ANOVA[2]	Linear[3]	Quadratic[3]
pH_{45min}[4]	6.04[b]	6.13[ab]	6.15[a]	6.20[a]	0.02	0.021	0.003	0.560
pH_{24h}[4]	5.72	5.72	5.70	5.79	0.02	0.345	0.261	0.243
Lightness (L*)	57.29	57.34	57.18	57.69	0.11	0.434	0.312	0.318
Redness (a*)	13.78	14.34	14.76	14.42	0.14	0.083	0.051	0.095
Yellowness (b*)	14.70	14.51	14.43	15.00	0.18	0.698	0.623	0.308
Drip loss[5] (%)	4.37[B]	4.24[A]	4.15[A]	4.20[A]	0.02	0.007	0.004	0.041

SEM, standard error of the mean; ANOVA, analysis of variance.
[1] n = 7 per treatment, with 2 birds per replicate. [2] One-way ANOVA of all treatment groups.
[3] Linear and quadratic effects of increasing inclusion levels of octacosanol. [4] pH_{45min}, 45 min postmortem; pH_{24h}, 24 h postmortem.
[5] Measured when hung at 48 h.
[a,b] Values within a row with different superscript letters are significantly different (p<0.05).
[A,B] Values within a row with different superscript letters differ significantly (p<0.01).

chicks and that the dietary addition of octacosanol at 24 mg/kg diet showed the best feed efficiency during the overall period. Xu and Shen (1997) obtained the same results by supplementing the diet of broiler chicks with 25 mg/kg octacosanol for 42 days. In addition, the study by Xu et al. (2007) reported that a 1% octacosanol (wt/wt) diet group of E-KO mice gained weight at a rate comparable with the control group, while the study by Xiang et al. (2012) revealed that moderate doses of octacosanol in rats effectively increase the ADFI and ADG, and improve feed efficiency. These results support the hypothesis that octacosanol may improve the growth performance of domestic animals.

Other studies have shown that dietary octacosanol supplementation effectively promotes the secretion of growth hormone (GH) in the blood of rats (Yang, 2012), which can promote protein deposition and bone growth by regulating metabolic processes and energy metabolism in the body (Brown-Borg and Bartke, 2012). Moreover, octacosanol is effective in modulating lipid metabolism, in enhancing glycogen storage, and in decreasing the rate of glycogen utilization in muscle (Kato et al., 1995; Kim et al., 2004), which is due in part to increased energy metabolism through the activation of adenosine monophosphate-activated protein kinase (AMPK) in the muscle (Oliaro-Bosso et al., 2009). These findings indicate that octacosanol may potentially be used as an ergogenic agent. Accordingly, one possible reason for the improvement in the broilers' growth performance with octacosanol is the resulting increase in GH levels in the chicks along with an enhanced rate of glycogen utilization for growth (Kim et al., 2004). Further research is needed to clarify the effect of octacosanol on GH in broilers' blood and the efficacy of octacosanol as an ergogenic compound.

To our knowledge, the effects of dietary supplementation of octacosanol on carcass characteristics have not yet been reported. In our current study, octacosanol improved the eviscerated yield and breast muscle yield of broilers. Moreover, it led to a decrease in abdominal fat content. Octacosanol was also shown to be effective in improving the carcass characteristics of broilers.

In fast-growing broilers, growth occurs primarily in the pectorals and feathers during the late stages of development (Scheele, 1997). The relative increase in growth in octacosanol-fed chicks in the grower phase may partly contribute to the comparable increase in breast muscle. Published studies have reported that octacosanol may stimulate the mRNA expression of glycogen synthase in muscle (Yang, 2012), which is effective in regulating glycogen supply and in maintaining the energy balance in muscle. The major metabolic consequences of adaptation to physical functions are a slower rate of muscle glycogen utilization and a greater reliance on fat oxidation in the muscle (Crowley et al., 1996). Thus, in our study, more energy and nutrients were directed towards growth, e.g. breast development, when octacosanol was added to the diet, which improved the feed utilization by broilers during the grower stage. In contrast, no effects on leg muscle yield were observed with dietary supplementation of octacosanol. This result may be explained by the different compositions of muscle fibre types, the rate of protein turnover (Baillie and Garlick, 1991) and the differences in chemical composition (Douris et al., 2006; Lavery et al., 2008) between the breast and leg muscles of animals. In addition, it has been reported that muscle growth factors are more effective in breast muscles than in leg muscles (Qiao et al., 2013). This effect of octacosanol on muscle tissues in broilers is worthy of follow-up investigation.

In the present study, the abdominal fat percentage decreased in response to the increase in octacosanol. Kato et al. (1995) also showed that dietary octacosanol supplementation significantly reduced the weight of perirenal adipose tissue in high-fat diet rats. Moreover, other studies have reported that the development of adipose tissue depends on the availability of serum lipids, which are principal substrates in lipid metabolism (Zhang et al., 2011; Liao et al., 2015). Accordingly, the decrease in abdominal fat content evoked by octacosanol supplementation may be associated with the effect of serum lipid metabolism and the inhibition of cholesterol synthesis (Shimura et al., 1987; Hernandez et al., 1992; Arruzazabala et al., 1994; Castano et al., 2000). Additional experiments are needed to confirm the effect of octacosanol on lipid metabolism.

Research into the influence of octacosanol on muscle qualities, such as fibre size, density, and transformation, remains limited. In this study, we investigated the effect of octacosanol on the meat quality of animals. It is demonstrated that the addition of octacosanol influences pH_{45min}, a* values and drip loss. Glycolysis—anaerobic metabolism initiated in post-mortem breast muscle—can produce lactic acid and lead to a reduction in meat pH (Cai et al., 2015), which will influence the colour and water-holding capacity of chicken. In addition, meat colour, which mainly depends on the content of the muscle myoglobin, is a comprehensive index reflecting muscle physiology and biochemical and microbiological changes. Drip loss is a sensitive index of muscle protein structure and charge change, which directly influence the flavour, tenderness, processing and storage of the muscle. An improvement in pH_{45min}, a* values and drip loss results in improved quality. The mechanisms for the improvement in meat quality concomitant with dietary octacosanol supplementation remain unclear, but it has been presumed that the effect may be attributed to the antioxidant properties of octacosanol.

Some studies have suggested that preventing lipid peroxidation in intramuscular fat—which is rich in unsaturated fatty acids and easily oxidized, and has phospholipids as its main component—is beneficial to maintaining the quality and flavour of meat (McCormick, 1999; Chikunya et al., 2004; Kamboh and Zhu, 2013). Studies have confirmed that octacosanol can inhibit both liver and blood lipid oxidation (Ohta et al., 2008; Long et al., 2015) and protect lipoprotein particles against lipid peroxidation in animals (Menendez et al., 1999; Yu, 2003). Therefore, one reason for the improvement in meat quality produced by octacosanol may be the antioxidant effect. The exact mechanism, however, needs further elucidation.

CONCLUSION

In summary, the results of the present study indicate that octacosanol is an effective and safe feed additive as it can improve feed efficiency, stimulate eviscerated yield and breast muscle yield, and contribute to good meat quality. In this study, dietary supplementation of octacosanol at 24 mg/kg was regarded as the recommended dosage in the diet of broilers.

ACKNOWLEDGMENTS

The financial support provided by the China Agriculture Research System-Beijing Team for Poultry Industry and the Agricultural Science and Technology Innovation Program (ASTIP) is gratefully appreciated.

REFERENCES

Aneiros, E., R. Mas, B. Calderon, J. Illnait, L. Fernandez, G. Castano, and J. C. Fernandez. 1995. Effects of policosanol in lowering cholesterol levels in patients with type II hypercholesterolemia. Curr. Ther. Res. 56:176-182.

Arruzazabala, M. L., D. Carbajal, R. Mas, V. Molina, S. Valdes, and A. Laguna. 1994. Cholesterol-lowering effects of policosanol in rabbits. Biol. Res. 27:205-208.

Aviagen Inc. 2014. Arbor Acres Broiler Management Handbook. Aviagen Inc., Huntsville, AL, USA.

Baillie, A. G. and P. J. Garlick. 1991. Responses of protein synthesis in different skeletal muscles to fasting and insulin in rats. Am. J. Physiol. 260: E891-E896.

Brown-Borg, H. M. and A. Bartke. 2012. GH and IGF1: roles in energy metabolism of long-living GH mutant mice. J. Gerontol. A. Biol. Sci. Med. Sci. 67A:652-660.

Cai, L., Y. S. Park, S. I. Seong, S. W. Yoo, and I. H. Kim. 2015. Effects of rare earth elements-enriched yeast on growth performance, nutrient digestibility, meat quality, relative organ weight, and excreta microflora in broiler chickens. Livest. Sci. 172:43-49.

Carbajal, D., V. Molina, S. Valdes, L. Arruzazabala, I. Rodeiro, R. Mas, and J. Magraner. 1996. Possible cytoprotective mechanism in rats of D-002, an anti-ulcerogenic product isolated from beeswax. J. Pharm. Pharmacol. 48:858-860.

Castano, G., R. Mas, L. Feranadez, J. C. Fernandez, J. Illnait, L. E. Lopez, and E. Alvarez. 2000. Effects of policosanol on postmenopausal women with type II hypercholesterolemia. Gynecol. Endocrinol. 14:187-195.

Chikunya. S., G. Demirel, M. Enser, J. D. Wood, R. G. Wilkinson, and L. A. Sinclair. 2004. Biohydrogenation of dietary n-3 PUFA and stability of ingested vitamin E in the rumen, and their effects on microbial activity in sheep. Br. J. Nutr. 91:539-550.

Crowley, M. A., W. T. Willis, K. S. Matt, and C. M. Donovan. 1996. A reduced lactate mass explains much of the glycogen sparing associated with training. J. Appl. Physiol. 81:362-367.

de Oliveira, A. M., L. M. Conserva, J. N. de Souza Ferro, F. de Almeida Brito, R. P. Lyra Lemos, and E. Barreto. 2012. Antinociceptive and anti-inflammatory effects of octacosanol from the leaves of Sabicea grisea var. grisea in mice. Int. J. Mol. Sci. 13:1598-1611.

Douris, P. C., B. P. White, R. R. Cullen, W. E. Keltz, J. Meli, D. M. Mondiello, and D. Wenger. 2006. The relationship between maximal repetition performance and muscle fiber type as estimated by noninvasive technique in the quadriceps of untrained women. J. Strength. Cond. Res. 20:699-703.

Gouni-Berthold, I. and H. K. Berthold, 2002. Policosanol: Clinical pharmacology and therapeutic significance of a new lipid-lowering agent. Am. Heart. J. 143:356-365.

Hernandez, F., J. Illait, R. Mas, G. Castano, L. Fernandez, M. Gonzalez, N. Cordovi, and J. C. Fernandez. 1992. Effects of policosanol on serum lipids and lipoproteins in healthy volunteers. Curr. Ther. Res. 51:568-575.

Honikel, K. O. 1998. Reference methods for the assessment of physical characteristics of meat. Meat Sci. 49:447-457.

Kamboh, A. A. and W. Y. Zhu. 2013. Effect of increasing levels of bioflavonoids in broiler feed on plasma anti-oxidative potential, lipid metabolites, and fatty acid composition of meat. Poult. Sci. 92:454-461.

Kato, S., K. I. Karino, S. Hasegawa, J. Nagasawa, A. Nagasaki, M. Eguchi, T. Ichinose, K. Tago, H. Okumori, K. Hamatani, M. Takahashi, J. Ogasawara, S. Masushige, and S. Masushige. 1995. Octacosanol affects lipid metabolism in rats fed on a high-fat diet. Br. J. Nutr. 73:433-441.

Kim, H., S. Park, D. S. Han, and T. Park. 2004. Octacosanol supplementation increases running endurance time and improves biochemical parameters after exhaustion in trained rats. J. Med. Food 6:345-351.

Lavery, G. G., E. A. Walker, N. Turan, D. Rogoff, J. W. Ryder, J. M. Shelton, J. A. Richardson, F. Falciani, P. C. White, P. M. Stewart, K. L. Parkers, and D. R. McMillan. 2008. Deletion of hexose-6-phosphate dehydrogenase activates the unfolded protein response pathway and induces skeletal myopathy. J. Biol. Chem. 283:8453-8461.

Liao, X. D., R. J. Wu, G. Ma, L. M. Zhao, Z. J. Zheng, and R. J.

Zhang. 2015. Effects of *Clostridium butyricum* on antioxidant properties, meat quality and fatty acid composition of broiler birds. Lipids Health Dis. 14:36.

Long, L., M. Z. Gao, K. Peng, J. Sun, and S. X. Wang. 2015. Effects of octacosanol extracted from rice bran on production performance and blood parameters in weanling piglets. J. Chinese Cereals Oils Assoc. 30:94-100.

McCormick, R. J. 1999. Extracellular modification to muscle collagen: Implication for meat quality. Poult. Sci. 78:785-791.

Menendez, R., V. Fraga, A. M. Amor, R. M. Gonzalez, and R. Mas. 1999. Oral administration of policosanol inhibits *in vitro* copper ion-induced rat lipoprotein peroxidation. Physiol. Behav. 67:1-7.

Menendez, R., R. Mas, A. M. Amor, R. M. Gonzalez, J. C. Fernandez, I. Rodeiro, M. Zayas, and S. Jimenez. 2000. Effects of policosanol treatment on the susceptibility of low density lipoprotein (LDL) isolated from healthy volunteers to oxidative modification *in vitro*. Br. J. Clin. Pharmacol. 50:255-262.

MAC (Ministry of Agriculture of China). 2004. Feeding Standard of Chicken. NY/T 33-2004. Standards Press of China, Beijing, China.

NRC (National Research Council). 1994. Nutrient Requirements of Poultry. 9th rev. edn. National Academy Press, Washington, DC, USA.

Ohta, Y., K. Ohashi, T. Matsura, K. Tokunaga, A. Kitagawa, and K. Yamada. 2008. Octacosanol attenuates disrupted hepatic reactive oxygen species metabolism associated with acute liver injury progression in rats intoxicated with carbon tetrachloride. J. Clin. Biochem. Nutr. 42:118-125.

Oliaro-Bosso, S., E. C. Gaudino, S. Mantegna, E. Giraudo, C. Meda, F. Viola, and G. Cravotto. 2009. Regulation of HMGCoA reductase activity by policosanol and octacosadienol, a new synthetic analogue of octacosanol. Lipids 44:907-916.

Pons, P., A. Jimenez, M. Rodrigues, J. Illnait, R. Mas, L. Fernandez, and J. C. Fernandez. 1993. Effects of policosanol in elderly hypercholesterolemic patients. Curr. Ther. Res. 53:265-269.

Qiao, X., H. J. Zhang, S. G. Wu, H. Y. Yue, J. J. Zuo, D. Y. Feng, and G. H. Qi. 2013. Effects of β-hydroxy-β-methylbutyrate calcium on growth, blood parameters and carcass qualities of broiler chickens. Poult. Sci. 92:753-759.

Saint-John, M. and L. McNaughton. 1986. Octacosanol ingestion and its effects on metabolic responses to submaximal cycle ergometry, reaction time and chest and grip strength. Int. Clin. Nutr. Rev. 6:81-87.

Scheele, C. W. 1997. Pathological changes in metabolism of poultry related to increasing production levels. Vet. Q. 19:127-130.

Shimura, S., T. Hasegawa, S. Takano, and T. Suzuki. 1987. Studies on the effect of octacosanol on motor endurance in mice. Nutr. Rep. Int. 36:1029-1038.

Singh, D. K., L. Li, and T. D. Porter. 2006. Policosanol inhibits cholesterol synthesis in hepatoma cells by activation of AMP-kinase. J. Pharmacol. Exp. Ther. 318:1020-1026.

Taylor, J. C., L. Rapport, and G. B. Lockwood. 2003. Octacosanol in human health. Nutrition 19:192-195.

Xiang, Y., H. Yang, X. Wu, and J. P. Liu. 2012. Research and application progress of energy metabolism of body controlled by octacosanol. J. Xinxiang Univ. 29:44-46.

Xu, R. P. and H. Shen. 1997. Application of octacosanol in broilers diet. Feed Res. 5:26. (in Chinese)

Xu, Z. Y., E. Fitz, N. Riediger, and M. H. Moghadasian. 2007. Dietary octacosanol reduces plasma triacylglycerol levels but not atherogenesis in apolipoprotein E-knockout mice. Nutr. Res. 27:212-217.

Yang, H. 2012. The Preparation of Octacosanol and the Mechanism of the Impact on Energy Metabolism. M. Sc. Thesis, Changsha University of Science and Technology, Changsha, China.

Yu, C. Q. 2003. Study on damage of myocardial mitochondria in rats after exhaustive exercise. China Food Additives 2:35-37 (in Chinese).

Zhang, W. H., F. Gao, Q. F. Zhu, C. Li, Y. Jiang, S. F. Dai, and G. H. Zhou. 2011. Dietary sodium butyrate alleviates the oxidative stress induced by corticosterone exposure and improves meat quality in broiler chickens. Poult. Sci. 90:2592-2599.

25

Effects of Starvation on Lipid Metabolism and Gluconeogenesis in Yak

Xiaoqiang Yu, Quanhui Peng, Xiaolin Luo[1], Tianwu An[1], Jiuqiang Guan[1], and Zhisheng Wang*

Animal Nutrition Institute, Key Laboratory of Low Carbon Culture and Safety Production in Cattle in Sichuan, Sichuan Agricultural University, Ya'an, Sichuan 625014, China

ABSTRACT: This research was conducted to investigate the physiological consequences of undernourished yak. Twelve Maiwa yak (110.3±5.85 kg) were randomly divided into two groups (baseline and starvation group). The yak of baseline group were slaughtered at day 0, while the other group of yak were kept in shed without feed but allowed free access to water, salt and free movement for 9 days. Blood samples of the starvation group were collected on day 0, 1, 2, 3, 5, 7, 9 and the starved yak were slaughtered after the final blood sample collection. The liver and muscle glycogen of the starvation group decreased ($p<0.01$), and the lipid content also decreased while the content of moisture and ash increased ($p<0.05$) both in *Longissimus dorsi* and liver compared with the baseline group. The plasma insulin and glucose of the starved yak decreased at first and then kept stable but at a relatively lower level during the following days ($p<0.01$). On the contrary, the non-esterified fatty acids was increased ($p<0.01$). Beyond our expectation, the ketone bodies of β-hydroxybutyric acid and acetoacetic acid decreased with prolonged starvation ($p<0.01$). Furthermore, the mRNA expression of lipogenetic enzyme fatty acid synthase and lipoprotein lipase in subcutaneous adipose tissue of starved yak were down-regulated ($p<0.01$), whereas the mRNA expression of lipolytic enzyme carnitine palmitoyltransferase-1 and hormone sensitive lipase were up-regulated ($p<0.01$) after 9 days of starvation. The phosphoenolpyruvate carboxykinase and pyruvate carboxylase, responsible for hepatic gluconeogenesis were up-regulated ($p<0.01$). It was concluded that yak derive energy by gluconeogenesis promotion and fat storage mobilization during starvation but without ketone body accumulation in the plasma. (**Key Words:** Starvation, Maiwa Yak, Fat Catabolism, Ketone Bodies, Gluconeogenesis)

INTRODUCTION

Yak (*Bos grunniens*) are the characteristic and important grazing livestock on the Qinghai-Tibetan Plateau. About 15 million or more than 90% of the world's total yak population are currently herded in Chinese territories, which are important for Tibetan herders in providing milk, meat and transport (Zhang et al., 2014). Yak have to survive during inadequate feed supply in the long cold season (October to May) due to herbage deficiency under traditional pure grazing farming system, which is a centuries old grassland yak production cycle, where the yak is expected to satiate in summer, fatten in fall, become thin in winter, and die in spring (Xue et al., 2005).

The biochemical profile of the livers of ruminant species shows striking differences from that of simple-stomached animals. Therefore, it may not always be valid to extrapolate from the situation in rat liver to that in ruminant liver in order to interpret experimental observations in ruminants (Zammit, 1990). A large number of experiments had been done using sheep and goats (Piccione et al., 2002), young bulls (Tveit and Almlid, 1980), Angus cattle (Smith et al., 1998) to investigate the effects of starvation on circadian modulation and metabolism in adipose tissues. The most recent research declared that yak have developed special regulating mechanisms in the kidney in terms of glomerular filtration rate and purine derivative excretion to help the animal to adapt the alpine environment (Wang et al., 2009). Another paper from the same research team showed that the yak might be more efficient at utilizing N in a harsh environment than are cattle (Guo et al., 2012).

* Corresponding Author: Zhisheng Wang.

E-mail: zswangsicau@126.com

[1] Grassland Science Academy of Sichuan Province, Chengdu, Sichuan 611731, China.

However, there is no information on yak' fat storage mobilization and metabolism during starvation. Therefore, this experiment was conducted to investigate the fat mobilization and gluconeogenesis in the liver of yak during starvation.

MATERIALS AND METHODS

The experimental procedures used in this research were approved by the Institutional Animal Care of Sichuan Agricultural University.

Site description

This experiment was conducted at Guozhong Yak Breeding Farm, Hongyuan County, Sichuan Province, China (101°51′-103°23′E, 31°50′-33°22′N), situated at the eastern edge of Qinghai-Tibetan Plateau. The altitude there was over 3,600 m. The extreme minimum temperature was −36°C during the experimental period (August, 2013).

Experimental design and management

Twelve healthy male Maiwa yak (3 yr old, body weight 110.3±5.85 kg) were randomly divided into baseline and starvation groups, with 6 replicates in each group respectively. The animals were reared in a controlled indoor environment. In consideration of the gastric emptying time of ruminants being more than 120 h, and previous experiment being conducted with steers (Pothoven and Beitz, 1975), the formal starvation experiment lasted 9 days. During this period, the animals were allowed free access to water and salt. The yak of baseline group were killed by electrical stunning coupled with exsanguination at 10:00 am on day 0, whereas the starvation yak were killed with the same method on the 10th day morning.

Sample collection and analysis

Blood samples were taken from jugular vein into heparinized evacuated tubes on day 0, 1, 2, 3, 5, 7, 9 at 0900 to 1000 h, and centrifuged (3,000 rpm) for 20 min, plasma aliquots were frozen at −20°C until the laboratory analysis. The concentration of plasma glucose, non-esterified fatty acids (NEFA), β-hydroxybutyric acid (BHBA) and acetoacetic acid (ACAC) were assayed using an automatic analyzer (Olympus AU 600, Diamond Diagnostics, Tokyo, Japan). The concentration of insulin was quantified using fluoroimmunoassay (AutoDELFIA; PerkinElmer Life and Analytical Sciences, Turku, Finland).

The liver and subcutaneous adipose tissues were cut into pieces and placed into 5 mL tubes and treated with liquid nitrogen immediately and transferred to −80°C freezer for further analysis. Another 100 g of liver and *Longissimus dorsi* sample was collected and stored at −20°C for proximate chemical composition analysis according to the Association of Official Analytical Chemists methods (AOAC, 1990).

Primer design and real time-polymerase chain reaction analysis of gene expression

The mRNA expression of pyruvate carboxylase (PC) and phosphoenolpyruvate carboxykinase (PEPCK) in the liver and fatty acid synthase (FAS), lipoprotein lipase (LPL), carnitine palmitoyltransferase-1 (CPT-1) and hormone sensitive lipase (HSL) in subcutaneous adipose tissue were detected by real time-polymerase chain reaction (PCR). Real time-PCR primers were designed using the software Beacon Designer 7.0 (Premier Biosoft International, Palo Alto, CA, USA) to avoid amplification of genomic DNA. The sequence of the primers used, the fragment size, annealing temperature and the sequence references of the expected PCR products are shown in Table 1. Total RNA

Table 1. Specific primers used for real-time quantitative polymerase chain reaction

Genes	Accession No.	Primer sequence, 5'→3'	Length (bp)	Annealing T (°C)
β-actin	NM_173979	CATCCGCAAGGACCTCTAC ATGCCAATCTCATCTCGTTTT	340	61.3
FAS	NM_001012669	ACCTCGTGAAGGCTGTGACTCA TGAGTCGAGGCCAAGGTCTGAA	196	59.5
LPL	NM_001075120	CTGGACGGTGACAGGAATGTAT CAGACACTGGATAATGCTGCTG	131	59.4
CPT-1	NM_001034349	CAAAACCATGTTGTACAGCTTCCA GCTTCCTTCATCAGAGGCTTCA	111	58.4
HSL	NM_001080220	GATGAGAGGGTAATTGCCG GGATGGCAGGTGTGAACT	100	62.0
PEPCK	AY145503.1	GATGGAAAGTAGAGTGTGTGGGTG GATGGAAAGTAGAGTGTGTGGGTG	146	55.7
PC	NM_177946.4	TGCGGTCCATCCTGGTCAA ACGCCAGGTAGGACCAGTT	87	63.3

FAS, fatty acid synthase; LPL, lipoprotein lipase; CPT-1, carnitine palmitoyltransferase-1; HSL, hormone-sensitive lipase; PEPCK, pyruvate carboxy kinase; PC, pyruvate carboxylase.

was isolated from samples using the Trizol reagent (TaKaRa, Dalian, China) following the manufacturer's instructions. The extracted RNA concentration was determined by a spectrophotometer at 260/280 nm. Subsequently, a 2 μg sample of total RNA was extracted for reverse transcription with DNase prior to reverse transcription to cDNA using a PrimeScript RT Reagent Kit (Takara, China), and it was performed with a Mastercycler (Eppendorf, Hamburg, Germany) at the following settings: 94°C for 2 min, followed by 45 cycles of 94°C for 15 s, specific annealing temperature for 30 s, and 72°C for 30 s. The expression changes of mRNA relative to housekeeping gene β-actin were validated by an SYBR-based High-Specificity miRNA qRT-PCR Detection Kit (TaKaRa, China) on the CFX96TM Real-Time PCR Detection System (Bio-Rad, Hercules, CA, USA). The volume of each PCR reaction was 20 μL. The $2^{-\Delta\Delta Ct}$ method was used to determine the expression level differences between surveyed samples (Dorak, 2006).

Statistical analysis

A student T-test was carried out on the data of body weight, liver and *Longissimus dorsi* proximate chemical composition and gene expression. The MIXED model of SAS (SAS Institute, 1999) was used to analyze the data of insulin, glucose and ketones in the plasma, the model used was $Y = \mu + T_i + e_{ij}$, where Y was the dependent variable, μ was the population mean for the variable; T_i was the fixed effect of time points (time points = 7), e_{ij} was the random error associated with the observation ij. Tukey-Kramer procedure was applied to do the multiple comparisons. Means with different superscripts were obtained with "pdmix 800 SAS macro" (Saxton, 1998). Less than 0.05 p value was considered as statistically significant.

RESULTS AND DISCUSSION

Body weight

The effects of starvation on body weight of yak are shown in Table 2. The body weight of the yak decreased by 9.84% after 9 days of starvation, and the carcass weight decreased by 9.71% compared with the baseline group (p< 0.05). Animals undergoing starvation decrease their metabolic rates and utilize stored energy such as glycogen, lipids and even protein to provide energy for survival, and weight loss is the most obvious and direct response (McCue,

Table 2. Effects of starvation on body weight of yak

| Item | Treatment | | SEM | p-value |
	Baseline	Starvation		
Initial body weight (kg)	108.9	111.8	3.27	0.92
Final body weight (kg)	-	100.8	-	-
Carcass weight (kg)	51.5	46.5	3.66	0.05

SEM, standard error of the mean.

2010). The body weight of yak decreased by only 9.84% after 9 days of starvation, while Kirton et al. (1972) reported that the beef cattle lost 10.2% of body weight after 4 days of pre-slaughter starvation. Chaiyabutr et al. (1980) reported that the body weight of lactating goats decreased by 16% after 2 days of starvation. This suggests that yak have a stronger tolerant capability for feed deficiency than other ruminants.

Chemical composition of liver and muscle

The effects of starvation on proximate chemical composition of muscle and liver are shown in Table 3. After 9 days of starvation, the moisture content increased by 1.47% and 2.65% in *Longissimus dorsi* and liver, respectively (p<0.01). Similarly, the ash content increased by 16.67% and 28.57% in *Longissimus dorsi* and liver, respectively (p<0.05). Conversely, the lipid content in *Longissimus dorsi* decreased by 58.33% and 2.50% (p<0.05). In addition, the protein content of *Longissimus dorsi* declined by 5.38% (p<0.01), but increased by 5.88% in the liver (p<0.01). Under normal circumstances, along with the continuous consumption of energy, body moisture and ash content increases gradually. In present study, the moisture and ash content increased at the expense of lipid content both in muscle and liver, indicating the body fat storage was being utilized for energy supply. This was in agreement with Chwalibog et al. (2004), who reported that pigs started to mobilize body fat after approximately 16 h fasting. Furthermore, the protein content of *Longissimus dorsi* decreased, implying protein mobilization was also initiated. This was consistent with findings reported by Ndibualonji et al. (1997), who observed an increase of plasma 3-methylhistidine after 1 day of fasting in dairy cattle, which was the index of muscle protein catabolism. The fat storage was mobilized, and the total amount of liver lipid decreased by 8.0% as the consequence of the decrease

Table 3. Effect of starvation on liver and muscle proximate chemical composition of yak

| Item | Treatment | | SEM | p-value |
	Baseline	Starvation		
*Longissimus dorsi**				
Moisture (%)	75.0	76.1	1.57	<0.01
Protein (%)	22.3	21.1	0.63	<0.01
Lipid (%)	1.2	0.5	0.17	0.01
Ash (%)	1.2	1.4	0.09	0.04
Liver				
Moisture (%)	71.8	73.7	1.43	<0.01
Protein (%)	17.0	18.0	0.57	<0.01
Lipid (%)	4.0	3.9	0.11	0.04
Ash (%)	0.7	0.9	0.04	0.03

SEM, standard error of the mean.
* Moisture was calculated on the wet weight basis, and the protein, lipid and ash were calculated on the dry weight basis.

of liver weight (Supplementary Table S1). This might mean to some extent the liver was no longer in a healthy state.

Glycogen content in liver and muscle

After 9 days of starvation, the liver glycogen decreased by 77.57% (p<0.01), while the muscle glycogen decreased by 58.70% (p<0.01) (Table 4). The amount of liver glycogen was obtained by glycogen concentration (Table 4) multiplied by liver weight (Supplementary Table S1). The amount of glycogen in the liver decreased by 79.16% (p<0.01). This was in line with former reports (Sugden et al., 1976; Meynial-Denis et al., 2005), which reported similar glycogen decrease of carcass and liver in rats. This, also indicated yak use liver glycogen first and subsequently the muscle glycogen. An unique phenomenon was documented indicating that starvation led to a depletion of liver glycogen during the first 48 h in golden hamsters. However, after 72 or 96 h of starvation a new glycogen accumulation was detected (Sasse, 1975). This implied that the gluconeogenesis function commenced when glycogen was exhausted.

Plasma metabolites and insulin concentrations

The effects of starvation on yak plasma insulin concentration are illustrated in Figure 1. Compared with day 0, the insulin concentration decreased by 15.53% (p< 0.01), and further decreased by 47.13% (p<0.01) after 2 days of starvation (Figure 1a). Compared with the data after 2 days of starvation, the insulin concentration increased by 26.88% after 3 days of starvation (p<0.01). Subsequently, the plasma insulin concentration kept stable, and no significant difference was observed (p>0.05). The plasma glucose decreased by 5.14% after 1 day of starvation (p< 0.05), and continued to decrease by 22.67% (p<0.01), compared with day 0. During the following days, the glucose concentration kept relatively stable (p>0.05) (Figure 1b). This was consistent with the trend of insulin. The trend of insulin and glucose in present study were also mirrored by the phenomenon observed in humans by Unger et al. (1963). These results suggested that yak could at least survive 9 days or even longer when feed was deprived with its plasma glucose kept stable though at a lower level.

The concentration of NEFA increased as the starvation period prolonged (p<0.01) (Figure 1c). The NEFA concentration after 9 days of starvation increased by 4.60-fold compared with day 0. The increased NEFA after starvation was also observed in Holstein steers by DiMarco et al. (1981), with a slight difference, as about 8-fold of NEFA increase was obtained in the steers after 9 days of starvation. Brown Swiss steers had a 3.50-fold increase in NEFA after 8 days of starvation (Pothoven and Beitz, 1975). Reasons for the differences between different species warrant further investigations.

Table 4. Effect of starvation on liver and muscle glycogen of yak

Item	Treatment		SEM	p-value
	Baseline	Starvation		
Liver glycogen (μmol/g tissue)	12.4	2.8	0.17	0.01
Muscle glycogen (μmol/g tissue)*	2.3	1.0	0.04	0.01

SEM, standard error of the mean.
* Muscle was cut from *Longissimus dorsi*.

Usually, ketone bodies, BHBA, ACAC and acetone accumulate in the body fluids during starvation. Surprisingly, the concentration of BHBA in plasma decreased from day 0 to day 3 after starvation, then to a steady level till the end of the experiment (p<0.01) (Figure 1d). As far as the ACAC was concerned, it decreased sharply from day 0 to day 3 after starvation (p<0.01), and then decreased gradually till the end of the experiment (p<0.01) (Figure 1e). Under normal conditions, it is regarded that when hepatic glycogen originating from carbohydrate sources disappear, both ketones and free fatty acids serve to meet the physiologic requirement for energy. It was speculated that under circumstances of starvation ketosis, glucose provides about 10%, ketone bodies about 32%, and free fatty acids about 50% of the caloric needs in humans (Garber et al., 1974). On the other hand, in dairy cattle, determination of BHBA is considered as the golden standard to diagnose subclinical ketosis. The threshold concentration of BHBA to be regarded as subclinical ketosis is more than 1.2 mmol/L but less than 2.0 mmol/L (Iwersen et al., 2009; McArt et al., 2011). In addition, it was reviewed that the blood/plasma concentrations of BHBA (5 to 12 mmol/L), ACAC (2 to 4 mmol/L) and acetone (3 to 5 mmol/L) accumulated after several weeks of total fasting (Owen et al., 1983) in human. Our results demonstrated that BHBA concentration (1.17 mmol/L) was almost 40-fold that of ACAC (0.03 mmol/L), and acetone was almost undetectable. Therefore, no acetone data was presented in this study. Acetone is mainly formed by spontaneous decarboxylation of ACAC. The relatively low concentration of ACAC and the undetectable acetone in plasma in present study indicating yak utilize ketone bodies more effective than other species. An unique ketone bodies metabolism mechanism may exist in the liver of yak. It might also be an advantage of yak against the harsh alpine environment gained through thousands of years evolution.

Gene expression in subcutaneous adipose tissue and liver

The effects of starvation on the key enzyme relative mRNA expression of fat metabolism and gluconeogenesis in subcutaneous adipose tissue and liver are shown in Figure 2a. The relative mRNA expression abundance of

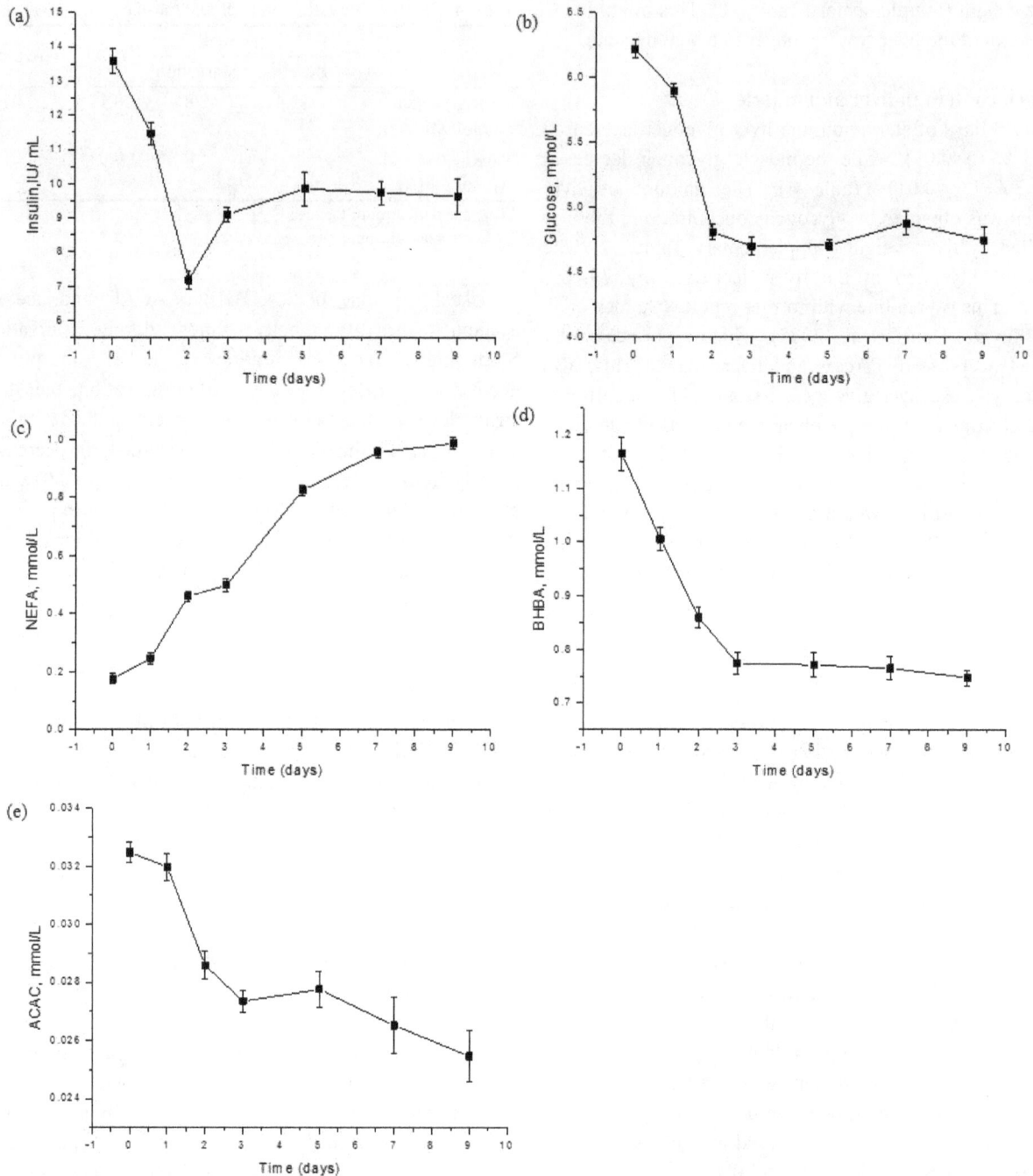

Figure 1. Effects of starvation on plasma insulin, glucose and ketone bodies concentration of yak. NEFA, non-esterified fatty acids; BHBA, β-hydroxybutyric acid; ACAC, acetoacetic acid. The vertical bars in figures are standard error of the mean.

lipogenic enzyme FAS and LPL in subcutaneous adipose tissue decreased by 86% and 83% (p<0.05), respectively. In the meanwhile, the expression abundance of lipolytic enzyme HSL and CPT-1 increased by 4.32 and 5.66-fold (p<0.01), respectively. It was reported by Pothoven and Beitz (1975) that FAS of Brown Swiss steers decreased 60% in the backfat after 1 day of fasting and 99% after 8 days. The decrease of FAS in subcutaneous and epididymal fat in mice with prolonged starvation (within 24 h) was also reported by Sponarova et al. (2005). These results indicated

the lipolysis was accelerated while lipogenesis was inhibited. The decreased LPL mRNA expression abundance was supported by DiMarco et al. (1981), who reported 63% decrease of LPL activity in subcutaneous adipose tissue in Holstein steers after 9 days of starvation. CPT-1 is the rate-limiting enzyme for fatty acid oxidation, catalyzing the first step in the transport of long-chain fatty acids from the cytosol into the mitochondrial matrix (McGarry and Foster, 1980). The increase mRNA expression of CPT-1 observed in present study was in agreement with Loor et al. (2005),

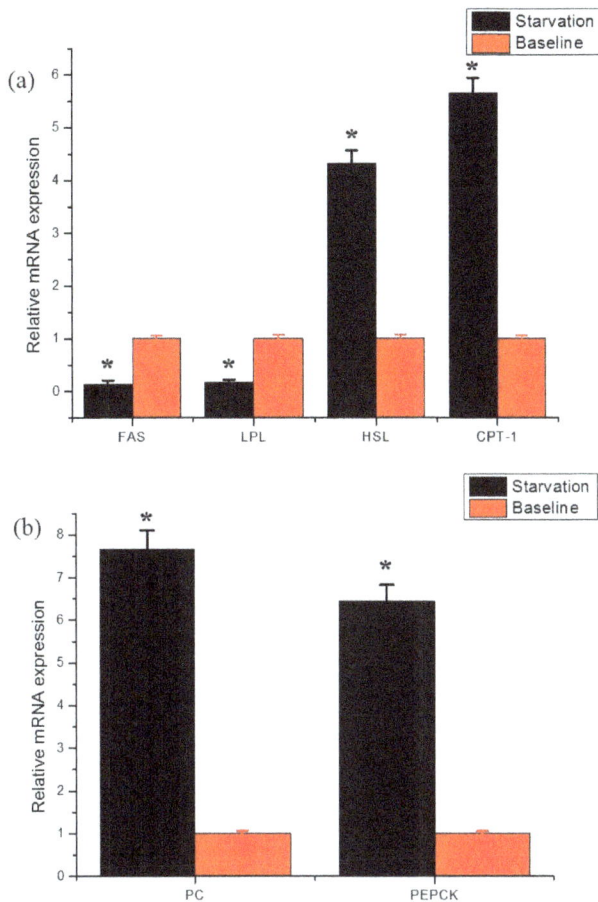

Figure 2. Relative abundance of mRNA expression of lipogenetic and lipolytic genes in subcutaneous adipose tissue and gluconeogenesis genes in the liver of yak. Values (±standard error of the mean) are expressed as mRNA quantity relative to the baseline group, which was given a value of one. FAS, fatty acid synthase; LPL, lipoprotein lipase; CPT, carnitine palmitoyltransferase-1; HSL, hormone sensitive lipase; PEPCK, phosphoenol pyruvate carboxy kinase; PC, pyruvate carboxylase. The asterisk indicates that means significantly different at $p < 0.01$ level.

who observed increased CPT mRNA expression in dairy cattle liver 1 day post-partum, when the animal was in the status of energy negative balance. Furthermore, the change trend of insulin and CPT-1 mRNA expression in present study was consistent with Li et al. (2013), who argued that high levels of insulin significantly inhibited the expression of genes related to fatty acid oxidation in calf hepatocytes. HSL is considered as a rate limiting enzyme in catabolism of fatty acids from intracellular triacylglycerol (Stralfors et al., 1987). Samra et al. (1996) studied the changes in lipid metabolism of adipose tissue in 24 healthy adults during early starvation (14 to 20 h). They declared the increased transcapillary efflux of NEFA reflected a significant increase in the action rate of HSL. In addition, they also

argued that insulin inhibits the activity of HSL. Both the NEFA and HSL mRNA expression increased while insulin decreased after starvation in present study. Bertile and Raclot (2011) also reported increased HSL mRNA expression and corresponding protein level (~1.3 fold) in fasted mice.

Genes that are important for hepatic adaptation are those encoding PEPCK and PC, which are described as rate-limiting enzymes for hepatic gluconeogenesis by Greenfield et al. (2000). They observed an increase in PC and PEPCK encoding mRNA abundance during the transition period when the dairy cattle are in negative energy balance condition. The mRNA expression of PC and PEPCK were increased 7.65 and 6.43-fold after starvation in present study (Figure 2b). However, Velez and Donkin (2005) reported that mRNA expression of PC was increased during feed restriction (50% of previously provided feed amount) and the capacity for gluconeogenesis from lactate was also increased while the mRNA expression of PEPKC was unchanged. This might indicate that different gluconeogenesis pathways are involved in different ruminants during starvation.

CONCLUSION

The liver and muscle glycogen of yak is almost exhausted and body fat storage is mobilized for energy supply after 9 days of starvation. On the other hand, the gluconeogenesis is up-regulated. These two aspects ensure the stability of blood glucose though at a relatvely lower level. During starvation, lipogenesis is depressed while the lipolysis is accelerated. This results in plasma non-esterified fatty acids accumulation. Unexpectedly, the ketone bodies BHBA and ACAC are decreased, whereas the actone is undetectable. This may be one of the physiological adaptations of yak after thousands of years selection to withstand the harsh alpine environment.

ACKNOWLEDGMENTS

This work was supported by the National Natural Science Foundation of China (Grant No.31470121) and the special key technologies of converting grass efficiently to livestock in Qinghai-Tibetan plateau community (201203008).

REFERENCES

AOAC (Association of Official Analytical Chemists). 1990.

Official Methods of Analysis, 15th edn. Association of Official Analytical Chemists, Arlington, VA, USA.

Bertile, F. and T. Raclot. 2011. ATGL and HSL are not coordinately regulated in response to fuel partitioning in fasted rats. J. Nutr. Biochem. 22:372-379.

Chaiyabutr, N., A. Faulkner, and M. Peaker. 1980. Effects of starvation on the cardiovascular system, water balance and milk secretion in lactating goats. Res. Vet. Sci. 28:291-295.

Chwalibog, A., A. H. Tauson, and G. Thorbek. 2004. Diurnal rhythm in heat production and oxidation of carbohydrate and fat in pigs during feeding, starvation and re-feeding. J. Anim. Physiol. Anim. Nutr. 88:266-274.

DiMarco, N. M., D. C. Beitz, and G. B. Whitehurst. 1981. Effect of fasting on free fatty acid, glycerol and cholesterol concentrations in blood plasma and lipoprotein lipase activity in adipose tissue of cattle. J. Anim. Sci. 52:75-82.

Dorak, M. T. 2006. Real-Time PCR, 1st edn, Taylor& Francis Group, New York, USA.

Garber, A. J., P. H. Menzel, G. Boden, and O. E. Owen. 1974. Hepatic ketogenesis and gluconeogenesis in humans. J. Clin. Invest. 54:981-989.

Greenfield, R. B., M. J. Cecava, and S. S. Donkin. 2000. Changes in mRNA expression for gluconeogenic enzymes in liver of dairy cattle during the transition to lactation. J. Dairy Sci. 83:1228-1236.

Guo, X. S., Y. Zhang, J. W. Zhou, R. J. Long, G. S. Xin, B. Qi, L. M. Ding, and H. C. Wang. 2012. Nitrogen metabolism and recycling in yaks (Bos grunniens) offered a forage-concentrate diet differing in N concentration. Anim. Prod. Sci. 52:287-296.

Iwersen, M., U. Falkenberg, R. Voigtsberger, D. Forderung, and W. Heuwieser. 2009. Evaluation of an electronic cowside test to detect subclinical ketosis in dairy cows. J. Dairy Sci. 92:2618-2624.

Kirton, A. H., D. J. Paterson, and D. M. Duganzich. 1972. Effect of pre-slaughter starvation in cattle. J. Anim. Sci. 34:555-559.

Li, P., C. C. Wu, M. Long, Y. Zhang, X. B. Li, J. B. He, Z. Wang, and G. W. Liu. 2013. Short communication: High insulin concentrations inhibit fatty acid oxidation-related gene expression in calf hepatocytes cultured in vitro. J. Dairy Sci. 96:3840-3844.

Loor, J. J., H. M. Dann, R. E. Everts, R. Oliveira, C. A. Green, N. A. J. Guretzky, S. L. Rodriguez-Zas, H. A. Lewin, and J. K. Drackley. 2005. Temporal gene expression profiling of liver from periparturient dairy cows reveals complex adaptive mechanisms in hepatic function. Physiol. Genomics 23:217-226.

McArt, J. A. A., D. V. Nydam, P. A. Ospina, and G. R. Oetzel. 2011. A field trial on the effect of propylene glycol on milk yield and resolution of ketosis in fresh cows diagnosed with subclinical ketosis. J. Dairy Sci. 94:6011-6020.

McCue, M. D. 2010. Starvation physiology: Reviewing the different strategies animals use to survive a common challenge. Comp. Biochem. Physiol. Part A Mol. Integr. Physiol. 156:1-18.

Meynial-Denis, D., A. Miri, G. Bielicki, M. Mignon, J. P. Renou, and J. Grizard. 2005. Insulin-dependent glycogen synthesis is delayed in onset in the skeletal muscle of food-deprived aged rats. J. Nutr. Biochem. 16:150-154.

McGarry, J. and D. Foster. 1980. Regulation of hepatic fatty acid oxidation and ketone body production. Annu. Rev. Biochem. 49:395-420.

Ndibualonji, B. B., D. Dehareng, and J. M. Godeau. 1997. Effects of starvation on plasma amino acids, urea and glucose in dairy cows. Ann. Zootech. 42:163-174.

Owen, O. E., S. Caprio, G. A. Reichard, M. A. Mozzoli, G. Boden, and R. S. Owen. 1983. 6 ketosis of starvation: A revisit and new perspectives. Clin. Endocrinol. Metab. 12:359-379.

Piccione, G., G. Caola, and R. Refinetti. 2002. Circadian modulation of starvation-induced hypothermia in sheep and goats. Chronobiol. Int. 19:531-541.

Pothoven, M. A. and A. C. Beitz. 1975. Changes in fatty acid synthesis and lipogenic enzymes in adipose tissue from fasted and fasted-refed steers. J. Nutr. 105:1055-1061.

SAS (Statistical Analysis System) Institute. 1999. User's Guide: Statistics. 8th edn. SAS Institute Inc., Cary, NC, USA.

Sasse, D. 1975. Dynamics of liver glycogen. Histochemistry 45:237-254.

Samra, J. S., M. L. Clark, S. M. Humphreys, I. A. Macdonald, and K. N. Frayn. 1996. Regulation of lipid metabolism in adipose tissue during early starvation. Am. J. Physiol. Endocrinol. Metab. 271:E541-E546.

Saxton, A. M. 1998. A macro for converting mean separation output to letter groupings in proc mixed. In: Proceedings of the 23rd SAS Users Group International. SAS Institute, Cary, NC. pp. 1243-1246.

Smith, S. B., K. C. Lin, J. J. Wilson, D. K. Lunt, and H. R. Cross. 1998. Starvation depresses acylglycerol biosynthesis in bovine subcutaneous but not intramuscular adipose tissue homogenates. Comp. Biochem. Physiol. Part B Biochem. Mol. Biol. 120:165-174.

Sponarova, J., K. J. Mustard, O. Horakova, P. Flachs, M. Rossmeisl, P. Brauner, K. Bardova, M. Thomason-Hughes, R. Braunerova, and P. Janovska et al. 2005. Involvement of AMP-activated protein kinase in fat depot-specific metabolic changes during starvation. FEBS. Lett. 579:6105-6110.

Stralfors, P., H. Olsson, and P. Belfrage. 1987. Hormone-sensitive lipase. Enzymes18:147-177.

Sugden, M. C., S. C. Sharples, and P. J. Randle. 1976. Carcass glycogen as a potential source of glucose during short-term starvation. Biochem. J. 160: 817-819.

Tveit, B. and T. Almlid. 1980. T4 degradation rate and plasma levels of TSH and thyroid hormones in ten young bulls during feeding conditions and 48 h of starvation. Acta Endocrinol. 93:435-439.

Unger, R. H., A. M. Eisentraut, and L. L. Madison. 1963. The effects of total starvation upon the levels of circulating glucagon and insulin in man. J. Clin. Invest. 42:1031-1039.

Velez, J. C. and S. S. Donkin. 2005. Feed restriction induces pyruvate carboxylase but not phosphoenolpyruvate carboxykinase in dairy cows. J. Dairy Sci. 88:2938-2948.

Wang, H., R. Long, W. Zhou, X. Li, J. Zhou, and X. Guo. 2009. A comparative study on urinary purine derivative excretion of yak (Bos grunniens), cattle (Bos taurus), and crossbred (Bos taurus×Bos grunniens) in the Qinghai-Tibetan plateau, China. J. Anim. Sci. 87:2355-2362.

26

Performance of Broiler Chickens Fed Low Protein, Limiting Amino Acid Supplemented Diets Formulated Either on Total or Standardized Ileal Digestible Amino Acid Basis

C. Basavanta Kumar*, R. G. Gloridoss[1], K. C. Singh[2], T. M. Prabhu, and B. N. Suresh[3]

Department of Animal Nutrition, Veterinary College, KVAFSU, Bangalore 560024, India

ABSTRACT: The aim of present experiment was to investigate the effect of protein reduction in commercial broiler chicken rations with incorporation of de-oiled rice bran (DORB) and supplementation of limiting amino acids (valine, isoleucine, and/or tryptophan) with ration formulation either on total amino acid (TAA) or standardized ileal digestible amino acids (SIDAA). The experimental design consisted of T_1, TAA control; T_2 and T_3, 0.75% and 1.5% protein reduction by 3% and 6% DORB incorporation, respectively by replacing soybean meal with supplemental limiting amino acids to meet TAA requirement; T_4, SIDAA control, T_5 and T_6, 0.75% and 1.5% protein reduction by DORB incorporation (3% and 6%) with supplemental limiting amino acids on SIDAA basis. A total of 360 d-old fast growing broiler chicks (Vencobb-400) were divided into 36 homogenous groups of ten chicks each, and six dietary treatments described were allocated randomly with six replications. During 42 days trial, the feed intake was significantly ($p < 0.05$) reduced by TAA factor compared to SIDAA factor and protein factor significantly ($p < 0.05$) reduced the feed intake at 1.5% reduction compared to normal protein group. This was observed only during pre-starter phase but not thereafter. The cumulative body weight gain (BWG) was significantly ($p < 0.05$) reduced in TAA formulations with protein step-down of 1.5% (T_3, 1,993 g) compared to control (T_1, 2,067 g), while under SIDAA formulations, BWG was not affected with protein reduction of 1.5% (T_6, 2,076 g) compared to T_4 (2,129 g). The feed conversion ratio (FCR) was significantly ($p < 0.05$) reduced in both TAA and SIDAA formulations with 1.5% protein step-down (T_3, 1.741; T_6, 1.704) compared to respective controls (T_1, 1.696; T_4, 1.663). The SIDAA formulation revealed significantly ($p < 0.05$) higher BWG (2,095 g) and better FCR (1.684) compared to TAA formulation (2,028 g; 1.721). Intake of crude protein and all limiting amino acids (SID basis) was higher in SIDAA group than TAA group with resultant higher nitrogen retention (4.438 vs 4.027 g/bird/d). The nitrogen excretion was minimized with 1.5% protein reduction (1.608 g/bird) compared to normal protein group (1.794 g/bird). The serum uric acid concentration was significantly reduced in T_3 (9.45 mg/dL) as compared to T_4 (10.75 mg/dL). All carcass parameters were significantly ($p < 0.05$) higher in SIDAA formulation over TAA formulation and 1.5% protein reduction significantly reduced carcass, breast and thigh yields. In conclusion, the dietary protein can be reduced by 0.75% with TAA formulation and 1.5% with SIDAA formulation through DORB incorporation and supplementation of limiting amino acids and among formulations, SIDAA formulation was better than TAA formulation. (**Key Words:** Broiler Chicken, De-oiled Rice Bran, Growth Performance, Limiting Amino Acids, Low Protein Diets, Standardized Ileal Digestible Amino Acids)

INTRODUCTION

Protein is considered as one of the major cost components in commercial poultry feed. Reduction of

* Corresponding Author: C. Basavanta Kumar.
E-mail: basavantac216@gmail.com
[1]Department of Instructional Livestock Farming Complex, Veterinary College, Bangalore 560024, India.
[2] Karnataka Veterinary, Animal and Fisheries Sciences University, Bidar, Karnataka 585401, India.
[3] Department of Instructional Livestock Farming Complex, Veterinary College, Hassan 573202, India.

dietary protein level and use of synthetic amino acid is suggested to reduce the feed cost and also to contain the environmental pollution of nitrogen (Corzo et al., 2009). For protein reduction, locally available feed ingredient such as de-oiled rice bran (DORB) can be thought of as substitute to partially replace soybean meal the costliest and widely used protein source. Piyaratne et al. (2009) used rice bran (with oil) up to 20% in broiler rations by balancing limiting amino acids. However, in this study, the limiting amino acids were not supplemented as crystalline amino acids, rather the soybean meal content was increased to balance the limiting amino acids and hence, the protein content of the diet was increased by 2.0% in the diet balanced for all limiting amino acids. Moreover, in none of the studies utilizing DORB (Das and Ghosh, 2000; Khan et al., 2002; Piyaratne et al., 2009), the protein reduction was addressed and most of the studies were conducted with a low total lysine level (<1.10%) to avoid deficiencies of other limiting amino acids, and the used lysine levels in fact were far below the recommended levels for present day fast growing commercial broilers. In this context, our previous study clearly demonstrated significant reduction of broiler performance in 6% DORB based low protein diets due to deficiencies of amino acid valine, isoleucine and tryptophan (Basavanta Kumar et al., 2015).

Feed formulation to balance amino acid *viz*., total amino acids (TAA), apparent/true digestible, apparent/true ileal digestible and standardized ileal digestible amino acid (SIDAA) basis were suggested and tried. Among these, formulation of diets based on SIDAA was suggested as a mean for better utilization of substitutable feed ingredients (Hoehler et al., 2006; Szczurek, 2010). However, some studies (Djouvinov et al., 2005; Mairoka et al., 2005; Ghaffari et al., 2007) did not show any advantage in feed formulation based on SID/ileal digestible amino acid basis compared to TAA basis. Keeping this concept in view, the study was undertaken to reduce protein in broiler rations by incorporation of DORB with supplementation of limiting amino acids and in addition, a comparison was made between TAA and SIDAA formulations.

MATERIALS AND METHODS

Ingredients and amino acid analysis

The feed ingredients procured were analyzed for moisture, crude protein (CP), ether extract, crude fiber (AOAC, 2005), nitrogen free extractives as difference and the amino acid composition was estimated at Evonik Laboratory. The experimental diets were formulated based on the analyzed amino acid composition of ingredients. The SIDAA profile of the experimental diets was arrived at by multiplying the amino acid profile of each raw material with digestibility coefficients suggested by Hoehler et al. (2006; Table 1), and the metabolizable energy content of the diet was estimated according to Rostagno (2011) based on the analyzed proximate composition and the reported digestibility coefficients for each nutrient from Rostagno (2011).

Experimental design and diets

Three levels of DORB (0, 3%, and 6%) were combined with two types of formulation (TAA or SIDAA) to yield a total of six treatments (Table 2). The TAA control diet (T_1) was a typical corn-soybean meal type commercial diet and was formulated in such a way to meet the fourth limiting amino acid (valine/isoleucine) requirement solely from feed ingredients and first three limiting amino acids (methionine, lysine and threonine) were supplemented to meet TAA requirement. In T_2 and T_3, the protein was reduced by 0.75% and 1.5% units by replacing soybean meal with 3% and 6% DORB, respectively and all limiting amino acids were balanced with supplementation of crystalline amino acid on TAA basis to meet ideal amino acid ratio recommended by Baker (1997). Parallelly, a SIDAA corn-soy control diet (T_4) was formulated to meet recommended ideal SIDAA ratio (Hoehler et al., 2006) in such a way to meet the requirement of fourth limiting amino acid (valine/isoleucine). In treatments T_5 and T_6, the soybean meal was replaced with 3% and 6% DORB, respectively and both treatments were formulated to meet SIDAA

Table 1. Ingredient amino acid profile and individual amino acids standardized ileal digestibility coefficients of ingredients employed in feed formulation

Ingredients	Lys	Met	M+C[1]	Thr	Val	Ile	Trp	Leu	Arg	Phe	Gly+Ser	His
Total amino acid composition (%, as is)[2]												
Maize	0.255	0.173	0.357	0.306	0.408	0.296	0.061	1.050	0.408	0.418	0.745	0.255
Soybean meal	2.727	0.577	1.214	1.712	2.080	2.00	0.577	3.364	3.334	2.279	4.656	1.184
Rapeseed meal	1.741	0.690	1.641	1.481	1.801	1.421	0.51	2.482	2.352	1.471	3.239	1.021
De-oiled rice bran	0.732	0.321	0.631	0.621	0.892	0.571	0.200	1.162	1.223	0.732	1.620	0.451
Meat and bone meal	2.121	0.523	0.898	1.292	1.815	1.115	0.207	2.457	3.187	1.391	8.320	0.681
Percentage standardized ileal digestibility coefficients[3]												
Maize	92	94	90	85	92	95	81	94	93	94	75/83	95
Soybean meal	90	91	86	85	88	89	89	89	93	89	80/84	92
Rapeseed meal	80	84	80	73	79	79	80	82	87	83	76/72	85
De-oiled rice bran	74	77	72	69	75	75	79	73	86	76	70/67	82
Meat and bone meal	69	72	62	62	70	69	53	71	77	70	69/52	71

[1] Methionine+cysteine.

[2] Based on the analyzed amino acid profile of ingredients.

[3] Adopted from Hoehler et al. (2006) except for Gly+Ser, which were adopted from Bryden et al. (2009), as apparent digestible coefficients.

Table 2. Ingredient composition (%, as is) of the experimental diets at pre-starter, starter and finisher phases

	Pre-starter phase (1-14 d)						Starter phase (15-28 d)						Finisher phase (29-42 d)					
	TAA formulation (%)			SIDAA formulation (%)			TAA formulation (%)			SIDAA formulation (%)			TAA formulation (%)			SIDAA formulation (%)		
DORB level	0	3	6	0	3	6	0	3	6	0	3	6	0	3	6	0	3	6
CP reduction	0	-0.75	-1.5	0	-0.75	-1.5	0	-0.75	-1.5	0	-0.75	-1.5	0	-0.75	-1.5	0	-0.75	-1.5
Treatment	T_1	T_2	T_3	T_4	T_5	T_6	T_1	T_2	T_3	T_4	T_5	T_6	T_1	T_2	T_3	T_4	T_5	T_6
Ingredient (%)																		
Maize	49.83	49.27	48.68	48.39	47.80	47.16	55.07	54.49	53.90	53.62	52.99	52.34	62.28	61.69	61.10	59.64	59.00	58.35
Soybean meal	40.93	37.93	34.93	42.11	39.11	36.11	33.85	30.85	27.85	35.07	32.07	29.07	26.00	23.00	20.00	28.21	25.21	22.21
DORB	-	3.00	6.00	-	3.00	6.00	-	3.00	6.00	-	3.00	6.00	-	3.00	6.00	-	3.00	6.00
Meat and bone meal	-	-	-	-	-	-	4.00	4.00	4.00	4.00	4.00	4.00	5.00	5.00	5.00	5.00	5.00	5.00
Rice bran oil	4.39	4.72	5.06	4.68	4.99	5.28	4.14	4.46	4.77	4.44	4.74	5.03	4.20	4.51	4.81	4.73	5.01	5.30
Calcite powder	1.25	1.28	1.30	1.25	1.27	1.29	0.54	0.56	0.58	0.53	0.55	0.57	0.44	0.46	0.48	0.43	0.45	0.47
L-lysine, 78%	0.116	0.157	0.280	0.082	0.181	0.282	0.136	0.218	0.300	0.100	0.200	0.300	0.169	0.250	0.332	0.105	0.205	0.305
DL-methionine, 99%	0.288	0.307	0.328	0.292	0.325	0.359	0.299	0.319	0.339	0.284	0.318	0.352	0.247	0.266	0.286	0.248	0.281	0.315
L-threonine, 98%	0.040	0.076	0.112	0.037	0.083	0.131	0.079	0.115	0.151	0.049	0.096	0.143	0.074	0.109	0.145	0.046	0.093	0.141
L-valine, 98%	-	0.036	0.076	-	0.062	0.122	-	0.039	0.079	-	0.062	0.122	-	0.034	0.074	-	0.053	0.113
L-isoleucine, 98%		-	0.016			0.049	-	0.024	0.071	-	0.035	0.093	-	0.046	0.092	-	0.058	0.116
L-Tryptophan, 98%	-	-	-	-	-	-	-	-	-	-	-	-	-	-	0.001	-	0.003	0.020
Potassium carbonate	-	0.036	0.072	-	0.035	0.073	-	0.037	0.075	-	0.039	0.076	-	0.036	0.072	-	0.037	0.074
Constant components[1]	3.15	3.15	3.15	3.15	3.15	3.15	1.90	1.90	1.90	1.90	1.90	1.90	1.60	1.60	1.60	1.60	1.60	1.60
Total	100.00	100.00	100.00	100.00	100.00	100.00	100.00	100.00	100.00	100.00	100.00	100.00	100.00	100.00	100.00	100.00	100.00	100.00
Calculated nutrient composition (as is)																		
CP (%)[2]	22.85	22.10	21.33	23.28	22.54	21.92	22.12	21.41	20.46	22.50	21.90	21.09	19.69	18.96	18.23	20.40	19.65	18.92
ME (kcal/kg)	3,000	3,000	3,000	3,000	3,000	3,000	3,100	3,100	3,100	3,100	3,100	3,100	3,200	3,200	3,200	3,200	3,200	3,200
Ca (%)	1.01	1.01	1.01	1.01	1.01	1.01	0.90	0.90	0.90	0.91	0.91	0.91	0.89	0.89	0.89	0.89	0.89	0.89
P_{av} (%)	0.47	0.47	0.48	0.47	0.48	0.48	0.45	0.45	0.46	0.45	0.45	0.46	0.44	0.44	0.45	0.44	0.45	0.45

TAA, total amino acid; SIDAA, standardized ileal digestible amino acid; DORB, de-oiled rice bran; CP, crude protein; ME, metabolizable energy.

[1] Contained Fe, 9,000 mg; I, 200 mg; Cu, 1,500 mg; Mn, 9,000mg; Zn, 8,000 mg; Se, 30 mg; vit A, 1 mIU; vit. D_3, 0.2 mIU; vit. E, 3.0 g; vit. C, 5.0 g; vit. B_1, 0.2 g; vit. B_2, 1.0 g; vit. B_6, 0.3 g; vit. B_{12}, 0.0015; niacin, 3.0 g; calcium-D-pantothenate, 1.5 g; biotin, 0.010 g; folic acid, 0.20 g; vit-K, 0.4 g; Di-calcium phosphate, 2,000g; salt, 400 g herbal liver stimulant, 170 g; semduramicin, 3.0 g; tetracyclin, 3.0 g; toxin binder, 200 g. During starter phase, as above, except Di-calcium phosphate, 750 g. During finisher phase, as in pre-starter phase, except Di-calcium phosphate, 450 g.

[2] Analyzed value.

requirement with supplementation of crystalline amino acids. In order to have a reasonable comparison between TAA and SIDAA formulated groups, all the diets were made iso-caloric and the SID lysine content of TAA control (T_1) was considered as a basal level for SIDAA based control (T_4). The other limiting amino acids levels were maintained to meet minimum ideal amino acid ratio recommended either for TAA or SIDAA. The diets were formulated using Microsoft Excel based program. All diets were fed in mash form during pre-starter (0 to 14 days), starter (15 to 28 days) and finisher (29 to 42 days) phases. The ingredient and nutrient composition of the experimental diets is presented in Table 2, and the amino acid composition of the experimental diets is presented in Table 3.

Experimental birds

The present experiment was carried out after approval for use of chicks and experimental procedures by Institutional Animal Ethics Committee under the guidelines of Committee for the Purpose of Control and Supervision of Experiments on Animals (CPCSEA), Ministry of Environment, Forests and Climate Change, Government of India. A total of 360 one d-old straight run fast growing commercial broiler chicks (Vencobb-400) were divided into thirty six homogenous groups with ten chicks in each pen.

The six experimental diets were randomly allocated to six pens each and each pen was considered as one replicate unit. All the chicks were reared under deep litter system in conventional open ventilated shed with standard vaccination program and uniform managemental practices throughout the 42 d period.

Parameters studied

Growth performance parameters: At weekly interval the replicate wise feed intake and body weight (BW) of individual birds were recorded. The mortality of the bird was recorded as and when occurred. The mortality corrected feed conversion ratio (FCR) was calculated as unit feed intake to the unit body weight gain (BWG) (Kumar et al., 2015).

Intake of metabolizable energy, crude protein and amino acids: Based on the feed intake, intake of metabolizable energy (ME) was calculated from the estimated value while the CP and amino acid intakes were calculated based on the analyzed CP and amino acid composition.

Serum biochemical profile: On 42nd d, blood was collected from two birds form each replicate for serum collection which was analyzed for uric acid using clinical auto-analyzer (BS-300, MINDRAY, ShenZen, China) as per the manufacture's specifications using a standard commercial kit.

Table 3. The total and standardized ileal digestible amino acid composition of experimental diets (%, as is) at pre-starter, starter and finisher phases

Amino acid	Pre-starter phase (0-14 d)							Starter phase (15-28 d)							Finisher phase (29-42 d)						
	Require-ments	TAA formulation			SIDAA formulation			Require-ments	TAA formulation			SIDAA formulation			Require-ments	TAA formulation			SIDAA formulation		
		T_1	T_2	T_3	T_4	T_5	T_6		T_1	T_2	T_3	T_4	T_5	T_6		T_1	T_2	T_3	T_4	T_5	T_6
Total amino acid content[1] (%, as is)																					
Lys	1.33	1.33	1.33	1.33	1.33	1.35	1.36	1.25	1.25	1.25	1.25	1.25	1.27	1.28	1.10	1.10	1.10	1.10	1.11	1.12	1.13
M+C	0.96	0.96	0.96	0.97	0.98	0.98	1.00	0.94	0.94	0.94	0.94	0.93	0.95	0.96	0.83	0.83	0.83	0.83	0.84	0.86	0.87
Thr	0.89	0.89	0.89	0.89	0.90	0.91	0.93	0.88	0.88	0.88	0.88	0.86	0.87	0.88	0.77	0.77	0.77	0.77	0.77	0.78	0.80
Val	1.05	1.05	1.05	1.05	1.08	1.09	1.14	1.00	1.00	1.00	1.00	1.02	1.04	1.06	0.88	0.89	0.88	0.88	0.92	0.93	0.95
Ile	0.89	0.97	0.92	0.89	0.99	0.94	0.94	0.86	0.88	0.86	0.86	0.91	0.89	0.90	0.76	0.76	0.76	0.76	0.80	0.81	0.82
Trp	0.21	0.27	0.26	0.24	0.27	0.26	0.25	0.21	0.24	0.23	0.21	0.24	0.23	0.22	0.18	0.20	0.19	0.18	0.21	0.20	0.21
Standardized ileal digestible amino acid content[2] (%, as is)																					
Lys	1.21	1.21	1.20	1.18	1.21	1.21	1.21	1.12	1.12	1.11	1.09	1.12	1.12	1.12	0.98	0.98	0.97	0.96	0.98	0.98	0.98
M+C	0.88	0.87	0.86	0.84	0.88	0.88	0.88	0.84	0.85	0.83	0.82	0.84	0.84	0.84	0.76	0.74	0.73	0.71	0.76	0.76	0.76
Thr	0.77	0.76	0.75	0.74	0.77	0.77	0.77	0.73	0.74	0.73	0.72	0.73	0.73	0.73	0.65	0.65	0.64	0.63	0.65	0.65	0.65
Val	0.95	0.94	0.91	0.90	0.95	0.95	0.95	0.90	0.88	0.86	0.84	0.90	0.90	0.90	0.80	0.77	0.75	0.73	0.80	0.80	0.80
Ile	0.82	0.87	0.81	0.77	0.89	0.83	0.82	0.78	0.79	0.76	0.75	0.81	0.78	0.78	0.71	0.68	0.67	0.65	0.71	0.71	0.71
Trp	0.19	0.24	0.22	0.20	0.24	0.23	0.21	0.18	0.21	0.19	0.17	0.21	0.20	0.18	0.17	0.17	0.15	0.14	0.18	0.17	0.17

TAA, total amino acid; SIDAA, standardized ileal digestible amino acid; M+C, methionine+cysteine.

* Requirements were calculated according to ideal amino acid ratio (Baker, 1997) except at pre-starter phase, where valine requirement was calculated according to ideal amino acid ratio of Hoehler et al. (2006), The diet amino acid composition was derived based on analyzed ingredient amino acid composition.

** Requirement were calculated according to ideal amino acid ratio (Hoehler et al., 2006) and diet composition was derived by multiplying amino acid composition with SID coefficients of amino acid of each ingredient (Hoehler et al., 2006).

Carcass characteristics: On d 42, two birds from each replicate were randomly selected, starved over night with the provision for *ad libitum* water and sacrificed by cervical dislocation. The dressing percentage was calculated as the per cent of the carcass weight to the BW after removing the feathers, neck, legs and internal viscera. Weights of different cuts *viz.,* breast, thigh, drumstick, and wing of the carcass were taken and each part was expressed as percentage of pre-slaughter BW (g/100 g).

Statistical analysis

The experimental data was statistically analyzed by two-way analysis of variance (ANOVA) with Bonferroni post test (p<0.05) to separate the factor and interaction effect and also by one-way ANOVA with Tukey's multiple range test (p<0.05) by using GraphPad Prism (GraphPad Prism 5.01 for windows, GraphPad Software, San Diego CA, USA, www.graphpad.com).

RESULTS AND DISCUSSION

Feed intake

The feed intake (Table 4) under TAA formulation was significantly reduced with 1.5% protein reduction (6% DORB; T_3) compared to TAA control (T_1) despite of limiting amino acid supplementation. This was only observed during pre-starter phase. Interestingly under SIDAA treatments, such type of negation was not seen. Factorial analysis revealed significantly (p<0.05) higher feed intake in SIDAA formulation and 1.5% unit protein reduction was found to significantly reduce the feed intake compared to normal protein group (0% DORB) during pre-

starter phase. The influence of dietary treatments or factors on the feed intake only during relatively younger age (<14 days) could be due to relatively under developed GIT (Batal and Parson, 2002) which undeniably resulted in poor amino acid digestibility, consequently resulting in still lesser SIDAA composition than estimated (Adedokun et al., 2008). The resultant amino acid imbalance might have severely depressed the feed intake during pre-starter phase. During rest of the phases, the feed intake remained similar among the treatments irrespective of differences in the amino acid profile of diets, which follows the "Theory of food intake and growth" proposed by Emmans (1981; 1989) which emphasizes that, birds attempt to grow at their genetic potential, for which they attempt to eat sufficient quantity feed (hence nutrient) required to grow at that rate unless and until the bulkiness of feed or the inability of birds to lose sufficient heat to environment constrains the feed intake. The non-significant pattern of feed intake observed between SIDAA and TAA formulations was also observed by Mairoka et al. (2005) and subsequently many studies also reported similar feed intake between TAA and digestible amino acid formulations (Szczurek, 2010; Nasr and Kheiri, 2012). Similar to findings of this study, non-significant effect of low protein limiting amino acid supplemented diets on the feed intake was also observed in previous studies (Narmond et al., 2008; Darsi et al., 2012).

Cumulative intake of metabolizable energy, crude protein and limiting amino acids

The cumulative intake of ME was not affected due to dietary treatments, while the CP intake (Table 4) tended to decrease (p<0.001) as the level of protein reduced by 1.5%

Table 4. Feed intake and cumulative intake of metabolizable energy, crude protein and amino acids in broiler chicken fed DORB based low protein, limiting amino acid supplemented diets

DORB level (%)	Formulation type	Treatment no.	Feed intake (g/bird)				ME (Mcal/bird)	Cumulative nutrient intake (g/bird)						
			0-14 d	15-28 d	29-42 d	0-42 d		CP	SID Lys	SID M+C	SID Thr	SID Val	SID Ile	SID Trp
0	TAA	T_1	328.1[b]	1,038	2,055	3,421	10.78	709.3[b]	35.81[ab]	26.90[abc]	23.59[ab]	28.06[bc]	24.92[bc]	6.40[d]
3	TAA	T_2	305.5[ab]	1,042	2,067	3,415	10.76	684.1[ab]	35.24[ab]	26.35[ab]	23.17[ab]	27.18[ab]	24.11[ab]	5.83[b]
6	TAA	T_3	277.0[a]	1,064	2,065	3,406	10.74	656.4[a]	34.64[a]	25.81[a]	22.73[a]	26.14[a]	23.57[a]	5.29[a]
0	SIDAA	T_4	344.8[b]	1,068	2,076	3,489	10.99	744.1[c]	36.56[b]	27.75[c]	23.94[b]	29.53[d]	26.36[d]	6.82[e]
3	SIDAA	T_5	324.1[b]	1,015	2,099	3,438	10.84	707.7[b]	35.94[ab]	27.30[bc]	23.54[ab]	28.92[cd]	25.53[cd]	6.21[cd]
6	SIDAA	T_6	305.9[ab]	1,057	2,128	3,492	11.01	692.8[b]	36.48[b]	27.71[c]	23.90[b]	29.37[cd]	25.87[cd]	6.10[bc]
		SEM	5.430	8.091	13.68	15.47	0.051	5.432	0.192	0.167	0.124	0.240	0.197	0.084
		p-value	0.001	0.431	0.733	0.492	0.498	<0.001	0.018	<0.001	0.028	<0.001	<0.001	<0.001
Factor effects														
Effect of formulation type														
TAA			303.5[a]	1,048	2,062	3,414	10.76	683.3[a]	35.23[a]	26.35[a]	23.16[a]	27.13[a]	24.20[a]	5.84[a]
SIDAA			324.9[b]	1,047	2,101	3,473	10.94	714.9[b]	36.33[b]	27.59[b]	23.79[b]	29.27[b]	25.92[b]	6.38[b]
		p-value	0.027	0.952	0.234	0.110	0.081	<0.001	0.003	<0.001	0.008	<0.001	<0.001	<0.001
Effect of protein reduction (%)														
0.00			336.4[b]	1,053	2,066	3,455	10.88	726.7[b]	36.18	27.32	23.76	28.80[b]	25.64[b]	6.61[b]
−0.75			314.8[ab]	1,029	2,083	3,426	10.80	695.9[a]	35.59	26.82	23.35	28.05[ab]	24.82[a]	6.02[a]
−1.50			291.5[a]	1,061	2,097	3,449	10.87	674.6[a]	35.56	26.76	23.31	27.75[a]	24.72[a]	5.69[a]
		p-value	0.001	0.252	0.804	0.579	0.762	<0.001	0.244	0.151	0.192	0.009	0.006	<0.001
Interaction effect (p value)			0.733	0.356	0.650	0.720	0.725	0.672	0.306	0.185	0.234	0.023	0.231	<0.009

DORB, de-oiled rice bran; ME, metabolizable energy; CP, crude protein; SID, standardized ileal digestible; M+C, methionine+cysteine; TAA, total amino acid; SIDAA, standardized ileal digestible amino acid.
[a-c] Within a column and within a group, means bearing different superscripts differ significantly (p<0.05).

unit (T_3) in TAA formulations and by 0.75% and above in SIDAA formulated treatments compared to respective controls. Results revealed significantly higher CP intake in SIDAA formulations over TAA formulations and step down of protein by 0.75% and 1.50% units significantly reduced the CP intake vis-à-vis normal protein group. The cumulative ME intake fairly remained similar among the treatments as an indication of adoptive behavior of birds to consume feed to meet their energy requirement for growth (Emmans, 1981; 1989). The deficiency of available amino acids (SIDAA) in T_2 perhaps resulted in increased feed intake to match for the limiting amino acid deficiency, which made the birds to consume similar CP as that of T_1. In contrast, under SIDAA treatments, since the diets were having same SIDAA levels, the birds probably did not increase their feed intake as the bird's amino acid requirement was met, subsequently resulting in lesser CP intake with 3% DORB inclusion itself. The higher CP intake due to SIDAA factor than TAA factor was rather a reflection of relatively higher dietary CP content in diet *per-se*. Inclusion of DORB to reduce CP with limiting amino acid supplementation resulted in significant reduction of CP intake both at 3% and 6% inclusions compared to no DORB group, justifying the importance of supplemental amino acids in reducing dietary protein levels (Narmond et al., 2008; Darsi et al., 2012). The significantly higher intakes of all limiting amino acids (SID basis) observed under SIDAA formulation over TAA formulation was a reflection of

higher SID amino acid content in SIDAA formulations than TAA formulation. The protein factor significantly reduced the intake of SID valine with protein reduction of 1.5% units, while SID isoleucine and tryptophan intakes were significantly reduced at and above 0.75% unit protein step down compared to normal protein group. The interaction effect of main factors was noticed only for SID valine and tryptophan intakes, where the SIDAA formulation significantly improved intakes of these two amino acids for same level of protein. In spite of protein reduction, the birds were able to meet the requirement of first three limiting amino acids probably by making minor adjustment in feed intake however; this adaptation was not adequate to increase the intake of subsequent limiting amino acids (valine, isoleucine, and tryptophan).

Body weight gain

The BWG (Table 5) under 1.5% unit low protein TAA formulation (T_3) was significantly (p<0.001) reduced during pre-starter and on cumulative basis compared to control (T_1). In contrast, under SIDAA formulation, 1.5% protein reduction (T_6) revealed significant BWG reduction *vis-à-vis* control (T_4) only during pre-starter phase however; similar performance was evident during rest of the phases and cumulatively irrespective of protein level. The BWG under SIDAA formulated 1.5% low protein diet (T_6) was comparable to TAA control (T_1) during all phases and cumulatively. The results revealed significantly (p<0.05)

Table 5. Body weight gain, feed conversion ratios, nitrogen excretion, retention (39 to 42 day) and serum uric acid levels (42nd day) of birds under different treatments

DORB level (%)	Formulation type	Treatments	Body weight gain (g/bird)				Feed conversion ratio (feed/gain)				Nitrogen (g/bird/d)		Serum uric acid (mg/dL)
			0-14 d	15-28 d	29-42 d	0-42 d	0-14 d	15-28 d	29-42 d	0-42 d	Excretion	Retention	
0	TAA	T$_1$	275.1^{bc}	739.6	1,052^{ab}	2,067^{bc}	1.193^{ab}	1.466^{bc}	1.990^{abc}	1.696^{bc}	1.747^{ab}	4.189^{ab}	10.15^{ab}
3	TAA	T$_2$	248.2^{ab}	733.8	1,043^{a}	2,025^{ab}	1.216^{c}	1.489^{cd}	2.011^{cd}	1.724^{cd}	1.736^{ab}	4.051^{ab}	9.94^{ab}
6	TAA	T$_3$	225.7^{a}	719.2	1,048^{a}	1,993^{a}	1.227^{c}	1.508^{d}	2.024^{d}	1.741^{d}	1.547^{a}	3.840^{a}	9.45^{a}
0	SIDAA	T$_4$	291.9^{c}	761.4	1,075^{ab}	2,129^{c}	1.181^{a}	1.433^{a}	1.964^{a}	1.663^{a}	1.840^{b}	4.338^{b}	10.75^{b}
3	SIDAA	T$_5$	272.1^{bc}	733.4	1,076^{ab}	2,081^{bc}	1.191^{ab}	1.444^{ab}	1.979^{ab}	1.686^{ab}	1.717^{ab}	4.491^{b}	10.62^{ab}
6	SIDAA	T$_6$	254.1^{ab}	729.5	1,092^{b}	2,076^{bc}	1.204^{bc}	1.461^{b}	1.995^{bc}	1.704^{bc}	1.668^{ab}	4.485^{b}	10.24^{ab}
		SEM	4.990	5.177	4.668	9.300	0.003	0.005	0.004	0.005	0.024	0.056	0.085
		p-value	<0.001	0.297	0.005	0.001	<0.001	<0.001	<0.001	<0.001	0.011	<0.001	0.036
Factor effects													
Effect of formulation type													
TAA			249.7^{a}	730.9	1047^{a}	2,028^{a}	1.212^{b}	1.488^{b}	2.008^{b}	1.721^{b}	1.677	4.027^{a}	9.85
SIDAA			272.7^{b}	741.4	1081^{b}	2,095^{b}	1.192^{a}	1.446^{a}	1.979^{a}	1.684^{a}	1.742	4.438^{b}	10.54
		p-value	0.004	0.308	<0.001	<0.001	<0.001	0.001	<0.001	<0.001	0.131	<0.001	0.875
Effect of protein reduction (%)													
0.00			283.5^{b}	750.5	1,064	2,098^{b}	1.187^{a}	1.450^{a}	1.977^{a}	1.680^{a}	1.794^{b}	4.263	10.45
−0.75			260.1^{ab}	733.6	1,059	2,053^{ab}	1.204^{b}	1.466^{ab}	1.995^{ab}	1.705^{b}	1.726^{ab}	4.271	10.28
−1.50			239.9^{b}	724.4	1,070	2,034^{a}	1.216^{c}	1.484^{b}	2.009^{b}	1.723^{b}	1.608^{a}	4.163	9.85
		p-value	<0.001	0.120	0.553	<0.001	<0.001	0.001	<0.001	<0.001	0.004	0.522	0.098
Interaction effect (p-value)			0.818	0.676	0.553	0.664	0.011	0.480	0.873	0.892	0.367	0.076	0.945

DORB, de-oiled rice bran; TAA, total amino acid; SIDAA, standardized ileal digestible amino acid.
[a-b] Within a column and within a group, means bearing different superscripts differ significantly (p<0.05).

improved BWG under SIDAA formulation over TAA formulation throughout the experimental period. The protein factor revealed a feasibility of protein reduction of 0.75% with limiting amino acid supplementation, while further protein reduction (1.5%) significantly reduced the BWG during pre-starter and on cumulative basis.

During the pre-starter phase, comparison within formulation type revealed significant growth retardation on 1.5% unit protein step down among both TAA and SIDAA based treatments which could be due to age related differences in digestibility coefficients of amino acids in ingredients (Adedokun et al., 2008). On cumulative basis, among TAA based treatments, 1.5% protein step down (T$_3$) resulted in significant depression of BWG despite of limiting amino acid supplementation. This can be attributed to differences in amino acid content on SID basis, which inevitably resulted in significantly reduced intake of all limiting amino acids. In contrast, among SIDAA formulations, the BWG was similar to that of control (T$_4$) despite of protein reduction. Significantly improved BWG due to SIDAA formulation type is a clear reflection of significantly higher intake of all limiting amino acids. The lower intake of lysine (3.03%), M+C (4.49%) and threonine (2.65%) to minor extent and valine (7.31%), isoleucine (6.64%) and tryptophan (8.46%) to a major extent in TAA formulation type over SIDAA type perhaps impeded protein accretion and subsequently reduced the growth performance. Among amino acid, methionine is primarily required for initiation of protein synthesis and has been revealed to influence myogenic gene expression in broilers (Wen et al., 2014) and moreover, lysine, methionine, threonine, valine and isoleucine are components of muscle protein and their deficiency invariably reduced the BWG. The growth retardation as a result of deficiency of either individual or various combinations of limiting amino acids lysine, methionine, threonine, valine, isoleucine and tryptophan is well noticed in previous studies (Corzo et al., 2009; Corzo et al., 2011; Basavanta Kumar et. al., 2015). This type of difference between TAA and DAA formulation was also reported by earlier studies (Szczurek, 2010; Nasr and Kheiri, 2012).

Feed conversion ratio

The FCR (Table 5) was influenced by six dietary treatments and both factors throughout the feeding phases and cumulatively. Among both TAA and SIDAA formulations, the FCR was found to significantly deteriorate with a protein step down of 1.5% units compared to respective controls during all phases and cumulatively. On cumulative basis, the better FCR observed in SIDAA control (T$_4$) was significantly superior to all TAA based treatments (T$_1$ to T$_3$) and 1.5% low protein SIDAA treatment (T$_6$). Factorial separation revealed significantly (p<0.001) improved FCR in SIDAA formulation over TAA formulation during all the three phases and on cumulative basis. The protein factor revealed significant deterioration of FCR with a protein reduction level of 0.75% and above during pre-starter phase and cumulatively, while during the starter and finisher phases, protein step down of 1.5% resulted in significant FCR depression compared to normal protein group (0% DORB) and protein reduction of 0.75% unit being mediocre remained similar to both normal protein group and 1.5% low protein group.

The significantly better feed efficiency noted on factorial approach under SIDAA groups can be traced back to a significantly higher intake of all limiting amino acids as described previously. The deficiencies of various amino acid viz., valine (Corzo et al., 2011), valine and isoleucine (Corzo et al., 2009) and valine, isoleucine and tryptophan on DORB based rations (Basavanta Kumar et al., 2015) have been shown to depress feed efficiency similar to present study. The protein factor with a reduction of 1.5% unit protein and supplementation of limiting amino acids resulted in significant depression of FCR at all production phases, while, 0.75% unit protein reduction significantly reduced FCR only during pre-starter phase. Our present findings contradicts previous studies (Waldroup et al., 2005; Darsi et al., 2012) which could be due to the age related differences in the SIDAA coefficients of ingredients (Adedokun et al., 2008) and due to this reason, even in spite of formulation on SIDAA basis, in reality still there might be moderate limiting amino acid deficiency and probably this deficiency was overcome by birds through slightly increasing feed intake to grow at their genetic potential (Emmans, 1981; 1989) consequently resulting in deterioration of feed efficiency despite of optimum BWG.

Nitrogen excretion, retention and serum uric acid levels

The nitrogen excretion (g/bird/d; Table 5) was significantly minimized in TAA formulated 1.5% low protein diet (T_3) than highest observation of SIDAA control (T_4). Formulation type had no influence on nitrogen excretion, while the dietary protein level significantly ($p < 0.05$) minimized nitrogen excretion at 1.5% protein reduction compared to normal protein diet. On the other hand, the nitrogen retention noticed in all SIDAA formulated treatments was significantly superior to T_3. This positive effect of SIDAA formulation on nitrogen retention was clearly separated on factorial analysis. The serum uric acid level was significantly reduced in TAA based 1.5% low protein diet (T_3) compared to SIDAA control (T_4). The serum uric acid level was significantly reduced in TAA based 1.5% low protein diet (T_3) compared to SIDAA control (T_4), where a protein reduction of more than 2% existed supporting the previous findings (Narmond et al., 2008; Darsi et al., 2012).

Carcass parameters

The results revealed significantly higher yields of all carcass parameters (Table 6) due to SIDAA formulation over TAA formulation. The protein factor significantly reduced the yields of carcass, breast and thigh at 1.5% protein reduction compared to normal protein group. The

Table 6. Carcass characteristics of birds fed low protein, limiting amino acid supplemented diets at the end of 42nd day trial

DORB level (%)	Formulation type	Treatment no.	Carcass parameter (% of pre-slaughter weight)				
			Carcass yield	Breast yield	Thigh yield	Drumstick yield	Wing yield
0	TAA	T_1	71.26[ab]	21.81[ab]	13.35[b]	09.66[abc]	7.702[abc]
3	TAA	T_2	71.32[ab]	21.20[ab]	12.75[ab]	09.12[ab]	7.180[ab]
6	TAA	T_3	70.60[a]	20.71[a]	12.25[a]	08.96[a]	7.016[a]
0	SIDAA	T_4	72.66[c]	22.06[b]	13.61[b]	10.33[c]	8.393[c]
3	SIDAA	T_5	72.54[bc]	21.97[ab]	13.52[b]	10.20[bc]	8.264[bc]
6	SIDAA	T_6	71.88[abc]	21.57[ab]	13.11[ab]	09.84[abc]	7.901[abc]
		SEM	0.151	0.135	0.117	0.119	0.114
		p-value	<0.001	0.026	0.003	0.001	0.001
Effect of formulation type							
TAA			71.06[a]	21.24[a]	12.78[a]	9.247[a]	7.299[a]
SIDAA			72.36[b]	21.87[b]	13.41[b]	10.12[b]	8.186[b]
		p-value	<0.001	0.016	0.004	<0.001	<0.001
Effect of protein reduction (%)							
0.00			71.96[b]	21.94[b]	13.48[b]	9.995	8.048
−0.75			71.93[ab]	21.59[ab]	13.14[ab]	9.660	7.722
−1.50			71.24[a]	21.14[a]	12.68[a]	9.400	7.459
		p-value	0.040	0.044	0.012	0.080	0.085
Interaction effect p-value			0.957	0.576	0.460	0.745	0.755

DORB, de-oiled rice bran; TAA, total amino acid; SIDAA, standardized ileal digestible amino acid.
[a-b] Within a column and within a group, means bearing different superscripts differ significantly (p<0.05).

abdominal fat percentage was significantly lower in SIDAA formulation over TAA formulation. Similar to our present findings, previous studies also reported higher carcass and breast meat yields (Szczurek, 2010), breast and thigh yields (Khaskar and Golian, 2009) in ileal digestible amino acid formulations over TAA formulations. The abdominal fat percentage was significantly lower in SIDAA formulation over TAA formulation, which perhaps better explains diversification of energy towards protein accretion in presence of balanced SIDAA profile (Szczurek, 2010).

CONCLUSION

The findings of the present study revealed possibility of protein reduction in broiler diets by 1.5% units with incorporation of DORB (6%) and supplemental limiting amino acids on SIDAA formulation, while on TAA formulation only 0.75% unit protein reduction (3% DORB) was found possible. Among formulation types, SIDAA formulation was found to significantly improve broiler performance than TAA based formulation. In addition, the protein step down also minimized the nitrogen excretion and SIDAA formulation improved the nitrogen retention over TAA formulation.

ACKNOWLEDGMENTS

The authors duly acknowledge Veterinary College, KVAFSU, Bangalore, and KVAFSU, Bidar for facilitating the work, Dr. B.V. Rao Poultry Research Foundation/WPSA (IB) for providing financial assistance for project and the Department of Animal Husbandry, GOK for providing deputation facility. We also acknowledge Evonik Pte. Ltd. Singapore for analysis of amino acid composition of the feed ingredients.

REFERENCES

Adedokun, S. A., O. Adeola, C. N. Parsons, M. S. Lilburn, and T. J. Applegate. 2008. Standardized ileal amino acid digestibility of plant feedstuffs in broiler chickens and turkey poults using a nitrogen-free or casein diet. Poult. Sci. 87:2535-2548.

AOAC (Association of Official Analytical Chemists). 2005. Official Methods of Analysis. 18th edn. Association of Official Analytical Chemists Int., Arlington, VA, USA.

Baker, D. H. 1997. Ideal Amino Acid Profiles for Swine and Poultry and Their Applications in Feed Formulations. Biokyowa Technical Review, Cape Girardeau, MO, USA. 9:1-24.

Basavanta Kumar, C., R. G. Gloridoss, K. C. Singh, T. M. Prabhu, Siddaramanna, B. N. Suresh, and G. A. Manegar, 2015. Impact of second line limiting amino acids' deficiency in broilers fed low protein diets with rapeseed meal and de-oiled rice bran. Vet. World 8:350-357.

Batal, A. B. and C. M. Parsons. 2002. Effects of age on nutrient

digestibility in chicks fed different diets. Poult. Sci. 81:400-407.

Bryden, W. L., X. Li, G. Ravindran, H. Li, and V. Ravindran. 2009. Ileal Digestible Amino Acid Values in Feedstuffs for Poultry. RIRDC Publication, Barton, Australia.

Corzo, A., W. A. Dozier III, L. Mejia, C. D. Zumwalt, M. T. Kidd, and P. B. Tillman. 2011. Nutritional feasibility of L-valine inclusion in commercial broiler diets. J. Appl. Poult. Res. 20:284-290.

Corzo, A., R. E. Loar II, and M. T. Kidd. 2009. Limitations of dietary isoleucine and valine in broiler chick diets. Poult. Sci. 88:1934-1938.

Darsi, E., M. Shivazad, M. Zaghari, N. F. Namroud, and R. Mohammadi, 2012. Effect of reduced dietary crude protein levels on growth performance, plasma uric acid and electrolyte concentration of male broiler chicks. J. Agric. Sci. Technol. 14:789-797.

Das, A. and S. K. Ghosh. 2000. Effect of feeding different levels of rice bran on performance of broilers. Indian J. Anim. Nutr. 17:333-335.

Djouvinov, D., M. Stefanov, S. Boicheva, and T. Vlaikova. 2005. Effect of diet formulation on basis of digestible amino acids and supplementation of probiotic on performance of broiler chicks. Trakia J. Sci. 3:61-69.

Emmans, G. C. 1981. A model of the growth and feed intake of ad libitum fed animals, particularly poultry. In: Computers in Animal Production (Eds. G. M. Hilllyer, C. T. Wittemore, and R. G. Gunn). British Society of Animal Production, London, UK. pp. 103-110.

Emmans, G. C. 1989. The growth of turkeys. In: Recent Advances in Turkey Science (Eds. C. Nixey and T. C. Grey). Butterworths, London, UK. pp. 135-166.

Ghaffari, M., M. Shivazad, M. Zaghari, and R. Taherkhani. 2007. Effect of energy levels of diets formulated on total or digestible amino acid basis on broiler performance and carcass traits. Asian J. Poult. Sci. 1:16-21.

GraphPad Prism. 2007. GraphPad Prism version 5.01 for Windows. GraphPad Software, San Diego, CA, USA.

Hoehler D., A. Lemme, V. Ravindran, W. L. Bryden, and H. S. Rostagno. 2006. Feed formulation in broiler chickens based on standardized ileal amino acid digestibility. In: Proceedings of Advances in Poultry Nutrition, VIII. International Symposium on Aquatic Nutrition. UANL, Monterrey, New Lion, Mexico. pp. 197-212.

Khan, M. Y., T. N. Pasha, A. Khalique, Z. Ali, and H. Rehman. 2002. Effect of feeding hydrogen peroxide treated defatted rice polishing on performance of broiler chicks. Int. J. Poult. Sci., 1:193-196.

Khaskar, V. and A. Golian, 2009. Comparison of ileal digestible versus total amino acid feed formulation on broiler performance. J. Anim. Vet. Adv. 8:1308-1311.

Kumar, C. B., B. S. V. Reddy, R. G. Gloridoss, T. M. Prabhu, B. N. Suresh, and S. N. Kumar. 2015. Amelioration of aflatoxicosis through a bio-technologically derived aflatoxin degrading commercial product in broilers. Pak. Vet. J. 35:217-221.

Mairoka, A., F. Dahlke, A. M. Penz Jr., and A. M. Kessler. 2005.

Diets formulated on total or digestible amino acid basis with different energy levels and physical form on broiler performance. Rev. Bras. Cienc. Avic. 7:47-50.

Namroud, N. F., M. Shivazad, and M. Zaghari. 2008. Effects of fortifying low crude protein diets with crystalline amino acids on performance, blood ammonia level, and excreta characteristics of broiler chicks. Poult. Sci. 87:2250-2258.

Nasr, J. and F. Kheiri. 2012. Effects of lysine levels of diets formulated based on total or digestible amino acids on broiler carcass composition. Rev. Bras. Cienc. Avic. 14:233-304.

Piyaratne, M. K. D. K., N. S. B. M. Atapattu, A. P. S. Mendis, and A. G. C. Amarasinghe. 2009. Effects of balancing rice bran based diets for up to four amino acids on growth performance of broilers. Trop. Agric. Res. Ext. 12:57-61.

Rostagno, H. S. 2011. Brazilian Tables for Poultry and Swine. 3rd edn. Federal University Viscosa, Viscosa, Brazil.

Szczurek, W. 2010. Practical validation of efficacy of the standardized ileal digestible amino acid values in diet formulation for broiler chickens. J. Anim. Feed Sci. 19:590-598.

Waldroup, P. W., Q. Jiang, and C. A. Fritts. 2005. Effect of supplementing broiler diets low in crude protein with essential and nonessential amino acids. Int. J. Poult. Sci. 4:425-431.

Wen, C., X. Chen, G. Y. Chen, P. Wu, Y. P. Chen, Y. M. Zhou, and T. Wang. 2014. Methionine improves breast muscle growth and alters myogenic gene expression in broilers. J. Anim. Sci. 92:1068-1073.

27

Effects of Physically Effective Neutral Detergent Fiber Content on Intake, Digestibility, and Chewing Activity in Fattening Heifer Fed Total Mixed Ration

Mi Rae Oh[a], Heeok Hong[1,a], Hong Liang Li, Byong Tae Jeon, Cheong Hee Choi, Yu Ling Ding, Yu Jiao Tang, Eun Kyung Kim, Se Young Jang, Hye Jin Seong, and Sang Ho Moon*

Korea Nokyong Research Center, Division of Food Bio Science, Konkuk University, Chungju 27478, Korea

ABSTRACT: The objective of this study was to determine the effects of physically effective neutral detergent fiber (peNDF) content in total mixed ration (TMR) on dry matter intake, digestibility, and chewing activity in fattening Hanwoo (*Bos taurus coreanae*) heifers. The experiment was designed as a replicated 3×3 Latin square using 12 heifers. Fattening heifers were offered one of three diets [high (T1), medium (T2), and low (T3) peNDF] obtained by different mixing times (3, 10, and 25 min) for the same TMR feed. The peNDF content of TMR was determined by multiplying the proportion of dry matter retained by a 1.18 mm-screen in a Penn State Particle Separator by the dietary NDF content. The $peNDF_{1.18}$ content was 30.36%, 29.20%, and 27.50% for the T1, T2, and T3 diets, respectively (p<0.05). Dry matter intake was not affected by peNDF content in TMR. Total weight gain in T1 group was significantly higher (p<0.05) than in T2 and T3 groups. However, weight gain did not differ between T2 and T3 groups. The feed conversion ratio decreased with an increase in the peNDF content (T1: 12.18, T2: 14.17, and T3: 14.01 g/g). An increase in the peNDF content of TMR was associated with a linear increase in the digestibility of dry matter, crude protein, crude fiber, neutral detergent fiber, and acid detergent fiber (p<0.05). Also, an increase in peNDF content of the TMR resulted in a linear increase in the number of chews in eating and ruminating (p<0.05), and consequently in the number of total chews (p<0.05). These results indicate that peNDF content affects digestibility and chewing activity. Consequently, the peNDF content of TMR should be considered for improving feed efficiency, digestibility, body weight gain, and performance in fattening heifers. (**Key Words:** Physically Effective Neutral Detergent Fiber, Chewing Activity, Digestibility, Dry Matter Intake, Fattening Heifer, Total Mixed Ration)

INTRODUCTION

Digestibility of fiber is important in ruminants such as cattle and sheep that evolved as forage consumers. Plant cell walls, which is measure as fiber, cannot be digested by animals but must be fermented by microorganisms (Mertens, 1988). Digestion of fiber requires more energy and complex mechanisms, and ruminants have developed attributes that result in efficient digestion. In addition, to perform well and to avoid metabolic disorders, ruminants need adequate dry matter (DM) intake, energy, and appropriate fiber content in their rations. Thus, it is necessary to accurately estimate how much dietary fiber cattle require, because forages and roughage are lower in digestibility and available energy than grains and other concentrates, and reducing fiber to minimum levels in the diet is often desirable for high-performance dairy cows and beef cattle (Mertens, 1997). Mertens (1997) defined physically effective neutral detergent fiber (peNDF) as the fraction of neutral detergent fiber (NDF) that stimulates rumination and contributes to a proper ruminal digesta mat consistency. The peNDF system is based on measuring feed particle size and NDF content of forage to accurately predict the chewing response (Grant and Colenbrander, 1990).

* Corresponding Author: Sang Ho Moon.
E-mail: moon0204@kku.ac.kr
[1] Department of Medical Science, School of Medicine, Konkuk University, Seoul 05029, Korea.
[a] These authors contributed equally to this work.

According to critical size theory, particles longer than 1.18 mm have the greatest resistance to passage and are largely responsible for stimulating chewing and rumination (Poppi et al., 1981). Digestibility is directly proportional to the digestible fraction of fiber and rate of fiber digestion, but it is inversely related to the rate of release of particles from the nonescapable to escapable fiber pool and the rate of escape. Particle size reduction increases the release rate from the nonescapable fraction which in turn results in reduced digestibility. Ruminal particulate matter is mostly below the threshold size for escape, and particle size reduction may not be the rate-limiting step in clearance from the reticulorumen. Therefore, an adequate particle size of fiber should be an integral part in the diet of ruminants to take advantage of their unique digestive capability (Mertens, 1997).

Although various methods are available to measure particle size in diets, the Penn State Particle Separator (PSPS) has become widely accepted as a quick and practical method for routine use on farms to evaluate the particle size of forages and total mixed ration (TMR) (Lammers et al., 1996). Mertens (1997) defined peNDF as the proportion of DM retained by a 1.18-mm screen multiplied by the dietary NDF (peNDF$_{>1.18}$) by using a dry-sieving technique. However, little information is available on the physical properties of TMR, which is widely used in dairy and beef cattle production in Korea. Therefore, an accurate evaluation of the physical characteristics of TMR is necessary to improve beef cattle production.

The objective of this study was to determine the effects of peNDF content on feed intake, digestibility, and chewing activity in fattening heifers fed TMR.

MATERIALS AND METHODS

Animals and diets

Fattening heifers of Hanwoo (*Bos taurus coreanae*, average 460.5±45.6 kg) were used to investigate the effects of the peNDF content of TMR on feed intake, digestibility, and chewing activity. Twelve heifers were assigned to treatments in replicated 3×3 Latin square design. Animals were housed in individual tie stalls and were fed TMR once a day at 0800 hours. This encouraged *ad libitum* intake. Animals had free access to fresh water and mineral blocks at all times. Each period consisted of 14 days of adaption to diets and 7 days of experimental measurements. The amount of feed offered and refused were measured and recorded daily during the last 7-d of the period to calculate feed intake and apparent nutrient digestibility. All animal-based procedures were approved by the Institutional Animal Care and Use Committee at Konkuk University (KU11058).

Each heifer was offered one of three diets based on same chemical formula but differing in peNDF content. The three dietary peNDF levels were obtained by different mixing times of TMR diets: 3 min (T1), 10 min (T2), and 25 min (T3). The experimental TMR diets which were mixed with perennial ryegrass hay (10% fresh matter [FM]), annual ryegrass hay (10% FM), tall fescue hay (5% FM), and other feedstuffs (75% FM) including corn gluten feed, wheat husk, ground corn grain, corn flake, cotton seed, and etc. were produced with the commercial formula for beef cattle recommended by the Institute of Bio Feed (Seoul, Korea). The chemical composition of the experimental diets is shown in Table 1.

Feed samples were collected once a week, and refusals were collected daily and DM intake determined. The collected samples were dried in an oven at 65°C for 48 h and ground through a 1-mm mesh screen (Model 4, Thomas Scientific, Swedesboro, NJ, USA) for analysis of chemical composition (AOAC, 1990). Particle size distributions for experimental TMR diet were determined using the PSPS (Lammers et al., 1996) containing three sieves (19, 8, and 1.18 mm) and a pan, and physically effectiveness factors (pef) were calculated as the sum of the DM portion retained on three sieves (19, 8, and 1.18 mm). The peNDF$_{1.18}$ content of TMR was calculated by multiplying NDF content of the TMR by pef$_{1.18}$ (Kononoff and Heinrichs, 2003a). The proportion of sample DM collected in a ≥1.18 mm sieve is commonly used as the physical effectiveness factor in the equation (Mertens, 1988).

Table 1. Chemical composition and particle size distribution of the total mixed ration (TMR) measured using the Penn State Particle Separator

Item	peNDF content[1]		
	T1	T2	T3
Chemical composition**			
DM (%)	85.62±0.14	86.44±0.18	86.58±0.24
CP (% DM)	13.38±0.37	13.64±0.26	13.48±0.26
EE (% DM)	3.97±0.05	3.90±0.06	3.69±0.13
Ash (% DM)	10.18±0.06	10.37±0.08	10.46±0.12
CF (% DM)	16.66±0.26	16.96±0.32	16.53±0.26
NDF (% DM)	40.54±0.41	40.32±0.06	40.47±0.27
ADF (% DM)	28.50±0.31	29.31±0.08	28.02±0.51
Particle size distribution (% DM)			
19 mm	14.15±1.18^{aD}	5.81±0.97^{bC}	1.81±0.97^{cD}
8 mm	22.09±1.57^{bC}	27.43±1.95^{aB}	23.81±3.67^{abC}
1.18 mm	38.66±1.73^{bA}	39.39±0.95^{bA}	42.28±1.97^{aA}
Pan	25.10±0.89^{bB}	27.59±1.56^{bB}	32.09±1.56^{aB}
peNDF (%)			
pef$_{>1.18}$	74.90±0.89^{a}	72.41±1.96^{a}	67.96±1.55^{b}
peNDF$_{>1.18}$	30.36±0.31^{a}	29.20±0.04^{b}	27.50±0.18^{c}

DM, dry matter; CP, crude protein; EE, ether extract; CF, crude fiber; NDF, neutral detergent fiber; ADF, acid detergent fiber; peNDF, physically effective neutral detergent fiber.
[1] T1: 3 min, T2: 10min, and T3: 25 min.
[a,b,c] Means with different superscript in the same row are different (p<0.05).
[A,B,C,D] Means with different superscript in the same column are different (p<0.05).

Table 2. Effects of peNDF content DM intake, feces excretion, average daily gain, and feed conversion ratio in fattening heifers

Item	peNDF content[1]		
	T1	T2	T3
DM intake (DMI, kg/d)	10.96±0.66[NS]	11.34±1.20	11.35±1.11
peNDF$_{>1.18}$ intake (kg/d)	3.32±0.32[NS]	3.31±0.67	3.08±0.32
NDF intake (kg/d)	4.44±0.26[NS]	4.57±0.48	4.54±0.44
Fecal excretion (kg/d)	3.63±0.32[NS]	3.87±0.94	4.01±0.43
Average daily gain (ADG, kg/d)	0.90±0.67[a]	0.80±0.39[b]	0.81±0.42[b]
Feed conversion ratio DMI/ADG (g/g)	12.18±5.85[NS]	14.17±5.79	14.01±8.07

peNDF, physically effective neutral detergent fiber; DM, dry matter; NS, not significant.
[1] T1: 3 min, T2: 10min, and T3: 25 min.
[a,b] Means with different superscript in the same row are different (p<0.05).

Apparent digestion and chewing activity

Total fecal collections were made for the apparent digestibility between days 15 and 21 of each period. Fecal sample of each heifer was collected immediately after every excretion, and each sample was mixed daily and then divided into three subsamples. One subsample from each sample used to form a combined sample and then dried at 65°C and ground before chemical analysis and digestion coefficient calculation.

Chewing activities were monitored continuously for 48 h. Six digital video cameras (SDC-435, Samsung, Seoul, Korea) and two multiplexers (PDR-XM3004, Egpis, Goyang, Korea) were used to monitor the behavior of each heifer. Cameras were carefully arranged to avoid blind spots.

Statistical analysis

The main effects of treatments were subjected to one-way analysis of variance using the general linear model procedure of SAS (Version 9.2, SAS Institute, Cary, NC, USA). When a statistically significant difference was recognized in the analysis, Duncan's multiple range test was used to detect statistical significance (p<0.05) among treatment groups.

RESULTS

Chemical composition, particle size distribution, and peNDF content

The chemical composition of the experimental diets was similar for high (T1), medium (T2), and low (T3) peNDF groups, because they had the same formulation and ingredients (Table 1), with only different mixing times. However, particle distribution was largely affected by mixing time. The proportion of particles retained on the 19-mm sieve decreased with increasing mixing time of TMR (T1: 14.15%, T2: 5.18%, and T3: 1.81%) and those retained on the 1.18-mm and pan sieves and the pan were significantly increased (T1: 38.66% and 25.10%, T2: 39.39% and 27.59%, and T3: 42.28% and 32.09%) (p<0.05). The peNDF$_{>1.18}$ content of TMR decreased (p<0.05) linearly with increasing mixing time of TMR. The peNDF$_{>1.18}$ content of the experimental

diets was 30.36%, 29.20% and 27.19% for T1, T2 and T3, respectively (p<0.05). The peNDF content calculated using actual chewing time (Mertens, 2002) was similar to these results.

Intake, average daily gain, and digestibility

The DM intake was 10.96±0.66, 11.34±1.20, and 11.35±1.11 kg/d for T1, T2, and T3, respectively, and there was no significant difference. The NDF and peNDF intakes were not affected by the peNDF content of the experimental diets (Table 2). Fecal excretion was 3.63±0.32, 3.87±0.94, and 4.01±0.43 kg/d for T1, T2, and T3, respectively, and there was no significantly difference.

Average daily gain (ADG) with T1 was significantly higher (p<0.05) than with T2 and T3. Feed conversion ratio was decreased by increasing the peNDF content (12.18±5.85, 14.17±5.79, and 14.01±8.07 in T1, T2, and T3, respectively).

The apparent digestibility of DM was significantly higher (p<0.05) in T1 than in T2 and T3 (Table 3). Nutrient digestibility except for the ether extract was significantly affected (p<0.05) by peNDF content. The apparent digestibility of crude protein (CP), crude fiber (CF), NDF, and acid detergent fiber (ADF) in T1 and T2 was significantly higher (p<0.05) than in T3.

Table 3. Effects of peNDF content on nutrient digestibility in fattening heifer

Item	peNDF content[1]		
	T1	T2	T3
DM (%)	66.85±0.02[a]	65.56±0.09[b]	64.42±0.04[b]
CP (% DM)	62.21±0.02[a]	62.40±0.09[a]	59.98±0.05[b]
EE (% DM)	90.59±0.01[a]	90.11±0.03[a]	90.45±0.02[a]
CF (% DM)	53.53±0.03[a]	52.69±0.11[a]	48.57±0.07[b]
NDF (% DM)	44.39±1.19[a]	43.79±1.33[a]	42.11±2.18[b]
ADF (% DM)	36.50±1.90[a]	35.70±2.40[a]	31.40±2.53[b]

peNDF, physically effective neutral detergent fiber; DM, dry matter; CP, crude protein; EE, ether extract; CF, crude fiber; NDF, neutral detergent fiber; ADF, acid detergent fiber.
[1] T1: 3 min, T2: 10 min, and T3: 25 min.
[a,b] Means with different superscript in the same row are different (p<0.05).

Chewing activities

The number of eating, rumination, and total chews increased linearly with increasing peNDF content (p<0.05) (Table 4). Ruminating time and total chewing time were also significantly affected (p<0.05) by the peNDF content of TMR. However, the eating time was not affected by peNDF content in T2 and T3. Eating rate, ruminating efficiency and chewing efficiency were significantly decreased (p<0.05) with an increase in the peNDF content.

DISCUSSION

The PSPS is based on the properties of Standard S424 of the American Society of Agricultural Engineers (ASAE, 1998) and has been proven to generate similar results to the vertical oscillating sieving method (Lammers et al., 1996; Buckmaster, 2000; Teimouri Yansari et al., 2004). It is generally understood by nutritionists that the physical characteristics of the diet, including forage chopping length, are important elements for reducing the risk of acidosis. In this study, the peNDF content of the TMR diets was reduced with increased mixing times of TMR. The peNDF$_{>1.18}$ values (30.36%, 29.20%, and 27.50% for T1, T2, and T3) of the TMR in the present study were higher than the recommended minimum ranges (19% to 21% and 18.9% to 21.2%) from Mertens (1997) and Einarson et al. (2004), probably because of a higher NDF content (above 40%) and the formulation based on various hays in this study. In addition, when measured using the PSPS with three sieves (19, 8, and 1.18 mm), the proportion of peNDF primarily depends on the mixing time of TMR and its particle size because the proportion of T3 was greatly decreased compared to other diets at the 19-mm sieve.

In this study, peNDF content did not affect DM intake and neutral detergent fiber intake. Based on the results from a previous study (Yang and Beauchemin, 2006), dairy cows may intentionally select long feed particles when ruminal pH is low due to low intake of peNDF. Other studies have reported no effects on DM intake and NDF intake from reduced peNDF (Soita et al., 2000; Beauchemin et al., 2003; Calberry et al., 2003). The DM intake can be regulated by many factors such as physical characteristics and ingredient and nutrient composition (Allen, 1997), and an interaction between dietary peNDF content and the forage-to-concentrate ratio may affect DM intake. For example, when high concentrate diets are fed to lactating dairy cows, a metabolic, rather than a physical factor places a constraint on feed intake. A reduction of peNDF content resulting in a higher particulate passage rate from the rumen would allow for a greater feed intake (Allen, 1997). A small increase in feed intake due to low peNDF content was also observed in cows fed a high concentrate diet (58%) when forage was coarse (Theoretical cut length = 19 mm; Einarson et al., 2004).

In this study, ADG of T1 was significantly higher than that of T2 and T3 (p<0.05). Feed conversion ratio was decreased by increasing the peNDF content (12.18, 14.17, and 14.01 g/g). The effects of peNDF content on body weight and feed conversion ratio have been substantially investigated, but there is limited information on the effects of peNDF content in TMR.

The apparent digestibility of DM, CP, CF, NDF and ADF was linearly increased with increasing dietary peNDF content (Table 3), and these results accord with some previous reports (Yang et al., 2002; Schwab et al., 2003; Kononoff and Heinrichs, 2003b). The difference in response between our study and that of Kononoff and Heinrichs (2003b) was probably related to the differences in peNDF

Table 4. Effects of peNDF content on chewing activity of fattening heifers

Item	peNDF content[1]		
	T1	T2	T3
	-- No --		
Total chews	35,361.92±1,059.21[a]	31,454.25±2,834.68[b]	27,980.67±1,433.12[c]
Eating	16,279.00±712.50[a]	13,467.83±1,933.71[b]	11,734.00±1,289.65[c]
Ruminating	19,082.92±554.29[a]	17,986.42±948.60[b]	16,246.67±579.72[c]
	-- min --		
Total chewing time	637.75±12.72[a]	586.83±39.35[b]	551.42±13.65[c]
Eating time	238.58±9.26[a]	213.25±23.14[b]	213.08±9.61[b]
Ruminating time	399.17±5.77[a]	373.58±17.19[b]	338.33±12.45[c]
	-- kg/h --		
Eating rate[2]	2.76±0.24[b]	3.55±0.64[a]	3.55±0.28[a]
Ruminating efficiency[3]	1.65±0.17[b]	1.83±0.40[b]	2.01±0.23[a]
Chewing efficiency[4]	1.03±0.09[b]	1.16±0.22[a]	1.23±0.11[a]

peNDF, physically effective neutral detergent fiber.
[1] T1: 3 min, T2: 10min, and T3: 25 min.
[2] Eating rate = intake (kg/d)/eating time (h/d). [3] Ruminating efficiency = intake (kg/d)/ruminating (h/d).
[4] Chewing efficiency = intake (kg/d)/total chewing.
[a,b,c] Means with different superscript in the same row are different (p<0.05).

contents of the TMR (range of 15.6% to 18.4% vs 30.36% to 27.50%). Whether forage particle length affects fiber digestibility appears to depend upon the concentration of peNDF in the diet. For example, fiber digestibility was observed to be reduced with smaller forage particle length of alfalfa silage when peNDF content was decreased to a minimum of 7.2% (Yang et al., 2002), but the effect was not observed when the peNDF content was 9.5% (Yang et al., 2001).

Reducing the peNDF content of TMR increases surface area available for microbial attack, thereby accelerating digestion, but fine particles often have a faster passage rate from the rumen such that digestibility is reduced. In addition, long particles are reduced in size during ingestive mastication, thereby eliminating the effects of particle length. Michalet-Doreau et al. (2002) reported that mean particle length markedly decreased after ingestive mastication.

Eating time in this experiment gradually declined with decreased peNDF content. Beauchemin et al. (2003) found that, although eating time (min/kg of DM) was similar, the number of chews per kilogram of DM intake during the eating period increased linearly (p<0.05) with increasing dietary peNDF; that is, the chewing rate increased. The authors suggested that cows could efficiently reduce long particles by increasing their chewing rate during the eating period. For example, Brouk and Belyea (1993) observed that chewing time was similar, but chewing rate was higher for cows fed long chopped alfalfa hay, and other researchers have reported both increased chewing rate and chewing time with increasing forage particle size (Kononoff and Heinrichs, 2003a). Therefore, cows likely increase their chewing activity through either chewing rate, chewing time, or both in relation to increased peNDF intake (Yang et al., 2006). In the present study, increased total chewing time with higher dietary peNDF content resulted in increased rumination time rather than increased eating time.

CONCLUSION

Physical characteristics of diets are critical for obtaining proper chewing activity as well as for animal production. In this study, reducing the peNDF content decreased total chewing activity, ruminating time, and total chewing time of fattening heifers. Average daily gain in T1 group was significantly higher than in T2 and T3 groups. Thus level of peNDF greater than 30% may be better for the optimal performance of fattening heifers. Feed conversion ratio was decreased by increasing the peNDF content. Digestibility in the total tract, especially for DM, CP, CF, NDF, and ADF was decreased with decreasing peNDF content. Interestingly, increased content of peNDF did not affect the DM intake significantly, but it did increase the digestibility. Therefore,

the PSPS is the most useful tool that can be used on the farm to evaluate fiber adequacy in forages and TMR. The values obtained can be used to determine the physical effectiveness of fiber, which is a good indicator of the ruminal potential of the feed.

ACKNOWLEDGMENTS

This work was carried out with support from the "Cooperative Research Program for Agriculture Science & Technology Development (project No. PJ011999)", Rural Development Administration, Republic of Korea.

REFERENCES

Allen, M. S. 1997. Relationship between fermentation acid production in the rumen and the requirement for physically effective fiber. J. Dairy Sci. 80:1447-1462.

AOAC (Association of Official Analytical Chemists) International. 1990. Official Methods of Analysis. 15th edn. Association of Official Analytical Chemists, Washington, DC, USA.

ASAE (American Society of Agricultural Engineers). 1998. S424: Method of Determining and Expressing Particle size of Chopped Forage Materials by Sieving Screening. ASAE Standards. Am. Soc. Agric. Eng., St. Joseph, MI, USA.

Beauchemin, K. A., W. Z. Yang, and L. M. Rode. 2003. Effects of particle size of alfalfa based dairy cow diets on chewing activity, ruminal fermentation, and milk production. J. Dairy Sci. 86:630-643.

Buckmaster, D. M. 2000. Particle size in dairy cows. In: Recent Advances in Animal Nutrition (Eds. P. C. Garnsworthy and J. Wiseman). Nottingham University Press, Nottingham, UK. pp. 109-128.

Brouk, M. and R. Belyea. 1993. Chewing activity and digestive responses of cows fed alfalfa forages. J. Dairy Sci. 76:175-182

Calberry, J. M., J. C. Plaizier, M. S. Einarson, and B. W. McBride. 2003. Effects of replacing chopped alfalfa hay with alfalfa silage in a total mixed ration on production and rumen conditions of lactating dairy cows. J. Dairy Sci. 86:3611-3619.

Einarson, M. S, J. C. Plaizier, and K. M. Wittenberg. 2004. Effects of barley silage chop length on productivity and rumen conditions of lactating dairy cows fed a total mixed ration. J. Dairy Sci. 87:2987-2996.

Grant, R. J. and V. F. Colenbrander. 1990. Milk fate depression in dairy cow: role of silage particle size. J. Dairy Sci. 73:1834-1842.

Kononoff, P. J. and A. J. Heinrichs. 2003a. The effect of reducing alfalfa haylage particle size on cows in early lactation. J. Dairy Sci. 86:1445-1457.

Kononoff, P. J. and A. J. Heinrichs. 2003b. The effect of corn silage particle size and cottonseed hulls on cows in early lactation. J. Dairy Sci. 86:2438-2451.

Lammers, B. P., D. R. Buckmaster, and A. J. Heinrichs. 1996. A simple method for the analysis of particle sizes of forage and total mixed rations. J. Dairy Sci. 79:922-928.

Mertens, D. R. 1988. Balancing carbohydrates in dairy rations. In: Proceedings of the Large Herd Dairy Management Conference, Syracuse, NY, USA. 150 p.

Mertens, D. R. 1997. Creating a system for meeting the fiber requirements of dairy cows. J. Dairy Sci. 80:1463-1481.

Mertens, D. R. 2002. Measuring fiber and its effectiveness in ruminant diets. Proceeding Plains Nutrition Council Spring Conference. San Antonio, TX, USA. pp. 40-66.

Michalet-Doreau, B., I. Fernandez, and G. Fonty. 2002. A comparison of enzymatic and molecular approaches to characterize the cellulolytic microbial ecosystems of the rumen and the cecum. J. Anim. Sci. 80:790-796.

Poppi, D. P., D. J. Minson, and J. H. Ternouth. 1981. Studies of cattle and sheep eating leaf and stem fractions of grasses. I. The voluntary intake, digestibility and retention time in the reticulo-rumen. Crop. Pasture Sci. 32:99-108.

Soita, H. W., D. A. Christensen, and J. J. McKinnon. 2000. Influence of particle size on the effectiveness of the fiber in barley silage. J. Dairy Sci. 83:2295-2300.

Schwab, E. C., R. D. Shaver, J. G. Lauer, and J. G. Coors. 2003. Estimating silage energy value and milk yield to rank corn hybrids. Anim. Feed Sci. Technol. 109:1-18.

Teimouri Yansari, A., R. Valizadeh, A. Naserian, D. A. Christensen, P. Yu, and F. Eftekhari Shahroodi. 2004. Effects of alfalfa particle size and specific gravity on chewing activity, digestibility, and performance of Holstein dairy cows. J. Dairy Sci. 87:3912-3924.

Yang, W. Z. and K. A. Beauchemin. 2006. Effects of physically effective fiber on chewing activity and ruminal pH of dairy cows fed diets based on barley silage. J. Dairy Sci. 89:217-228.

Yang, W. Z., K. A. Beauchemin, and L. M. Rode. 2001. Effects of grain processing, forage to concentrate ratio, and forage particle size on rumen pH and digestion by dairy cows. J. Dairy Sci. 84:2203-2216

Yang, W. Z., K. A. Beauchemin and L. M. Rode. 2002. Effects of particle size of alfalfa-based dairy cow diets on site and extent of digestion. J. Dairy Sci. 85:1958-1968.

28

Dietary Niacin Supplementation Suppressed Hepatic Lipid Accumulation in Rabbits

Lei Liu, Chunyan Li, Chunyan Fu, and Fuchang Li*

Department of Animal Science, Shandong Agricultural University, Taian, Shandong 271018, China

ABSTRACT: An experiment was conducted to investigate the effect of niacin supplementation on hepatic lipid metabolism in rabbits. Rex Rabbits (90 d, n = 32) were allocated to two equal treatment groups: Fed basal diet (control) or fed basal diet with additional 200 mg/kg niacin supplementation (niacin). The results show that niacin significantly increased the levels of plasma adiponectin, hepatic apoprotein B and hepatic leptin receptors mRNA (p<0.05), but significantly decreased the hepatic fatty acid synthase activity and adiponectin receptor 2, insulin receptor and acetyl-CoA carboxylase mRNA levels (p<0.05). Plasma insulin had a decreasing tendency in the niacin treatment group compared with control (p = 0.067). Plasma very low density lipoproteins, leptin levels and the hepatic adiponectin receptor 1 and carnitine palmitoyl transferase 1 genes expression were not significantly altered with niacin addition to the diet (p>0.05). However, niacin treatment significantly inhibited the hepatocytes lipid accumulation compared with the control group (p<0.05). In conclusion, niacin treatment can decrease hepatic fatty acids synthesis, but does not alter fatty acids oxidation and triacylglycerol export. And this whole process attenuates lipid accumulation in liver. Besides, the hormones of insulin, leptin and adiponectin are associated with the regulation of niacin in hepatic lipid metabolism in rabbits. (**Key Words:** Niacin, Liver, Lipid Metabolism, Rabbits)

INTRODUCTION

Liver plays a key role in lipid metabolism and is the hub of fatty acid synthesis and lipid circulation through lipoprotein synthesis in mammals. Triacylglycerols may accumulate in hepatocytes or are exported as constituents of very low density lipoproteins (VLDL) (Nguyen et al., 2008). The excessive accumulation of lipid droplets into the hepatocytes can result in hepatic steatosis. The triacylglycerol content of hepatocytes is regulated by the activity of cellular molecules that facilitates hepatic fatty acid synthesis and esterification ('input'), and hepatic fatty acid oxidation and triacylglycerol export ('output'). The genes encoding levels of Acetyl-CoA carboxylase (ACC), fatty acid synthase (FAS) and carnitine palmitoyltransferase (CPT) 1 were respectively involved in the regulation of FAS and oxidation. Besides, the lipid metabolism in liver in mammals can be regulated by hormones (e.g., insulin, leptin, and adiponectin) (Saltiel and Kahn, 2001; Berg et al., 2002; Gallardo et al., 2007; Peng et al., 2009) and nutritional state (carbohydrates, B vitamins) (Havel, 2004; Ganji et al., 2015).

Emerging findings indicate that niacin decreases plasma VLDL level by inhibiting hepatocyte diacylglycerol acyltransferase 2 (Kamanna et al., 2013). Lamon-Fava et al. (2008) found that niacin treatment significantly increased high density lipoprotein (HDL) apo-protein A-I (ApoA-I) concentrations and production, as well as enhanced the clearance of triglyceride-rich lipoproteins. G-protein-coupled receptor 109 A (GPR109A) was identified as receptors for niacin (Wise et al., 2003). Hepatic-specific over-expression of GPR109A in mice reduced plasma HDL cholesterol accompanied by reduction in hepatic cholesterol efflux to ApoA-I (Li et al., 2010). Generally, the studies of niacin were focused on relative lipid regulation in plasma for the treatment of dyslipidemia and atherosclerosis. But the effect of niacin on hepatic fatty acid metabolism is still unclear.

* Corresponding Author: Fuchang Li.
E-mail: chlf@sdau.edu.cn

Obesity and obesity-associated fatty liver disease are becoming global health problems in adults as well as children (Browning et al., 2004; Hamaguchi et al., 2005). Hepatic steatosis is encountered in about 20% to 35% of the general adult population in the United States (Reddy and Rao, 2006). The prevention or treatment of hepatic steatosis is difficult to implement, especially by nutrition intervention methods. In the present study, we investigated the effect of dietary niacin addition on hepatic fatty acid metabolism in rabbits, and determined the possibility of niacin attenuates hepatic lipid accumulation.

MATERIALS AND METHODS

Animals

Three-month-old Rex-rabbits were individually housed in locally made cages ($60\times40\times40$ cm). Temperature and lighting were maintained according to commercial conditions. The diets were formulated according to the values for growing rabbits from NRC (1977) and de Blas and Mateos (1998). The diets were pressure pelleted with the diameter of the pellets being 4 mm. All rabbits received a starter diet containing 16% crude protein, 14% crude fiber and 11 MJ/kg of digestible energy. All rabbits had free access to food and water during the rearing period. This study was approved by the Shandong Agricultural University and conducted in accordance with the "Guidelines for Experimental Animals" of the Ministry of Science and Technology (Beijing, China).

Experimental protocol and sample collection

At 90 days of age, 32 rabbits of similar body weight ($1,860\pm100$ g) were divided into 2 groups (16 replicates per group and 1 rabbit per replicate): Fed basal diet (control, measured niacin content was 28 mg/kg) or fed basal diet with 200 mg/kg niacin addition (niacin, measured niacin content was 230 mg/kg). The experiment lasted for 8 wks which included 1 wk adaptation period and 7 wks experimental period. At the end of the trial, 16 rabbits (8 rabbits per group, 4 male and 4 female, with the average body weight of the 8 rabbits equal to the average body weight of entire treatment group) were electrically stunned (70 V, pulsed direct current, 50 Hz for 5 s), and 10 mL blood was collected immediately from the heart. Plasma was obtained following centrifugation at 400 g for 10 min at 4°C and stored at −20°C for subsequent analysis. The rabbits were sacrificed by cervical dislocation, and the liver was collected. After being snap-frozen in liquid nitrogen, the liver tissue samples were stored at −80°C.

Measurements

Plasma VLDL concentration was determined using the method described by Barter and Lally (1978). Plasma adiponectin and leptin concentrations were determined using a validated sandwich enzyme-linked immunosorbent assay (Uscn Life Science Inc., Wuhan, China) with an adiponectin- or leptin-specific antibody. Intra-assay coefficient of variation of adiponectin was 2.3%, and leptin was 3.0%.

Plasma insulin was measured by radioimmunoassay with a guinea pig anti-porcine insulin serum (3 V Bio-engineering group Co., Weifang, China). In this procedure, ^{125}I-labelled porcine insulin competes with chicken insulin for sites on anti-porcine insulin antibodies that are immobilised on the wall of a polypropylene tube. Significant cross-reactivity has been observed between chicken insulin and guinea pig anti-porcine sera (Simon et al., 1974). The insulin in this study was referred to as immunoreactive insulin. The sensitivity of the assay was 1 μIU/mL, and all samples were included in the same assay to avoid inter-assay variability. The intra-assay coefficient of variation was 2.01%.

Liver samples were ground with saline (saline volume/sample weight = 4:1) to measure apo-protein B (Apo B) concentration and FAS activity. Apo B concentrations in liver were measured using immunoturbidimetric assays. The FAS activity in liver was measured using ultraviolet chromatometry. All the commercial diagnostic kits were from Jiancheng Bioengineering Institute (Nanjing, China).

The accumulation of cytoplasmic lipid droplets was visualised by Oil Red O staining according to the protocol of Lillie and Fullmer (1976). Briefly, tissues were immediately frozen in liquid nitrogen and cut with a Leica CM-1850 cryostat microtome (Leica, Wetzlar, Germany). Then 16-mm thick sections were fixed in 4% formaldehyde for 10 min and stained with filtered 0.5% Oil Red O (Sigma-Aldrich, St Louis, MO, USA), which was made by dissolving in isopropyl alcohol for 15 min at room temperature. Morphometric analysis was performed on 10 randomly chosen fields containing transverse sections of liver from each rabbit. The selected fields were photographed using an Olympus CX41 phase contrast microscope (Olympus, Tokyo, Japan). The volume density of each Oil Red O positive liver tissue was determined using the point-counting method described by Weibel and Bolender (1973).

Total RNA extraction and quantitative real-time polymerase chain reaction (PCR) were performed as described previously (Li et al., 2014). Sequences of primers are shown in Table 1. The PCR data were analyzed with the $2^{-\Delta\Delta CT}$ method (Livak and Schmittgen, 2001). The mRNA levels of target genes were normalised to glyceraldehyde 3-phosphate dehydrogenase (GAPDH) mRNA (ΔCT). On the

Table 1. Gene-specific primers used for the analysis of rabbit gene expression

Gene	GenBank accession no.	Primer sequences (5'-3')	Product size (bp)
Adpn R1	AM886136	F: CGGCTCATCTACCTCTCCAT R: ACACACCTGCTCTTGTCTGC	109
Adpn R2	AM886135	F: CTGGCTCAAGGATAACGACTT R: AATGTTGCCTGTCTCTGTGTG	109
INSR	XM_008249177	F: CGCTACCAATCCTTCTGTCC R: TAGTGCGTGATGTTGCCATT	111
LEPR	XM_008265107	F:AAGAACAGAGATGAGGTGGTGC R:CCAGTGTGGCGTATTTCACG	187
ACC	XM_008271160	F: TGGCTGTATCCATTATGTCAAGCG R: TGAAGAAAGGGTCAGGAAGGCAGTA	235
CPT1	XM_002724092	F: AGGTGCTCCTCTCCTACCACGG R: GTTGCTGTTCACCATCAGTGGC	379
GAPDH	NM_001082253	F: TGCCACCCACTCCTCTACCTTCG R: CCGGTGGTTTGAGGGCTCTTACT	163

Adpn R, adiponectin receptor; *INSR*, insulin receptor; *LEPR*, leptin receptor; *ACC*, Acetyl-CoA carboxylase; *CPT1*, carnitine palmitoyltransferase 1; *GAPDH*, glyceraldehyde 3-phosphate dehydrogenase.

basis of the Ct values, GAPDH mRNA expression was stable across the treatments in this study ($p>0.1$).

Statistical analysis

The data are presented as the means±standard error of the mean, with n = 8 for plasma metabolites, hormones, hepatic FAS activity, Oil Red O staining and all mRNA levels analysis. All of the data collected were subjected to one-way analysis of variance analysis with the Statistical Analysis Systems statistical software package (Version 8e, SAS Institute, Cary, NC, USA), and the primary effect of niacin treatment was evaluated. Homogeneity of variances among groups was confirmed using Bartlett's test (SAS Institute, USA). When the primary effect of treatment was significant, differences between means were assessed by unpaired T-Test. $p<0.05$ was statistically significant, and $0.05<p<0.1$ was changed tendency.

RESULTS

The relative hormones and VLDL concentration in plasma were determined, and the results are shown in Figure 1. Although no significant difference was observed in plasma levels of VLDL (Figure 1A) and leptin (Figure 1B) in niacin group rabbits compared with the control, plasma adiponectin was significantly increased with 200 mg/kg niacin addition ($p<0.05$, Figure 1C). In addition, plasma insulin had a decreasing tendency in the niacin treatment group compared with the control group (p = 0.067, Figure 1D).

Though determining the relative genes expression or enzyme activity-related lipid metabolism, we found that dietary addition of niacin significantly decreased the FAS activity (Figure 2A) and adiponectin receptor 2 (adpn R2)

(Figure 2B), insulin receptor (INSR) (Figure 2C) and ACC (Figure 2D) mRNA levels ($p<0.05$) but increased the Apo B (Figure 2E) content and leptin receptor (LEPR) gene expression (Figure 2F) ($p<0.05$). Besides, the *adpn R1* (Figure 2G) and *CPT1* (Figure 2H) genes expression were not significantly altered when niacin addition in diet.

From Figure 3, we can determine that dietary niacin treatment had a significant inhibition on cytoplasmic lipid accumulation compared with control.

DISCUSSION

Dietary niacin can regulate lipid metabolism in kidney, adipose tissue and lipid transportation in plasma (Rubic et al., 2004; Cho et al., 2010; Fabbrini et al., 2010). Recent researches show that niacin increases HDL by reducing hepatic expression cholesteryl transfer protein and reducing the uptake of HDL (Jin et al., 1997; van der Hoorn et al., 2008). However, the effect of niacin on fatty acids metabolism in liver is still unclear. Our study shows that dietary niacin supplementation inhibited hepatic fatty acids synthesis and lipid accumulation.

Lipid metabolism in liver is key for energy homeostasis in rabbits. Fatty acids synthesis in liver is tightly controlled by nutritional conditions (Asai and Miyazawa, 2001). In our study, the *ACC* gene and FAS activity were significantly decreased in liver after niacin supplementation, which suggest that the hepatic fatty acids synthesis process is significantly inhibited by niacin. The results are consistent with the study in rats, as niacin administration relieved the increased ACC and FAS protein due to chronic renal failure in kidney (Cho et al., 2010). But the niacin treatment didn't change the hepatic *CPT1* gene expression in normal rabbit liver or human primary hepatocytes (Carling et al., 2008),

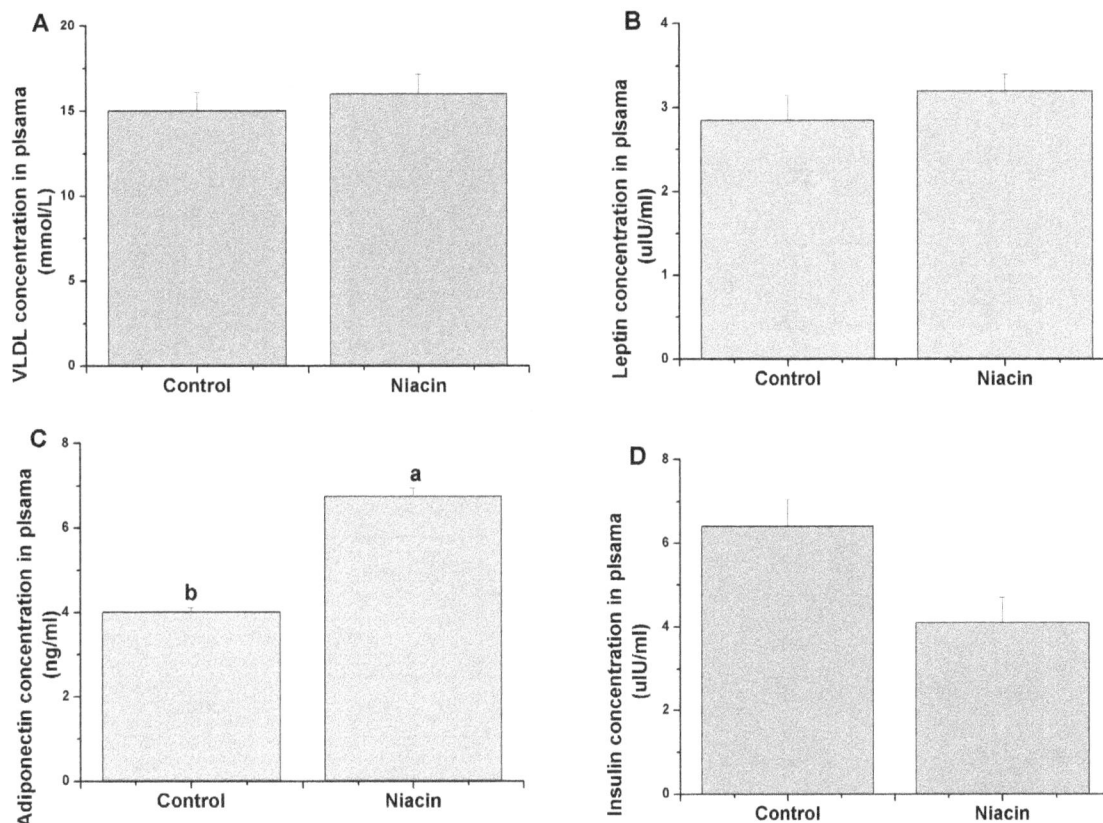

Figure 1. Effects of dietary niacin on plasma very low density lipoproteins (VLDL) (A), leptin (B), adiponectin (C), and insulin (D) concentration. Values are shown as the mean±standard error of the mean. [a,b] Means with different superscripts are significantly different (p<0.05).

which is the rate-limiting enzyme of fatty acid catabolic pathway. The results suggest that the regulation of niacin on fatty acid metabolism is only on the anabolic pathway, but not on the catabolic pathway.

Hepatic lipid components that are synthesized in smooth endoplasmic reticulum, are glycosylated in golgi apparatus, and released into blood in the form of VLDL. Apo B is the key component whose rate of synthesis in the rough endoplasmic reticulum controls the overall rate of VLDL production (Cruz-Bautista et al., 2015). Results from liver slice studies has suggested that species with limited hepatic lipogenesis have less ability to secrete VLDL from the liver compared with species in which the liver is a major or moderate source of lipogenesis (Pullen et al., 1990). But the plasma VLDL concentration was not altered after niacin administration in our experiment. The major reason might be related to the increased hepatic Apo B level. Although there was less lipogenesis, niacin can stimulate the other synthetic component (Apo B) production. Which can result in a unaltered VLDL level.

Hepatic fatty acids metabolism is also affected by hormones (e.g., insulin, leptin, and adiponectin). Many studies have found the transcription rate of the *FAS* gene is

quickly increased after insulin administration in diabetic mouse and rats (Paulauskis and Sul, 1988; Ruzzin et al., 2010). Besides, insulin can also inhibit Apo B secretion in isolated human hepatocytes (Salhanick et al., 1991). These effects are achieved by the binding of insulin to the INSR at the cell surface (Nakae and Accili, 1999). In the present experiment, the decreased tendency of plasma insulin and significant inhibition of hepatic *INSR* gene in niacin group suggest that the insulin signal is associated with the regulation of niacin on hepatic lipid metabolism. But these results are inconsistent with a previous observation, which found that niacin had no effect on insulin action in human (Fabbrini et al., 2010). Perhaps the effect of niacin on insulin can be species-dependent.

Leptin is another hormone that may be involved in lipogenesis. There is a growing consensus that leptin limits fat storage not only by inhibiting food intake but also by affecting specific metabolic pathways in liver and adipose (Anania, 2002; Kitamura et al., 2006). The previous studies show that leptin can stimulate fatty acid oxidation and inhibit lipogenesis by down-regulating the expression of genes involved in fatty acid and triglyceride synthesis in liver or adipose (Bai et al., 1996; Matsusue et al., 2003;

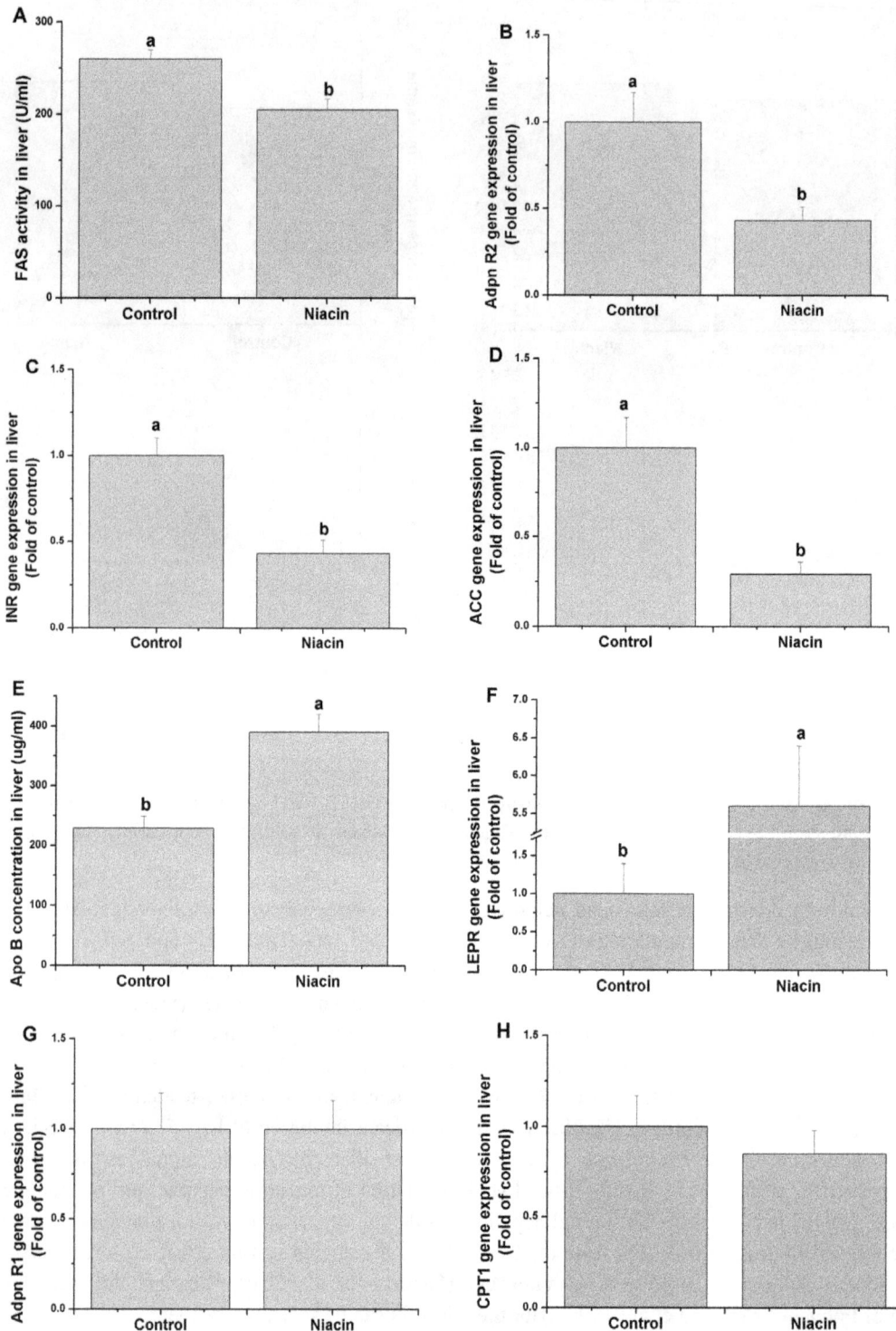

Figure 2. Effects of dietary niacin on hepatic fatty acid synthase (FAS) activity (A), adiponectin receptor 2 (*Adpn R2*) gene expression (B) insulin receptor (*INSR*) gene expression (C), acetyl-CoA carboxylase (*ACC*) gene expression (D), Apo B concentration (E), leptin receptor (*LEPR*) gene expression (F), adiponectin receptor 1 (*Adpn R1*) gene expression (G) and carnitine palmitoyltransferase (*CPT*) 1 gene expression (H). Values are shown as the mean±standard error of the mean. [a,b] Means with different superscripts are significantly different ($p<0.05$).

Buettner et al., 2008). Although the plasma leptin concentration was not affected by the niacin, the hepatic LEPR was increased 5-fold than control. The result shows that the hepatic leptin signal is involved to the niacin regulation on lipid metabolism.

Adiponectin, a protein exclusively secreted from

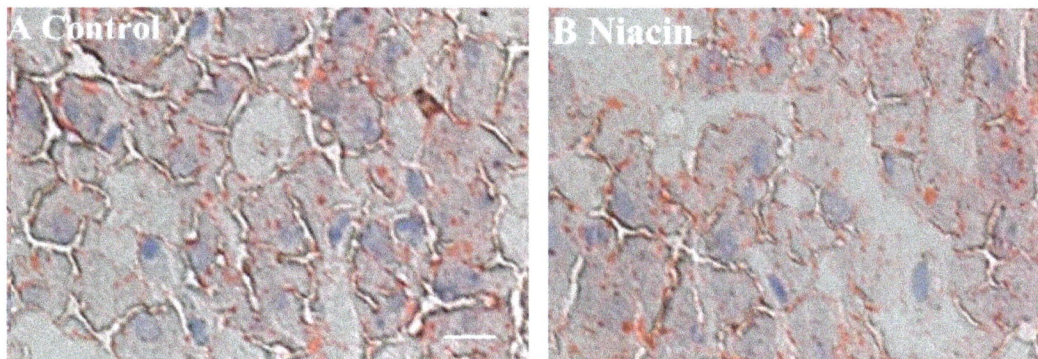

Figure 3. Oil Red O staining of liver for cytoplasmic lipid droplets (indicated by arrows) showing effect of dietary niacin treatment on lipid accumulation in hepatocyte (A: control, B: niacin) from rabbit. The scale bar in (A) represents 50 mm.

adipose tissue, and its agonists could be promising candidates for the treatment of obesity-associated metabolic syndromes (Berg et al., 2002). Recent studies suggest that adiponectin can act directly on hepatic lipid metabolism (Li et al., 2014; Lin et al., 2015). Adiponectin can regulate hepatic fatty acid metabolism *in vivo* by attenuating the activities of FAS and ACC (Xu et al., 2003). It can also improve dyslipidemia and decrease lipid accumulation in muscle, thus enhancing insulin sensitivity in this tissue (Yamauchi et al., 2001). In present study, niacin significantly increased the plasma adiponectin level, which counters the hepatic *ACC* gene expression and FAS activity. It is therefore hypothesized that niacin affects lipogenesis by affecting the adiponectin secretion. Adpn R2, not Adpn R1, may be the crosslink between adiponectin signaling and nonalcoholic fatty liver disease in obese mice (Peng et al., 2009). In present study, hepatic Adpn R1 and Adpn R2 mRNA levels did not show the same tendency with plasma adiponectin level after niacin treatment. Therefore, the relationship of Adipo R with adiponectin regulating hepatic lipogenesis in niacin treated-rabbits remains unclear and warrants further investigation.

In conclusion, niacin treatment can decrease hepatic fatty acids synthesis, but not alter fatty acids oxidation and triacylglycerol export, which attenuates lipid accumulation in liver. Besides, the hormones of insulin, leptin and adiponectin are associated with the regulation of niacin on the hepatic lipid metabolism in rabbits.

ACKNOWLEDGMENTS

This work was supported by the Modern Agro-industry Technology Research system (CARS-44-B-1), Agricultural Major Application Technology Innovation Project of Shandong Province (2015-2016), Project Funded by China Postdoctoral Science Foundation (2015M580601) and Youth Science and Technology Innovation Fund of Shandong Agricultural University (2015-2016).

REFERENCES

Anania, F. A. 2002. Leptin, liver, and obese mice--fibrosis in the fat lane. Hepatology 36:246-248.

Asai, A. and T. Miyazawa. 2001. Dietary curcuminoids prevent high-fat diet-induced lipid accumulation in rat liver and epididymal adipose tissue. J. Nutr. 131:2932-2935.

Bai, Y., S. Zhang, K. S. Kim, J. K. Lee, and K. H. Kim. 1996. Obese gene expression alters the ability of 30A5 preadipocytes to respond to lipogenic hormones. J. Biol. Chem. 271:13939-13942.

Barter, P. J. and J. I. Lally. 1978. Metabolism of esterified cholesterol in the plasma very low density lipoproteins of the rabbit. Atherosclerosis 31:355-364.

Berg, A. H., T. P. Combs, and P. E. Scherer. 2002. ACRP30/adiponectin: an adipokine regulating glucose and lipid metabolism. Trends Endocrinol. Metab. 13:84-89.

Browning, J. D., L. S. Szczepaniak, R. Dobbins, P. Nuremberg, J. D. Horton, J. C. Cohen, S. M. Grundy, and H. H. Hobbs. 2004. Prevalence of hepatic steatosis in an urban population in the United States: impact of ethnicity. Hepatology 40:1387-1395.

Buettner, C., E. D. Muse, A. Cheng, L. Chen, T. Scherer, A. Pocai, K. Su, B. Cheng, X. Li, and J. Harvey-White et al. 2008. Leptin controls adipose tissue lipogenesis via central, STAT3-independent mechanisms. Nat. Med. 14:667-675.

Carling, D., M. J. Sanders, and A. Woods. 2008. The regulation of AMP-activated protein kinase by upstream kinases. Int. J. Obes. 32:S55-S59.

Cho, K. H., H. J. Kim, V. S. Kamanna, and N. D. Vaziri. 2010. Niacin improves renal lipid metabolism and slows progression in chronic kidney disease. Biochim. Biophys. Acta 1800:6-15.

Cruz-Bautista, I., R. Mehta, J. Cabiedes, C. García-Ulloa, L. E. Guillen-Pineda, P. Almeda-Valdés, D. Cuevas-Ramos, and C. A. Aguilar-Salinas. 2015. Determinants of VLDL composition and apo B-containing particles in familial combined hyperlipidemia. Clin. Chim. Acta 438:160-165.

De blas, C. and G. G. Mateos. 1998. Feed formulation. In: Nutrition of the Rabbit (Eds. C. de Blas and J. Wiseman). CAB International, Wallingford, UK. pp. 222-232.

Fabbrini, E., B. S. Mohammed, K. M. Korenblat, F. Magkos, J. McCrea, B. W. Patterson, and S. Klein. 2010. Effect of fenofibrate and niacin on intrahepatic triglyceride content, very low-density lipoprotein kinetics, and insulin action in obese subjects with nonalcoholic fatty liver disease. J. Clin. Endocrinol. Metab. 95:2727-2735.

Gallardo, N., E. Bonzón-Kulichenko, T. Fernández-Agulló, E. Moltó, S. Gómez-Alonso, P. Blanco, J. M. Carrascosa, M. Ros, and A. Andrés. 2007. Tissue-specific effects of central leptin on the expression of genes involved in lipid metabolism in liver and white adipose tissue. Endocrinology 148:5604-5610.

Ganji, S. H., M. L. Kashyap, and V. S. Kamanna. 2015. Niacin inhibits fat accumulation, oxidative stress, and inflammatory cytokine IL-8 in cultured hepatocytes: Impact on non-alcoholic fatty liver disease. Metabolism 64:982-990.

Hamaguchi, M., T. Kojima, N. Takeda, T. Nakagawa, H. Taniguchi, K. Fujii, T. Omatsu, T. Nakajima, H. Sarui, and M. Shimazaki et al. 2005. The metabolic syndrome as a predictor of nonalcoholic fatty liver disease. Ann. Intern. Med. 143:722-728.

Havel, P. J. 2004. Update on adipocyte hormones: regulation of energy balance and carbohydrate/lipid metabolism. Diabetes 53:143-151.

Jin, F. Y., V. S. Kamanna, and M. L. Kashyap. 1997. Niacin decreases removal of high-density lipoprotein apolipoprotein AI but not cholesterol ester by Hep G2 cells. Implication for reverse cholesterol transport. Arterioscler. Thromb. Vasc. Biol. 17:2020-2028.

Kamanna, V. S., S. H. Ganji, and M. L. Kashyap. 2013. Recent advances in niacin and lipid metabolism. Curr. Opin. Lipidol. 24:239-245.

Kitamura, T., Y. Feng, Y. I. Kitamura, S. C. Chua Jr., A. W. Xu, G. S. Barsh, L. Rossetti, and D. Accili. 2006. Forkhead protein FoxO1 mediates Agrp-dependent effects of leptin on food intake. Nat. Med. 12:534-540.

Lamon-Fava, S., M. R. Diffenderfer, P. H. R. Barrett, A. Buchsbaum, M. Nyaku, K. V. Horvath, B. F. Asztalos, S. Otokozawa, M. Ai, and N. R. Matthan et al. 2008. Extended-release niacin alters the metabolism of plasma apolipoprotein (Apo) A-I and ApoB-containing lipoproteins. Arterioscler. Thromb. Vasc. Biol. 28:1672-1678.

Lillie, R. D. and H. M. Fullmer. 1976. Histopathologic Technic and Practical Histochemistry. 4th edn. McGraw-Hill, London, UK.

Li, X., J. S. Millar, N. Brownell, F. Briand, and D. J. Rader. 2010. Modulation of HDL metabolism by the niacin receptor GPR109A in mouse hepatocytes. Biochem. Pharmacol. 80:1450-1457.

Li, Y., G. Qin, J. Liu, L. Mao, Z. Zhang, and J. Shang. 2014. Adipose tissue regulates hepatic cholesterol metabolism via adiponectin. Life Sci. 118:27-33.

Lin, Z., X. Pan, F. Wu, D. Ye, Y. Zhang, Y. Wang, L. Jin, Q. Lian, Y. Huang, and H. Ding et al. 2015. Fibroblast growth factor 21 prevents atherosclerosis by suppression of hepatic sterol regulatory element-binding protein-2 and induction of adiponectin in mice. Circulation 131:1861-1871.

Livak, K. J. and T. D. Schmittgen. 2001. Analysis of relative gene expression data using real-time quantitative PCR and the 2 (-Delta Delta C (T)) method. Methods 25:402-408.

Matsusue, K., M. Haluzik, G. Lambert, S. H. Yim, O. Gavrilova, J. M. Ward, B. Brewer Jr., M. L. Reitman, and F. J. Gonzalez. 2003. liver-specific disruption of ppary in leptin-deficient mice improves fatty liver but aggravates diabetic phenotypes. J. Clin. Invest. 111:737-747.

Nakae, J. and D. Accili. 1999. The mechanism of insulin action. J. Pediatr. Endocrinol. Metab. 12:721-731.

Nguyen, P., V. Leray, M. Diez, S. Serisier, J. Le Bloc'h, B. Siliart, and H. Dumon. 2008. Liver lipid metabolism. J. Anim. Physiol. Anim. Nutr. 92:272-283.

Paulauskis, J. D. and H. S. Sul. 1989. Hormonal regulation of mouse fatty acid synthase gene transcription in liver. J. Biol. Chem. 264:574-577.

Peng, Y., D. Rideout, S. Rakita, M. Sajan, R. Farese, M. You, and M. M. Murr. 2009. Downregulation of adiponectin/AdipoR2 is associated with steatohepatitis in obese mice. J. Gastrointest. Surg. 13:2043-2049.

Pullen, D. L., J. S. Liesman, and R. S. Emery. 1990. A species comparison of liver slice synthesis and secretion of triacylglycerol from nonesterified fatty acids in media. J. Anim. Sci. 68:1395-1399.

Reddy, J. K. and M. S. Rao. 2006. Lipid metabolism and liver inflammation. II. Fatty liver disease and fatty acid oxidation. Am. J. Physiol. Gastrointest. Liver Physiol. 290:G852-G858.

Rubic, T., M. Trottmann, and R. L. Lorenz. 2004. Stimulation of CD36 and the key effector of reverse cholesterol transport ATP binding cassette A1 in monocytoid cells by niacin. Biochem. Pharmacol. 67:411-419.

Ruzzin, J., R. Petersen, E. Meugnier, L. Madsen, E. J. Lock, H. Lillefosse, T. Ma, S. Pesenti, S. B. Sonne, T. T. Marstrand, and M. K. Malde et al. 2010. Persistent organic pollutant exposure leads to insulin resistance syndrome. Environ. Health Perspect. 118:465-471.

Salhanick, A. I., S. I. Schwartz, J. M. Amatruda. 1991. Insulin inhibits apolipoprotein B secretion in isolated human hepatocytes. Metabolism 40:275-279.

Saltiel, A. R. and C. R. Kahn. 2001. Insulin signalling and the regulation of glucose and lipid metabolism. Nature 414:799-806.

Simon, J., P. Freychet, and G. Rosselin. 1974. Chicken insulin: radioimmunological characterization and enhanced activity in rat fat cells and liver plasma membranes. Endocrinology 95:1439-1449.

van der Hoorn, J. W., W. de Haan, J. F. Berbée, L. M. Havekes, J. W. Jukema, P. C. Rensen, and H. M. Princen. 2008. Niacin increases HDL by reducing hepatic expression and plasma levels of cholesteryl ester transfer protein in APOE* 3Leiden. CETP mice. Arterioscler Thromb. Vasc. Biol. 28:2016-2022.

Weibel, E. R. and R. P. Bolender. 1973. Stereological techniques for electron microscopic morphometry. In: Principles and Techniques of Electron Microscopy (Ed. M. A. Hayat). Van Nostrand Rheinhold Company, New York, pp. 237-296.

Wise, A., S. M. Foord, N. J. Fraser, A. A. Barnes, N. Elshourbagy, M. Eilert, D. M. Ignar, P. R. Murdock, K. Steplewski, and A. Green et al. 2003. Molecular identification of high and low affinity receptors for nicotinic acid. J. Biol. Chem. 278:9869-9874.

Permissions

All chapters in this book were first published in AJAS, by Asian-Australasian Association of Animal Production Societies; hereby published with permission under the Creative Commons Attribution License or equivalent. Every chapter published in this book has been scrutinized by our experts. Their significance has been extensively debated. The topics covered herein carry significant findings which will fuel the growth of the discipline. They may even be implemented as practical applications or may be referred to as a beginning point for another development.

The contributors of this book come from diverse backgrounds, making this book a truly international effort. This book will bring forth new frontiers with its revolutionizing research information and detailed analysis of the nascent developments around the world.

We would like to thank all the contributing authors for lending their expertise to make the book truly unique. They have played a crucial role in the development of this book. Without their invaluable contributions this book wouldn't have been possible. They have made vital efforts to compile up to date information on the varied aspects of this subject to make this book a valuable addition to the collection of many professionals and students.

This book was conceptualized with the vision of imparting up-to-date information and advanced data in this field. To ensure the same, a matchless editorial board was set up. Every individual on the board went through rigorous rounds of assessment to prove their worth. After which they invested a large part of their time researching and compiling the most relevant data for our readers.

The editorial board has been involved in producing this book since its inception. They have spent rigorous hours researching and exploring the diverse topics which have resulted in the successful publishing of this book. They have passed on their knowledge of decades through this book. To expedite this challenging task, the publisher supported the team at every step. A small team of assistant editors was also appointed to further simplify the editing procedure and attain best results for the readers.

Apart from the editorial board, the designing team has also invested a significant amount of their time in understanding the subject and creating the most relevant covers. They scrutinized every image to scout for the most suitable representation of the subject and create an appropriate cover for the book.

The publishing team has been an ardent support to the editorial, designing and production team. Their endless efforts to recruit the best for this project, has resulted in the accomplishment of this book. They are a veteran in the field of academics and their pool of knowledge is as vast as their experience in printing. Their expertise and guidance has proved useful at every step. Their uncompromising quality standards have made this book an exceptional effort. Their encouragement from time to time has been an inspiration for everyone.

The publisher and the editorial board hope that this book will prove to be a valuable piece of knowledge for researchers, students, practitioners and scholars across the globe.

List of Contributors

Nadia Musco, Raffaella Tudisco, Monica I. Cutrignelli, Federico Infascelli, Serena Calabrò, Oscar Ruiz, Claudio Arzola, Eduviges Burrola and Agustín Corral
Department of Veterinary Medicine and Animal Production, University of Napoli Federico II, Napoli 80137, Italy

Ivan B. Koura, Ghislain Awadjihè, Sebastien Adjolohoun and Marcel Houinato
Department of Animal Sciences, Faculty of Agricultural Sciences, University of Abomey-Calavi, Cotonou 526, Benin

Maria Pina Mollica
Department of Biology, University of Napoli Federico II, Napoli 80126, Italy

Oscar Ruiz, Claudio Arzola, Eduviges Burrola and Agustín Corral
College of Animal Science and Ecology, Autonomous University of Chihuahua, Chihuahua, Chih. 31000, Mexico

Yamicela Castillo and Mateo Itza
Department of Veterinary Medicine, Multidisciplinary Division, Autonomous University of Juarez City, Nuevo Casas Grandes, Chih. 31803, México

Jaime Salinas
College of Veterinary Medicine and Animal Science, Autonomous University of Tamaulipas, Cd. Victoria, Tamps. 87000, México

Michael E. Hume
Agricultural Research Service, Southern Plains Research Center, Food and Feed Safety Research Unit, United States Department of Agriculture, College Station, TX 77843, USA

Manuel Murillo
College of Veterinary Medicine and Animal Science, Juarez University of Durango State, Durango, Dgo. 34000, Mexico

Lizhi Wang and Bai Xue
Institute of Animal Nutrition, Sichuan Agricultural University, Yaan 625014, Sichuan, China

Y. Wang, N. K. Kopparapu, J. Liu and X. L. Liu
College of Food and Biological Engineering, Qiqihar University, Qiqihar 161006, China

L. Jin and Q. N. Wen
Agricultural Machinery Research Institute of Liaoning, Shenyang 110036, China

Y. G. Zhang
Animal Science and Technology Institute, Northeast Agriculture University, Harbin 150030, China

Chen Wei, Shixin Lin and Jinlong Wu
DSM China Animal Nutrition Centre, Bazhou, Hebei 065700, China

Guangyong Zhao, Tingting Zhang and Wensi Zheng
State Key Laboratory of Animal Nutrition, College of Animal Science and Technology, China Agricultural University, Beijing 100193, China

Sun Hee Moon, Xi Feng, Hyun Yong Lee and Jihee Kim
Department of Animal Science, Iowa State University, Ames, IA 50011, USA

Inyoung Lee
Naturence Co., Ltd., Sejong 339-824, Korea

Dong Uk Ahn
Department of Animal Science, Iowa State University, Ames, IA 50011, USA
Department of Animal Science and Technology, Sunchon National University, Sunchon 540-742, Korea

T. A. Woyengo, I. A. Emiola and C. M. Nyachoti
Department of Animal Science, University of Manitoba, Winnipeg, MB R3T 2N2, Canada

I. H. Kim
Department of Animal Resources and Science, Dankook University, Cheonan 330-714, Korea

S. B. AL-Suwaiegh
Animal and Fish Production Department, King Faisal University, Al-Ahsa, 31982, Kingdom of Saudi Arabia

Chang-Won Kang and Byoung-Ki An
Laboratory of Poultry Science, Department of Animal Science and Technology, College of Animal Bioscience and Technology, Konkuk University, Seoul 05029, Korea

Sang-Jin Kim
Laboratory of Poultry Science, Department of Animal Science and Technology, College of Animal Bioscience and Technology, Konkuk University, Seoul 05029, Korea Yuhan Corp, Seoul 06927, Korea

Kyung-Woo Lee
Laboratory of Poultry Science, Department of Animal Science and Technology, College of Animal Bioscience and Technology, Konkuk University, Seoul 05029, Korea

K. Cheng, Y. Niu, X. C. Zheng, H. Zhang, Y. P. Chen, L. L. Zhang, Y. M. Zhou and T. Wang
College of Animal Science and Technology, Nanjing Agricultural University, Nanjing 210095, China

M. Zhang and X. X. Huang
Jiangsu Wilmar Spring Fruit Nutrition Products Co., Ltd. Taixing 225434, China

W. Zhu, B. X. Zhang, K. Y. Yao and J. K. Wang
Institute of Dairy Science, College of Animal Sciences, Zhejiang University, Hangzhou 310058, China

I. Yoon and Y. H. Chung
Diamond V, Cedar Rapids, IA 52405, USA

B. Zhang, Z. H. Wei, H. Z. Sun, J. X. Liu and H. Y. Liu
Institute of Dairy Science, College of Animal Science, Zhejiang University, Hangzhou 310058, China

C. Wang
College of Animal Science and Technology, Zhejiang A & F University, Hangzhou, 311300, China

G. Z. Xu
Institute of Shanghai Dairy Science, Shanghai 200032, China

C. E. Oltramari, G. G. O. Nápoles, M. R. De Paula, J. T. Silva, M. P. C. Gallo, M. H. O. Pasetti and C. M. M. Bittar
Animal Science Department, University of São Paulo – USP/ESALQ, Piracicaba, SP 13418-900, Brazil

X. H. Jin, P. S. Heo, J. S. Hong and Y. Y. Kim
Department of Agricultural Biotechnology, College of Animal Life Sciences, Seoul National University, Seoul 151-921, Korea

N. J. Kim
National Academy of Agricultural Science, Wanju 565-851, Korea

S. S. Jin, S. W. Jung, J. C. Jang, W. L. Chung, J. H. Jeong and Y. Y. Kim
School of Agricultural Biotechnology and Research Institute for Agriculture and Life Science, Seoul National University, Seoul 151-921, Korea

Ísis Lazzarini, Edenio Detmann, Sebastião de Campos Valadares Filho, Mário Fonseca Paulino, Erick Darlisson Batista, Luana Marta de Almeida Rufino, William Lima Santiago dos Reis and Marcia de Oliveira Franco
Department of Animal Science, Universidade Federal de Viçosa, Viçosa, MG CEP 36570-000, Brazil

Mingmei Bai, Guixin Qin and Zewei Sun
Animal Production and Product Quality and Security Key Lab, Ministry of Education, Jilin Agricultural University, Changchun 130118, China
College of Animal Science and Technology, Jilin Agricultural University, Changchun 130118, China

Guohui Long
College of Life Science, Jilin Agricultural University, Changchun 130118, China

Dong Hyeon Kim, Hyuk Jun Lee, Young Ho Joo and Sam Churl Kim
Division of Applied Life Science (BK21Plus, Institute of Agriculture and Life Science), Gyeongsang National University, Jinju 52828, Korea

Sardar M. Amanullah
Division of Applied Life Science (BK21Plus, Institute of Agriculture and Life Science), Gyeongsang National University, Jinju 52828, Korea
Bangladesh Livestock Research Institute, Savar, Dhaka-1341, Bangladesh

Ouk Kyu Han
National Institute of Crop Science, RDA, Suwon 16429, Korea

Adegbola T. Adesogan
Department of Animal Sciences, IFAS, University of Florida, Gainesville, FL 32608, USA

S. K. Kim, T. H. Kim, S. K. Lee, K. W. Lee and B. K. An
Department of Animal Science and Technology, Konkuk University, Seoul 143-701, Korea

K. H. Chang and S. J. Cho,
CJ Cheiljedang Ltd., Seoul 100-400, Korea

Han Lin Li, Pin Yao Zhao, Yan Lei, Md Manik Hossain and In Ho Kim
Department of Animal Resource and Science, Dankook University, Cheonan 330-714, Korea

Jungsun Kang
Genebiotech. Co. Ltd., Seoul 06774, Korea

Wisitiporn Suksombat, Chayapol Meeprom and Rattakorn Mirattanaphrai
School of Animal Production Technology, Institute of Agricultural Technology, Suranaree University of Technology, Nakhon Ratchasima, 30000, Thailand

Lihuai Yu, Shunan Wang, Luoyang Ding, Mengzhi Wang, Li Dong and Hongrong Wang
The College of Animal Science and Technology, Yangzhou University, Yangzhou 225009, China

Xianghuan Liang
Yangzhou Kangyuan Dairy CO., LTD, Yangzhou, 215009, China

Shunxi Wang, Kai Peng and Lei Long
College of Engineering, China Agricultural University, Beijing 100083, China

Yuxi Wang
Agriculture and Agri-Food Canada, Lethbridge Research and Development Centre, Lethbridge, AB T1J 4B1, Canada

J. Wang, H. J. Zhang and G. H. Qi
Key Laboratory of Feed Biotechnology of Ministry of Agriculture, Feed Research Institute, Chinese Academy of Agricultural Sciences, Beijing 100081, China

S. G. Wu and F. Yuan,
Tianjin Naer Biotechnology Co., Ltd., Tianjin 300457, China

L. Long
Key Laboratory of Feed Biotechnology of Ministry of Agriculture, Feed Research Institute, Chinese Academy of Agricultural Sciences, Beijing 100081, China
Tianjin Naer Biotechnology Co., Ltd., Tianjin 300457, China

Xiaoqiang Yu, Quanhui Peng and Zhisheng Wang
Animal Nutrition Institute, Key Laboratory of Low Carbon Culture and Safety Production in Cattle in Sichuan, Sichuan Agricultural University, Ya'an, Sichuan 625014, China

Xiaolin Luo, Tianwu An and Jiuqiang Guan
Grassland Science Academy of Sichuan Province, Chengdu, Sichuan 611731, China

C. Basavanta Kumar and T. M. Prabhu
Department of Animal Nutrition, Veterinary College, KVAFSU, Bangalore 560024, India

R. G. Gloridoss
Department of Instructional Livestock Farming Complex, Veterinary College, Bangalore 560024, India

K. C. Singh
Karnataka Veterinary, Animal and Fisheries Sciences University, Bidar, Karnataka 585401, India

B. N. Suresh
Department of Instructional Livestock Farming Complex, Veterinary College, Hassan 573202, India

Mi Rae Oha, Hong Liang Li, Byong Tae Jeon, Cheong Hee Choi, Yu Ling Ding, Yu Jiao Tang, Eun Kyung Kim, Se Young Jang, Hye Jin Seong and Sang Ho Moon
Korea Nokyong Research Center, Division of Food Bio Science, Konkuk University, Chungju 27478, Korea

Heeok Hong
Department of Medical Science, School of Medicine, Konkuk University, Seoul 05029, Korea

Lei Liu, Chunyan Li, Chunyan Fu and Fuchang Li
Department of Animal Science, Shandong Agricultural University, Taian, Shandong 271018, China

Index